World Survey of Climatology Volume 6

CLIMATES OF CENTRAL AND SOUTHERN EUROPE

World Survey of Climatology

Editor in Chief:

H. E. LANDSBERG, College Park, Md. (U.S.A.)

Editors:

H. ARAKAWA, Tokyo (Japan)
R. A. BRYSON, Madison, Wisc. (U.S.A.)
H. FLOHN, Bonn (Germany)
J. GENTILLI, Nedlands, W.A. (Australia)
J. F. GRIFFITHS, College Station, Texas (U.S.A.)
F. K. HARE, Ottawa, Ont. (Canada)
P. E. LYDOLPH, Milwaukee, Wisc. (U.S.A.)
S. ORVIG, Montreal, Que. (Canada)
D. F. REX, Boulder, Colo. (U.S.A.)
W. SCHWERDTFEGER, Madison, Wisc. (U.S.A.)
K. TAKAHASHI, Tokyo (Japan)
H. VAN LOON, Boulder, Colo. (U.S.A.)
C. C. WALLÉN, Geneva (Switzerland)

World Survey of Climatology Volume 6

Climates of Central and Southern Europe

edited by

C. C. WALLÉN

World Meteorological Organisation
Geneva (Switzerland)

ELSEVIER SCIENTIFIC PUBLISHING COMPANY
Amsterdam-Oxford-New York 1977

ELSEVIER SCIENTIFIC PUBLISHING COMPANY
335 Jan van Galenstraat
P.O. Box 211, Amsterdam, The Netherlands

Distributors for the United States and Canada:

ELSEVIER NORTH-HOLLAND INC.
52 Vanderbilt Avenue
New York, N.Y. 10017

Library of Congress Cataloging in Publication Data
Main entry under title:

Climates of central and southern Europe.

   (World survey of climatology ; v. 6)
   Bibliography: p.
   Includes indexes.
   1. Central Europe--Climate. 2. Europe, Southern--
Climate. I. Wallén, Carl Christian. II. Series.
QC981.W67 vol. 6 [QC989.C45]   551.6'08s [551.6'9'43]
ISBN 0-444-41336-7   76-46572

ISBN: 0-444-40734-0 (series)
ISBN: 0-444-41336-7 (vol. 6)

With 84 illustrations and 126 tables

Printed in The Netherlands

World Survey of Climatology

List of Contributors to this Volume

V. Cantú
Osservatorio Centrale Meteorologia
e Fisico dello Spazio
Vigna di Valle (Italy)

D. Furlan
Hydrometeorological Institute of the S.R. Slovenija
Ljubljana (Yugoslavia)

W. Okolowicz
Geographical Institute
Climatological Branch
University of Warsaw
Warsaw (Poland)

H. Schirmer
Deutscher Wetterdienst
Offenbach a/M. (Bundesrepublik Deutschland)

M. Schüepp
Schweizerische Meteorologische Zentralanstalt
Zürich (Switzerland)

Contents

Contents

Chapter 4. THE CLIMATE OF ITALY
by V. CANTÚ

Chapter 5. THE CLIMATE OF SOUTHEAST EUROPE
by D. Furlan

Chapter 1

Introduction

C. C. WALLÉN

In Volume 5 of the *World Survey of Climatology* the climates of some western European countries are discussed in relation to the territories concerned. The volume also includes an Introduction which aims at presenting the basic features of climates of all western and central Europe. In the present volume the climates of the central and southern parts of Europe are presented again according to the territories concerned and there seems to be no reason to repeat the overall picture of the general circulation and the resulting climatic features as given in Volume 5. It may be argued that the review given in the introduction to Volume 5 is now somewhat old and therefore ought to be revised in the light of recent developments. In the editor's mind, however, the best way of considering the new developments that may have taken place is to give adequate attention to the most recent papers presented in the reference lists of the chapters in the present volume. The contributions have all been fairly well updated as far as references are concerned. In general, the recent developments add little to the overall picture although they often have contributed considerably to the development of new thinking in relation to the climatic conditions of certain areas of Europe.

The finalization of Volumes 5 and 6 of the *World Survey* has taken considerable time. This has been due to many circumstances which it is not possible to dwell on in this context. Let me instead express my sincere thanks to all the authors and other persons who, over a period of more than ten years, have been involved in the preparation of these two volumes. As well-known Europe is not only a very much split-up continent from the political point of view, the climates of the continent are also very much diversified and in many ways it has been both necessary and useful to have so many different persons with their own experience of various regions involved in the preparation of these volumes.

It should be noted, however, that there is both an advantage and a disadvantage in having a variety of authors presenting their ideas of the climates of different regions and territories. Some may argue that what has been gained through this personal presentation of views may have been lost in the overall co-ordinated picture of the European climates. However, in a time when conformity and equality are given preference in many contexts I am happy to have been the editor of a group of contributions which everyone in itself reflects the opinion of the particular author rather than of a dictatorial editor. It is my hope that the reader will appreciate this attitude of the editor even though he may have preferred a larger degree of conformity to have been reflected in this work. Finally, my special and very sincere thanks go to Dr. G. Manley, Dr. Thor Werner-Johannessen, Dr. R. Arléry, Dr. A. Lines Escardó, Dr. V. Cantú, Dr. V. Okołowicz,

Dr. D. Furlan, and Drs. M. Schüepp and H. Schirmer for their countless efforts to finalize these volumes in spite of various difficulties and many unforeseen circumstances.

Chapter 2

Climates of Central Europe

M. SCHÜEPP AND H. SCHIRMER

Historical account

Today's meteorological instruments did not originate in any of the four countries of the area under consideration; nevertheless, there are many long-term meteorological records from this area. Thermometers and barometers, which were in use in Italy, The Netherlands and France, were introduced to central Europe in the 17th century. Particularly in old-established university towns scientists pursued the reading of these instruments with great enthusiasm. Unfortunately, most of the observations which were taken in the 17th century have been lost. However, in the 18th century quite a few meteorological stations were established whose long-term records could be continued to the present time after the necessary instrument corrections and reductions have been applied. Von Rudloff (1967) gives a detailed history of the stations as well as a discussion of the climatic fluctuations seen on these long-term records.

In the brief survey given here it is worth mentioning that the stations with the longest meteorological records, which could be pursued until this day, are those at Berlin (from 1719), Basel (from 1755), Geneva (from 1768), Vienna (from 1775), Hohenpeissenberg (from 1781) and a station in the high Alps, namely Great St. Bernard (from 1818). These meteorological observations show a period of record minimum temperatures coinciding with the maximum glacier advances (about 1820 and 1855); the most striking features are the summer of 1816 (the year 1816 was called "the year without a summer", and was followed by a famine) and the cold winter 1829/30.

In the first half of the 19th century, mainly sporadic and incomplete meteorological records were obtained because the governmental authorities were not interested in sponsoring any meteorological research activity. Not until the years after 1860 was there a turn for the better. A first impetus for a uniform observational network had materialized through the so-called "Societas Meteorologica Palatina" by initiative of Johann Jacob Hemmer under the elector of the Palatinate, Carl Theodor; this network of 38 stations covered the whole of Europe from 1780 to 1792, but it could not be maintained in the chaos after the French Revolution. The meteorological stations, however, which were reestablished in the second half of the 19th century, continued to operate. During the following period a great number of qualified scientists worked in the area under consideration and—among others—the climatologists Hann, Hellmann, Köppen and Wild have become known all over the world. The last hundred years hence have been characterized by an increasing amount of climatic observations being processed for publication in the yearbooks of various countries as well as by many efforts to investigate regional climatic differences by comparison of these observations.

The establishment of synoptic weather observations between 1870 and 1880 enabled the scientists to find physical explanations for many of these regional differences. Furthermore, the establishment of an aerological network in the years after 1930 for the purpose of aviation weather forecast enabled the scientists to correlate surface and upper-air observations. Then, additional tools came to the scientists' aid, as for instance computers, satellite pictures and radar. At the same time, however, a certain handicap was introduced in the field of meteorological observations: more and more have man-made meteorological observations been replaced by the automatic meteorological observations and this often endangers the necessary continuity of the long-term climatological records. At the end of this chapter a list of references is attached which includes selected papers and books of maps that are deemed to furnish some helpful additional illustrative material for the reader. Of course, it cannot be expected that this list is complete, if so it would fill a voluminous reference book. This selection might appear somewhat subjective; however, this might be outweighed by the value of guidance given to a reader otherwise flooded with publications.

General geographical and climatological features

The area under consideration, namely the Federal Republic of Germany, the Democratic Republic of Germany, Austria and Switzerland, forms a broad strip extending from the coastal areas of the North Sea and the Baltic at a latitude of approximately 55°N to the southern foothills of the Alps at a latitude of approximately 46°N. The variety of climates in the individual parts of this area, however, is not only due to the difference in latitude but also and essentially to the influence of the individual mountain ranges. There are for instance the highlands of central Germany which do not reach great heights above mean sea level (maximum height 1,051 m above m.s.l. in the Fichtel Mountains besides the height of 1,142 m above m.s.l. in the Harz Mountains, an isolated massif in the north German lowlands); nevertheless, they act as a noteworthy climatic divide. The German lowlands to the north of these highlands are easily crossed by the Atlantic depressions moving from the British Isles toward the east, because there is comparatively little surface friction in this region. Thus, the normal life cycle of a depression is hardly interfered with, in contrast to Scandinavia where the Norwegian mountain range in the Skagerrak area may enhance wave disturbances and therefore cyclogenesis (see *World Survey of Climatology*, Vol.5, pp.26–27). Thus, the lowlands in the north of the Federal Republic of Germany are subject to the intrusions of original maritime air from the North Sea and the great number of inlets and islands along the sea shore are subject to an everlasting battle between the water and the land, mainly due to severe gales in winter in connection with spring tides: large parts of the North Frisian Islands (Schleswig-Holstein) for instance were lost in severe weather of this kind (Nordstrand, 1634: a third of the island was submerged in high waves). Recently, here, as in The Netherlands, man has begun fighting this brute, natural force by struggling, not only to prevent these breaks, but also to reclaim the lost land by means of modern technology.

The lowlands in the Democratic Republic of Germany extend much farther to the south;

they are much less subject to the maritime influence and they are characterized by an increase in continentality toward the east.

The region of central and southern Germany has a rather different weather regime. The diversified topography of the highlands, manifested by a great number of, in many cases, winding valleys, brings about a variety of natural subregions with climatic features of their own. Anticyclonic weather is therefore generally characterized by a stagnant layer of air near the ground which, especially in the long winter nights, is strongly cooled when the earth's surface suffers a net loss of heat due to terrestrial radiation. Thus, surface inversions occur very frequently. In summer they are mainly restricted to the early morning hours, in winter, however, they may last for days. This departure from the usual decrease of temperature with altitude creates a low temperature in the valleys and a high temperature on the top of hills and mountains. In the warm season typical mountain and valley winds will develop: the system normally consists of a weak downvalley component or mountain wind at night and a strong upvalley component or valley wind during daytime which reaches its greatest strength in the afternoon. The climatic tables (pp.61–73) do not show this autochthonic wind system, which might deviate considerably from the general gradient wind direction, because the mountain and valley wind system develops mostly in calm, clear weather without noticeable large-scale air advection. The prevailing wind direction indicated in Fig.27 (p.56) mainly coincides with the general surface flow. Hence a comparison of the surface wind roses of the central German and Alpine area on the one hand and the north German lowlands on the other hand will reveal significant differences in the prevailing wind direction. These differences are due to a strong modification of the general surface flow by the terrain as the surface wind undergoes a considerable channeling effect in the valleys.

The mountain ranges bordering the valleys gradually increase in height when approaching the Alps, the main mountain range of the area under discussion. The Vosges Mountains (highest peak 1,423 m above m.s.l.) and the Black Forest (highest peak 1,495 m above m.s.l.) border the upper Rhine Valley. The Swabian Alb, with a little lower peak height of 1,015 m above m.s.l., separates the Rhine and the Danube valleys. The Upper Palatine Forest, the Bavarian Forest and above all the Bohemian Forest (highest peak 1,457 m above m.s.l.) constitute the eastern partition from the Czech area of the Elbe Valley. The climatic subregions will be described in detail on pp.28–41.

The sweeping chain of the Alps is situated in Switzerland and Austria; it runs from west-southwest towards east-northeast together with the Jura Mountains (highest peak 1,723 m above m.s.l.) situated to the west of it as well as parts of the northern and southern Alpine foothills. The considerable height of the Alpine Massif (in its western part: Mont Blanc, 4,810 m, Monte Rosa, 4,638 m and Bernina, 4,051 m above m.s.l.; and in its eastern part: Gross Glockner, 3,798 m above m.s.l.) not only affects the development of local wind systems as described for the highlands, but also causes a significant modification of the general circulation: the enhancement of wave disturbances due to this mountain barrier is even stronger than that due to the Norwegian Mountains. Cold air advection from the North Atlantic effects the formation of a secondary cyclone in the vicinity of the Gulf of Genoa because the wall of the Alps protects the warm air of the Po Valley from being removed. This warm air is removed from the south Alpine foothills only after the tongue of cold air protrudes first through the lower Rhône Valley and later through the gap in the Danube Valley near Vienna, unless the layer of the cold air is so thick that it can directly overflow the chain of the Alps. In winter, such an overflowing often takes place resulting in the so-called north föhn, a strong and gusty katabatic flow of cold air along the southern slopes of the Alps; it will be discussed later on p.23, Fig.12.

The reverse of the north föhn is the south föhn of the Alps caused by an Atlantic cyclone moving from the west towards the Alps. Under these circumstances a flow is created which forces the air of the Gulf of Genoa to cross the Apennines, then to form an upper flow across the Po Valley and finally to cross the Alpine chain in a south–northerly direction. Heavy orographic precipitation is then produced on the windward side of the Alps, whereas warm dry katabatic winds and broken to scattered cloudiness result on the lee side. Later in Fig.11, a conspicuous example of this orographic flow will be given. Besides this modification of the general flow by the Alpine chain the steep Alpine valleys also create the formation of surface inversions and the development of mountain and valley wind systems as already outlined for the central European highlands. In the Alps, however, these effects are much more powerful. Broad valleys as for instance in the Valais in Switzerland are well known for their strong valley winds which in some places overtop the ridges (Maloja wind in the upper Engadine).

The general circulation over central Europe

The strong horizontal temperature gradient between the subtropics and the subpolar region creates the prevalence of westerly winds in the troposphere over central Europe in all months. Thus, the main air advection is from the Atlantic Ocean, which results in the predominantly maritime type of climate in the coastal areas of northern Germany with the above-mentioned regional differentiation of decreasing maritime influence and increasing continentality toward the east within the lowlands of the Democratic Republic of Germany.

However, southern Germany, Switzerland, and Austria in particular, lie at a distance of several hundreds of kilometres from the maritime west-European coast; therefore, this region experiences a gradually increasing continental influence on its climate. The large areas of high elevation in this region have a continuous winter snow cover of long duration from which the terrestrial radiation greatly intensifies the continental influence. In individual years, this local, radiational cooling can be enhanced by the advection of very cold air from Russia. When in pre-winter such a weather type (high pressure over Scandinavia and Finland causing easterly flow over central Europe) becomes persistent for quite some time, it is very unlikely that there will be a change of weather conditions due to the advance of warm air from the Atlantic: the upper low associated with the continental cold dome deflects the winds from their original westerly direction to veer to a northwesterly flow which steers the Atlantic depressions into the Mediterranean. The cold air in the rear of these depressions encounters very warm subtropical air from Africa and, at the interface of these air masses of rather different temperature, wave formation takes place. The easterly winds to the north of these cyclones enhance the advection of cold air from Russia or even Siberia into central Europe, i.e. they favour the persistence or even the intensification of the preceding weather situation. This is the physical explanation for the severe cold of individual winters (e.g., in this century in the winters 1928/29, 1941/42, 1946/47 and 1962/63), whereas the long-term records show that generally the rather mild Atlantic air dominates central Europe and may advance as far eastwards as the Oder or Vistula rivers (compare the European weather maps in *World Survey of Climatology*, Vol.5, p.4).

In the warm season, the intensity of the maritime influence upon central Europe for individual years also varies greatly, because in this season, too, central Europe is influenced by a climatic divide. In summer, a strong large-scale horizontal temperature gradient develops between the polar ice cap and the Mediterranean which enjoys sunny and warm weather influenced by the subtropical high-pressure belt. The polar front, which separates these air masses, then has a position far more to the north than in winter: its cloudiness and rainfall are mainly felt in the region of the British Isles where the "English lawn" is watered sufficiently, whereas only a special flora can stand the summer aridity of the Mediterranean. The central European area in most cases tends to have unsettled weather like that predominating over the British Isles.

The Atlantic depressions often bring thunder showers which are augmented by orographic rainfall so that the area under discussion—except for some marginal zones—receives its main amount of precipitation in summer time. Of course, it happens that in individual years the Azores high for a long period of the summer extends far to the north and northeast covering central Europe and causing dry spells as for instance in

TABLE I

MEAN TEMPERATURES (°C) 1951–1960

	Press. level (mbar)	Jan.	Feb.	Mar.	Apr.	May	June	July	Aug.	Sept.	Oct.	Nov.	Dec.	Annual
Schleswig	850	− 4.1	− 5.2	− 2.3	− 1.3	3.0	6.1	8.3	7.8	5.9	3.7	0.4	− 1.9	1.7
	700	−11.8	−12.7	−10.0	− 9.4	− 5.5	− 1.9	0.1	− 0.2	− 1.8	− 3.8	− 7.1	− 9.6	− 6.1
	500	−28.3	−29.1	−26.6	−25.5	−21.7	−17.6	−15.7	−16.2	−17.6	−19.9	−23.4	−25.7	−22.3
	300	−52.3	−52.8	−51.2	−50.2	−47.6	−44.3	−42.1	−42.9	−43.9	−45.8	−49.0	−50.8	−47.7
Hannover	850	− 3.6	− 4.6	− 1.6	− 0.5	3.7	6.8	9.0	8.6	6.4	4.2	0.6	− 1.4	2.3
	700	−11.4	−12.3	− 9.7	− 8.8	− 5.0	− 1.6	0.5	0.0	− 1.0	− 3.4	− 6.5	− 8.8	− 5.7
	500	−27.8	−28.7	−26.5	−25.0	−21.2	−17.1	−14.9	−15.5	−16.8	−19.4	−22.9	−25.3	−21.8
	300	−52.1	−52.5	−51.1	−50.2	−47.4	−43.8	−41.7	−42.4	−43.4	−45.7	−48.5	−50.8	−47.5
Lindenberg near Berlin	850	− 3.9	− 5.0	− 2.0	0.1	4.4	8.0	9.9	9.5	6.7	4.1	0.6	− 1.6	2.6
	700	−11.3	−12.3	− 9.5	− 8.3	− 4.5	− 0.8	1.0	0.5	− 1.1	− 3.1	− 6.5	− 8.7	− 5.3
	500	−27.4	−28.1	−25.8	−24.2	−20.3	−15.9	−13.8	−14.6	−16.3	−18.8	−22.3	−24.7	−21.0
	300	−51.5	−51.8	−50.6	−49.3	−46.4	−42.7	−40.2	−41.1	−42.5	−44.9	−48.4	−50.1	−46.6
Payerne	850	− 2.2	− 2.7	0.7	2.3	6.9	9.5	11.7	11.4	9.4	5.4	1.6	− 0.1	4.4
	700	− 9.4	−10.0	− 8.2	− 6.8	− 3.3	− 0.5	1.7	1.6	0.3	− 2.7	− 5.5	− 7.4	− 4.2
	500	−25.4	−26.6	−24.3	−22.8	−19.2	−15.8	−13.5	−13.8	−14.8	−18.2	−21.6	−23.4	−19.9
	300	−50.9	−51.2	−49.9	−48.9	−46.0	−42.8	−40.2	−40.2	−41.2	−44.8	−47.8	−49.4	−46.1
Munich	850	− 3.0	− 3.4	− 0.1	1.8	6.2	9.3	11.3	11.2	9.0	5.7	1.9	− 0.2	4.1
	700	−10.8	−11.5	− 9.2	− 7.7	− 3.8	− 0.6	1.3	1.3	0.0	− 2.8	− 5.9	− 8.1	− 4.8
	500	−27.0	−27.8	−25.6	−24.1	−20.0	−16.2	−13.9	−14.3	−15.4	−18.8	−22.1	−24.4	−20.8
	300	−52.2	−52.6	−51.4	−50.4	−47.3	−43.7	−41.0	−41.3	−42.4	−45.7	−48.8	−50.7	−47.3
Vienna	850	− 3.9	− 4.2	− 1.9	1.2	5.8	9.4	11.2	11.1	8.3	5.0	0.8	− 1.1	3.5
	700	−11.2	−11.5	− 9.6	− 8.1	− 3.8	− 0.2	1.4	1.5	− 0.4	− 2.8	− 6.0	− 8.4	− 4.9
	500	−27.5	−27.9	−26.4	−24.3	−20.2	−16.1	−14.1	−14.3	−16.0	−19.0	−22.4	−24.9	−21.1
	300	−52.4	−52.6	−52.3	−50.8	−47.6	−43.7	−41.4	−41.6	−42.9	−46.0	−49.0	−51.0	−47.6

the good wine years of this century 1904, 1911, 1921, 1947, 1949, 1959, 1964 and 1971. The dense aerological network which has been established in central Europe since World War II has furnished valid records of the mean temperature and wind conditions in the troposphere as well as in the lower stratosphere. Tables I–III list the mean values of the period 1951–1960 in the same format as has been used for western Europe in *W.S.C.*, Vol.5 on pp.139–141.

Fig.1 represents the annual variation of temperature for various long-term periods of record for three stations in central Europe. It can easily be seen that in the northeasterly region (Prague and Berlin) the winter cold persists until about mid-February followed by a continuous temperature rise until the beginning of June; in earlier times, however, this rise was often interrupted by the temperature drop of the "ice saints" on 11/12 May (evident from curve *a* for Berlin). In more recent time (periods of the *b* curves) this cold spell does not appear any more, whereas the so-called "sheep cold" is well marked in June (evident from the *b* curves for Berlin, Prague, and Karlsruhe). On the average, the maximum of summer heat occurs as late as about one month after the summer

TABLE II

MEAN WINDS AT THE 500-MBAR LEVEL, 1951–1960

Station		Jan.	Feb.	Mar.	Apr.	May	June	July	Aug.	Sept.	Oct.	Nov.	Dec.
Schleswig	$dd*$	296	299	298	286	270	255	257	257	270	264	271	278
	ff	10	8	5	8	6	8	9	9	10	9	8	12
	v_m	19	18	16	17	15	14	15	14	16	16	17	19
	S	52	44	32	47	39	57	60	63	62	55	48	64
Berlin	dd	294	292	295	287	275	258	266	257	278	270	276	283
	ff	10	8	5	7	7	8	9	10	10	8	7	10
	v_m	18	17	15	15	14	13	14	15	16	15	16	17
	S	56	47	35	48	52	62	63	69	63	52	43	59
Hannover	dd	294	296	289	287	270	259	260	255	271	266	274	282
	ff	11	8	5	7	6	7	10	11	10	8	7	11
	v_m	20	18	16	16	14	13	15	15	16	16	17	18
	S	54	44	32	44	42	52	65	73	61	50	42	61
Munich	dd	304	287	287	280	272	265	273	250	268	258	276	277
	ff	9	10	6	6	7	7	9	10	9	6	6	8
	v_m	18	16	15	13	13	12	13	14	15	14	15	15
	S	51	61	41	47	55	59	71	70	62	42	41	53
Payerne (1951–1960)	dd	298	298	298	296	296	268	272	264	262	266	276	280
	ff	10	7	6	5	6	7	8	10	9	6	5	7
(1961–1970)	dd	307	300	297	271	266	269	278	270	268	270	260	302
	ff	9	8	8	7	8	6	9	10	8	5	9	9
	v_m	18	18	16	15	15	12	14	14	13	13	17	17
	S	53	47	50	46	51	51	68	69	57	37	55	51
Vienna (1952–1966)	dd	295	288	290	287	280	265	274	267	277	264	278	281
	ff	8	10	6	7	6	6	7	10	10	8	6	9
	v_m	19	19	14	15	12	12	11	14	16	18	12	18
	S	44	54	43	44	48	49	65	69	61	42	46	52

* dd = direction of the resultant wind, in degrees; ff = resultant wind speed (m/sec); v_m = mean wind speed (m/sec); S = constancy, $100\,ff/v_m$ (lower than S of individual months).

Remarks. Schleswig, Hannover and Munich: 12h00–15h00 G.M.T. Berlin: midday ascents 1.1.1951–15.4.1956; morning ascents (07 G.M.T.) 16.4.1956–31.12.1960.

solstice, in the older periods of record it is even delayed to the end of July or beginning of August.

Tables I–III refer to the 500-mbar level which corresponds to a height of about 5.5 km above m.s.l. and which balances half the weight of the atmosphere; this level is particularly appropriate for representing flow conditions, for the following reasons: at this height the orographic influence of the Alps has weakened considerably and above this level there is hardly any change in wind direction up to the tropopause, whereas the wind speed increases. The powerful polar-front jet stream is usually found slightly above the 300-mbar level (about 9,000 m above m.s.l.), in winter imbedded in the mid-latitude westerlies.

The differences in continentality within the area of central Europe (Table I) are measured by the mean annual range of temperature, i.e. the difference between the mean temper-

TABLE III

RELATIVE FREQUENCIES (%) OF THE DOMINANT COMPONENTS OF WINDS*

Seasons	Wind comp.	Schleswig		Berlin		Hannover		Munich		Payerne		Vienna	
Spring	N	26	45	24	43	25	44	23	42	28	48	25	43
	S	19		19		19		19		20		18	
	W	37	48	38	48	38	49	39	48	42	52	44	46
	E	11		10		11		9		10		12	
	calm**	7		9		7		10		0.5		0.7	
Summer	N	12	34	12	29	13	31	16	31	21	35	21	37
	S	22		17		18		15		14		16	
	W	46	57	52	60	52	61	53	58	60	65	55	62
	E	11		8		9		5		5		7	
	calm	9		11		8		11		0.4		0.6	
Autumn	N	20	42	22	42	20	39	19	38	21	43	21	42
	S	22		20		19		19		22		21	
	W	46	52	45	52	47	55	43	52	46	57	49	58
	E	6		7		8		9		11		9	
	calm	6		6		6		10		0.3		0.4	
Winter	N	24	38	24	38	23	37	24	39	29	43	26	43
	S	14		14		14		15		14		17	
	W	49	57	48	57	49	58	45	53	45	57	48	57
	E	8		9		9		8		12		9	
	calm	5		5		5		8		0.4		0.2	

* At 500 mbar 00h–03h and 12h–15h G.M.T., 1951–1960.
** Calm: ≤ 4 m/sec for German stations, < 0.3 m/sec for Payerne and Vienna.

ature of the warmest and coldest months. Reference is made to stations in western Europe (Vol.5, p.139, table I) for comparison. At the 850-mbar level, the annual range of temperature obviously increases from west to east (Brest 10.1°C, Vienna 15.4°C), whereas in the south and the north, the range is less extreme (Nîmes 13.9°C, Schleswig 13.5°C).

In the Mediterranean, as already mentioned, there is a considerable change in the climatic regime from winter to summer, whereas the region of the North Sea and the Baltic usually remain within the influence of the westerlies. However, the northern region also experiences a certain annual variation of wind direction insofar as winds from the west are predominant in summer and from the northwest in winter, somewhat similar to the conditions in southern Europe but there is no thorough change in cloudiness and precipitation which is characteristic of the Mediterranean climate. This variation in surface flow is hardly noticeable at the 500-mbar level and therefore the detailed discussion of the annual wind variation is based upon observations taken on top of the Säntis at a height of 2,500 m above m.s.l. at 47°30'N at the northern edge of the Alps (Fig.2). In Fig.2 the frequency of wind directions is divided according to three main types: westerly wind (SW–WNW), föhn (SE–SSW) and bise (NE–ESE); the cases falling in between the Beaufort-force interval 1–4 were weighted once, those in the interval 5–8 twofold and those in the interval 9–12 threefold. Thus, the wind directions were related with the wind speed, the mean values of which have been plotted in the upper part of the figure as a smoothed curve.

There are two spells which are characterized by a lull in the westerlies; these spells occur at the end of May and at the end of September and separate summer and winter. About mid-June, a sudden onset and prevalence of west winds is typical; it represents

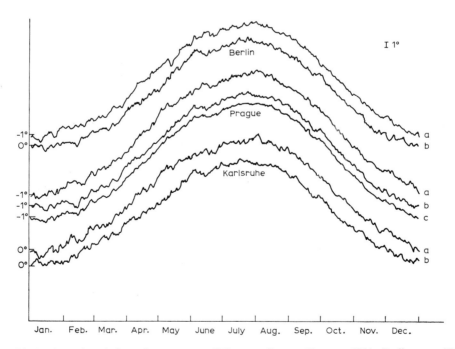

Fig.1. Annual variation of temperature (°C), according to FLOHN, 1954. Berlin: *a* = 1719–1837; *b* = 1848–1927. Prague: *a* = 1775–1839; *b* = 1840–1934; *c* = 1775–1934. Karlsruhe: *a* = 1779–1830; *b* = 1871–1920.

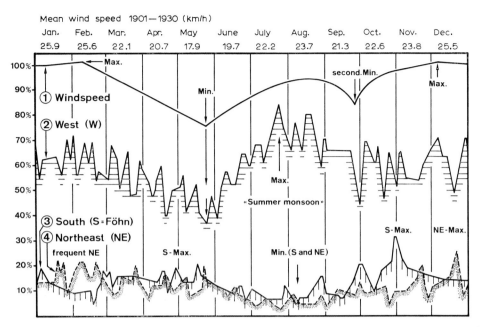

Mean wind speed 1901—1930 (km/h)

	Jan.	Feb.	Mar.	Apr.	May	June	July	Aug.	Sep.	Oct.	Nov.	Dec.
	25.9	25.6	22.1	20.7	17.9	19.7	22.2	23.7	21.3	22.6	23.8	25.5

Fig.2. Annual variation of wind regime on the Säntis, 1883–1940 (according to Scherhag, 1948).

the beginning of the so-called "European summer monsoon", a bad-weather singularity characterized by a drop in temperature (Fig.1); details of this phenomenon will be given in the section on "Monthly characteristics" (p.18).

The constancy[1] of winds in summer, after the onset of the summer monsoon, is much greater than in the winter months, especially in the southern regions (Payerne VII 70%); in the cold season of the year the frequency drops to values around 55%. The constancy of the winds is considerably less in the lower layers of the atmosphere, mainly and particularly so in places which lack any orographic channelization of the surface wind. This is especially true in the season when slack pressure gradients occur very frequently, i.e. in May and June. At Hannover the wind constancy at the 850-mbar level in that season is only about 40%, whereas in the other months it is similar to that of the free atmosphere, namely between 55 and 60%. Table III shows the seasonal distribution of the four main wind directions at the 500-mbar level. They can be related to three main types of circulation patterns, the basic features of which are outlined in Fig.3 according to advection type. The details are presented in the next section.

The meridional-circulation type has a great bearing on the weather regime of the winter season: depending on the longitudinal position of the ridges and troughs, there may be a severe winter in central Europe due to a persistent cold northerly flow (case I), or there may be abnormal winter warmth in central Europe due to a persistent warm southerly flow (case *II*). The statistics of Table III prove that case *I* is much more frequent in mid-winter than is case *II*, this being so because the cold land mass of the continent is often correlated with an upper low. Also in the remainder of the seasons, northerly upper winds occur with greater frequency at most stations than southerly

---

[1] See the definition in *World Survey of Climatology*, Vol.5, p.140: "It is convenient to add a parameter of dispersion to the vector mean wind: the most commonly used parameter is the 'constancy'. This is defined as the ratio, multiplied by 100, of the modulus of the vector mean wind to the mean speed of the wind. The latter quantity is the arithmetical mean of wind speeds regardless of direction."

upper winds. Furthermore, it is not surprising that the westerly zonal flow by far exceeds the easterly flow due to blocking. Certainly in winter the frequency of blocking situations amounts to a quarter of the frequency of the westerly zonal flow, but with only a twelfth in summer the amount of east winds is negligible.

The annual variation of atmospheric conditions

Classification of extended weather types

Before entering a discussion of the annual variation of the central-European climate the climatic components, in the form of the individual, extended weather types, must be explained.

The great interdiurnal variability of the weather in the temperate latitudes is well known; this interdiurnal variability could refer either to a simultaneous change of all the essential factors such as cloudiness, precipitation, humidity, temperature, wind and visibility or to a change of specific, single factors. Nevertheless, there are both short and extended spells which are found to have more or less uniform features so that a series of individual daily weather types may be merged into an extended weather type. In this connection and advancing from our discussion of the local and mesoscale conditions to the sub-continental central-European scale, BAUR's (1947) term "Grosswetterlage" (large-scale weather type) would be appropriate. Properly speaking, this expression should be worded: "large-scale extended weather type" (Grosswitterungslage), because both space and time are integrated. As for the time scale, the averaging covers a period of several days, the minimum duration of extended weather types in central Europe being 3 days and the mean duration $5\frac{1}{2}$ days, according to HESS and BREZOWSKY (1969).

These authors list 28 large-scale weather types; this large list has been condensed into a systematic skeleton scheme with a world-wide applicability. The weather types as well as the extended weather types can be divided into the following two basic categories.

(*a*) *Advective types* (Fig.3). The horizontal large-scale motions of the atmosphere are predominant, so that in flat terrain the vertical wind components are unimportant. However, the orography may add vertical components to this air flow when mountain barriers effect well marked upslope and lee phenomena. The advective type is frequent in higher latitudes, especially in the cold season, which is characterized by a rather

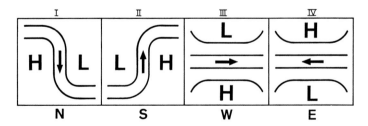

Fig.3. Schematic pressure pattern of the European weather map, central Europe being the central region of the map. Main circulation patterns with their types of advection: *I* and *II* = meridional circulation (upper trough); *III* = zonal circulation (subpolar trough–subtropical high-pressure belt); *IV* = blocking (subpolar high).

small temperature lapse rate and a great static stability in the atmosphere, so that vertical displacements are suppressed.

(*b*) *Convective types* (Fig.4). The vertical air motions predominantly influence the weather, either as a single effect or in connection with the effects of horizontal motions. The term convection is used here in a broad sense, not only with reference to the cellular convection in convective clouds in summer, which results in thunderstorm activity, but also including generally any lifting or subsidence of air during weather where the atmospheric motions are predominantly vertical. In central Europe, these occur mainly in the warm season, in lower latitudes, however, all the year round.

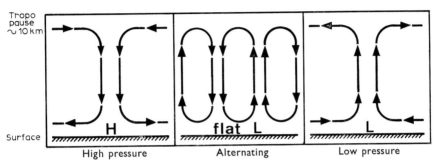

Fig.4. Flow pattern (vertical cross-sections of ageostrophic component) for convective weather types with marked vertical motions.

Both these two basic categories can be further divided into 4 and 3 extended weather types, respectively.

The basic criteria for the classification of the advective types (*a*) have been dealt with in the previous section (p.12). The classification is based upon the four main wind directions: north (for the weather pattern see Fig.12, p.23) and south (Fig.11, p.23) representing the meridional type of circulation; west (Fig.9, p.20) representing the zonal circulation; and east (Fig.10, p.20) representing the blocking situation.

In the case of vertical wind shear it is debatable whether the surface or the upper flow is the controlling factor. Baur as well as Hess and Brezowsky based their classifications mainly on the upper flow (500 mbar; compare Table III, p.9); however, in special cases, especially in the case of southeasterly surface flow, they had to account also for the low-level flow.

The basic criteria for the convective types (*b*) are the following. The general conditions for the occurrence of convective types (*b*) are small horizontal components of wind speed (less than about 15 knots) or strong winds of differing direction (cyclonic or anticyclonic curvature); then, the division into 3 extended weather types can be presented as in Fig.4. The classification of these 3 extended weather types of convective motion is based upon the sign of the vertical motion.

(*1*) Subsidence is predominant due to frictional divergence in a high-pressure system (anticyclonic flow resulting in cloud dissipation above the planetary boundary layer).

(*2*) Alternating small-scale updrafts and downdrafts due to terrain effects in the case of slack pressure gradients, i.e. weak horizontal flow (air-mass thunderstorms in summer).

(*3*) Ascent of air is predominant due to frictional convergence in a low-pressure system (cyclonic flow resulting in the formation of dense cloudiness and precipitation).

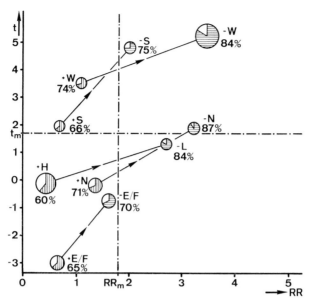

Fig.5. Temperature and precipitation characteristics for the seven basic extended weather types in winter (Dec.–Feb.) at Karlsruhe. For explanation see text.

Seasonal characteristics

The following analysis of the seasonal variation of weather in central Europe is based upon the above-described seven basic extended weather types, the evaluation yielding the frequencies with which these extended weather types occur as well as their effects on the temperature and precipitation regimes in central Europe.

Bürger's (1958) tabulations have been used for characterizing both the winter and the summer season in Figs.5 and 7, respectively, following Fliri's (1962) manner of representation. Each weather type is represented by an open circle in a rectangular coordinate system which has the precipitation amount RR (mm) as abscissa, and as ordinate the temperature t (°C). (The average temperature value of all weather types is t_m; RR_m is the average precipitation amount of all weather types.) The size of the circle, i.e. its surface area, represents the frequency with which the weather type under consideration occurs in the given season. The hatched portion of the circle represents the mean cloudiness similar to the basic symbols on the weather maps; in addition, the clouded portions of the sky have been given as percentages. The diagrams refer to the meteorological observations taken at Karlsruhe, from 1890 through 1950. Easterly flow and flat low situations have been combined (E and F), while according to the common practice at the German Weather Service (Deutscher Wetterdienst) (Hess and Brezowsky, 1969) the classification of the individual weather types has been divided into anticyclonic ($+$) and cyclonic ($-$) flow following the central-European weather conditions. Of course, cyclonic weather types have a high probability of precipitation and, in combination with the frequent westerly flow ($-W$), they have also mild temperatures. Cold and dry weather, however, are typical for anticyclonic easterly flow and flat low-pressure situations ($+E/F$) as well as high-pressure situations ($+H$).

Winter

In winter, the weather conditions in central Europe result from the interaction of three principal centres of action in northern latitudes which alternately influence the weather.

The Azores high directs mild, moist Atlantic air masses on its northern side towards the east.

The Icelandic low, with its rapid sequence of frontal passages, also steers alternating warm and cold air towards the east in the direction of the British Isles.

The climatological connotation of an "Icelandic low" (i.e., as it appears in the mean monthly synoptic charts) implies an extended weather type the main components of which are serial wave cyclones, often in the form of a "cyclone family", with a final outbreak of very cold arctic air from Greenland, Iceland or Spitsbergen entering central Europe by way of the North Sea.

The Siberian cold high is normally centred in Asia east of the Ural; it may happen, however, that a ridge extends towards the west into Scandinavia and central Europe thus blocking the joint westerly flow from the Azores high and the Icelandic low over western Europe, so that central Europe experiences a cold easterly flow.

The interaction between these three centres of action results normally in an alternation between zonal (westerly) and meridional (north–south) circulation, which in severe winters is interrupted by extended spells of blocking. In central Europe, the north component of the meridional flow is unusually pronounced, i.e. the related trough line is to the east of the area under consideration. In these spells, therefore, the temperature is below normal due to the air advection from the pole. In a blocking situation in midwinter the cold is further intensified as the Siberian high-pressure system becomes the source region of extremely cold air. On the other hand, in periods of persistent zonal circulation the temperature rises because the surface layer of cold air which forms due to nightly radiational cooling over the continent is forced away towards the east and replaced by mild maritime air which has been warmed from below by the relatively warm surface water of the ocean.

Cyclogenesis in the Mediterranean occurs when cold air from the polar region or northern Russia advances towards the Alps and enters the Mediterranean Basin by way of southern France; over the sea this cold air encounters subtropical air which even in winter is still rather warm. The channelization of the cold air through the corridor bounded by the Pyrenees on one hand and the southwestern Alps on the other and the propulsion of this northerly air through the lower Rhône Valley are favourable for the development of vortex disturbances at the interface between the cold and the warm air mass. This cyclogenesis is of dynamical origin and takes place in the vicinity of the Gulf of Genoa even before the cold air approaches (Chapter 4). In their progress eastward the Genoa cyclones enter the Adriatic Sea and not infrequently move towards the northeast and enter Hungary or Czechoslovakia. Following Van Bebber in the delineation of the depression tracks this would be a "Vb"-track which ordinarily implies heavy precipitation in eastern Europe. Nevertheless, the precipitation amounts differ due to influence of orography. The widespread precipitation areas of these Vb-depressions cover large parts of the Federal Republic of Germany and the Democratic Republic of Germany; however, they seldom affect the region west of the line formed by the Black

Forest via the Harz Mountains to the mouth of the Oder River; the dreadful summer floods of the Oder River are mostly due to disturbances of the Vb-type.

Spring and autumn

The two transitional seasons are dealt with jointly because they have many features in common. Nevertheless, there are decisive differences mainly in the vertical temperature distribution and the annual distribution of temperature due to the phase lag between the continent and the ocean (compare Fig.6A: Klagenfurt = continental; Emden = North Sea coastal area; Valentia, Ireland = Atlantic coast; North Sea = open sea).

The consequences of these differences are the following. The rather high sun in spring-time heats the land considerably thus effecting a steep vertical lapse rate, whereas the rather low sun in autumn does not compensate for the radiational cooling of the atmos-pheric layer near the ground so that a static stability develops. The mean annual varia-tion of the vertical lapse rate can be seen from Fig.6B (Vienna = continental; De Bilt = North Sea coastal area). Therefore, in spring the convective weather types are prevailing, i.e. characterized by low- or flat low-pressure systems due to the heating of the ground, whereas the frequent formation of high-pressure cells is an intrinsic feature of autumn often resulting in temperature inversions, characteristic of anticyclonic situations.

Spring and summer favour the formation of cumulus-type clouds which under the influence of increased heating turn into cumulonimbus and thunderstorms. Spring is the "Storm and Stress" period of the early year, whereas autumn gives the mellow calm

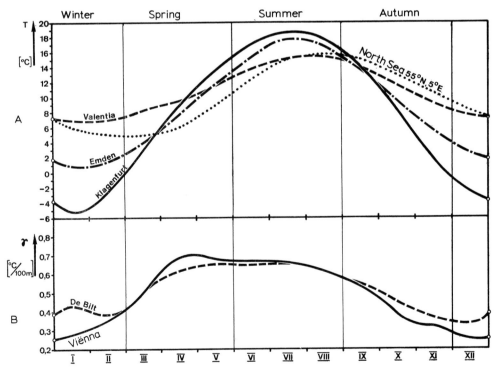

Fig.6. Mean annual march of: A. air temperature at Valentia, Emden and Klagenfurt, as well as surface-water temperature of the North Sea; B. environmental lapse rate.

of an Indian summer with the pleasant sunny and mild daylight hours after the dissipation of the frequent morning fog.

The phase lag between the annual temperature distribution over the ocean and over the land is not only observed in connection with the vertical lapse rate but also with the air and water temperature; this can be seen from the dotted curve in Fig.6A. The inland continentality generally results in almost equal temperatures in spring and autumn, whereas in the autumn season the oceanity is of particular importance to the coastal area: Fig.6 clearly shows the influence of the increasing distance from the sea shore.

After these comments on the differences of the climatic features of spring and autumn the following discussion will deal with their similarities. As far as changes of the basic centres of action are concerned, the Siberian high, which forms over Siberia in winter due to the winterly radiational cooling of the vast Eurasian continent, is weaker in both autumn and spring. The temperatures of the land and the sea approximate to each other. The subtropical high-pressure belt which in winter generally is restricted to the vicinity of the Azores, in spring and autumn frequently extends as a ridge over northern Spain to the Alps and farther into southern Russia; this synoptic pattern predominates around both mid-March and mid-October.

The Icelandic low-pressure system continues from winter into spring but it tends to weaken. Its splitting into two or more centres makes way for an extension of the Azores high towards the north in late spring, often resulting in a blocking situation. This "block" does not remain stationary over Scandinavia as it is generally the case in winter; in spring the block persists over the eastern North Atlantic so that on its eastern margin cold arctic air is steered into central Europe. This is the time of springtime cold-air outbreaks; from the North Sea coast to the Alps, exposed sites are endangered by the potential occurrence of late frost. The last third of October and all of November also experience this type of cold-air outbreaks.

In addition to the characteristic feature of these cold-air outbreaks spring and autumn offer a variety of other weather patterns. The zonal circulation with winds from westerly direction predominates both in April and November; April showers and November gales are well-known phenomena. When a series of wave cyclones travels eastward, the first strong cyclone generally initiates southerly flow over central Europe giving rise to the south föhn in the Alpine valleys. This weather fury is a typical symptom associated with the spring and autumn months. In spring, it melts the snow; in autumn it gives the full-flavoured ripeness to the grapes in the vineyards in the valleys. Its occurrence is irregular: sometimes it fails to appear for months, in other years it is frequent; thus it will continue to be a problem to the forecasters in the years to come.

Summer

When the polar area experiences 24 hours of sunshine, the horizontal temperature gradient between north and south weakens. The Icelandic low-pressure system lacks in intensity and the westerly flow over the Atlantic calms down. Fig.2 (p.11) shows that about the end of May there is a minimum in the frequency of occurrence of westerly winds as well as in the general wind speed, due to the predominance of weak pressure gradients. The sporadic occurrence of flat highs or lows inhibits the development of

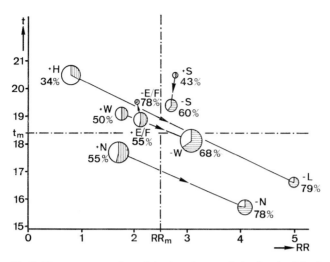

Fig.7. Temperature and precipitation characteristics for the 7 basic extended weather types in summer (June–Aug.) at Karlsruhe. The temperature regime of the cyclonic westerly flow ($-W$) has changed from winter to summer insofar as the relatively high temperatures of winter have dropped to just medium temperatures in summer. Now, the low-pressure situations ($-L$) and the cyclonic northerly flow ($-N$) are damp and cool, whereas the high-pressure situations ($+H$) meet the expectation of being warm.

any large-scale flow. The heating of the continent turns the weak south–north temperature gradient into a heat decline from the warm continent to the cool ocean. The weather development is comparable to the process in an electric circuit equipped with a generator and a spark gap. The generator gradually builds up an electric potential without an instantaneous effect. However, as soon as the electric field strength reaches a critical limit, sparking occurs; the electric current results in the compensation of voltage, and a fresh start can be made. In figurative use with regard to the weather, the temperature gradient between the continent and the ocean represents the generator and the inertia of the air represents the spark gap. From time to time, generally in intervals of one to two weeks, there is an intrusion of cool maritime air into central Europe, often accompanied by thunderstorm activity. Then, another heating period in the continental area begins. Through this process the zonal flow is more or less regularly restored in the summer months (see Fig.2). However, extended spells of heat or of unsettled weather are not to be expected as indicated in Fig.7 where the prevailing westerly flow is characterized by a broken cloud cover. Nevertheless, in individual years, a ridge of the Azores high-pressure system extends into central Europe resulting in subsidence and—at least in the southern parts of central Europe—dry and warm weather (high-pressure situations in Fig.7). On other occasions, the high does not extend eastwards, but towards the north in the direction of the British Isles and Scandinavia so that the northern parts of central Europe enjoy fair weather while the southern parts suffer from disturbances. The saying "the exception proves the rule" can be applied in particular to the field of meteorology.

Monthly characteristics (December through November)

The normal annual weather distribution has been described in detail for central Europe (FLOHN, 1954). However, this paper is intended to outline the basic features without presenting the exact sequence of the extended weather types, since "normal" weather will hardly last for some months in a row.

It is thought appropriate to begin this round trip through the year with the beginning of winter in December. At that time in most years the area of central Europe, with the exception of the Alps, is not yet clad in snow, while any low ground is covered with wide-spread ground fog or low stratus for days or even weeks. In the second third of December, the development of a high-pressure situation (pre-winter) often occurs as shown in Fig.8, resulting in a noteworthy temperature drop especially in the lowlands. Towards the end of the year this settled weather is often interrupted by a spell of wester-lies with warm air advection (Fig.9). The lowlands experience thaw (Christmas low) whereas in the Alps—often just before the winter-sport fans arrive for their Christmas vacation—at heights greater than 1,000 m above m.s.l., it continuously snows.

At a later time—especially around the 20th of January (mid-winter)—the high-pressure regime prevails again; this spell ordinarily brings the temperature minima of the winter season. A secondary frequency maximum of anticyclonic weather occurs in the last third of February (late winter). In between, a variety of individual deep westerly cyclone waves or short-term föhn situations (southerly flow), may occur. However, in mid-winter the föhn is not strong enough to force the surface layer of cold air out of the Alpine foreland so that the warm southerly flow is restricted to a partial area of the Alps.

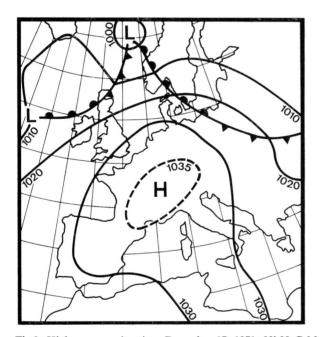

Fig.8. High-pressure situation, December 17, 1971, 00h00 G.M.T. Dry pre-winter type weather with sunshine in the highlands; high mountains mild, lowlands cold.

Northeasterly and easterly flow reach their maximum frequency of occurrence in February, for the wide-spread snowcover, which in the second half of January covers most of eastern, central and northern Europe, favours the formation of high-pressure centres over Scandinavia and Finland. Fig.10 gives an example of such a pressure pattern with severe cold-air advection from the northeast; the extremely low temperature minima during this synoptic situation often represent the long-term record temperature values, e.g., at Berlin −26.0°C on 11 Feb. 1929, at Frankfurt on the Main −21.5°C on 12 Feb. 1929, at Hüll (lower Bavaria) −37.8°C on 12 Feb. 1929 (the absolute temperature minimum of the Federal Republic of Germany until 1973), at Zürich −24.2°C on 12 Feb. 1929 and in Stift Zwettl (lower Austria) −36.6°C on 11 Feb. 1929.

Strong regional air currents often originate from this pressure pattern. North of the

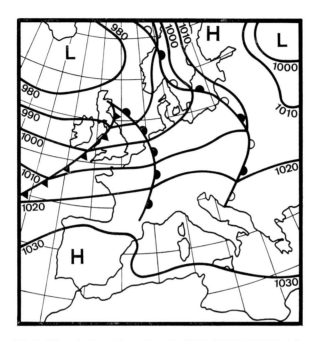

Fig.9. Westerly flow, December 24, 1967, 06h00 G.M.T. Advection of mild maritime air into central Europe; unsettled weather giving precipitation; strong mid-latitude westerlies.

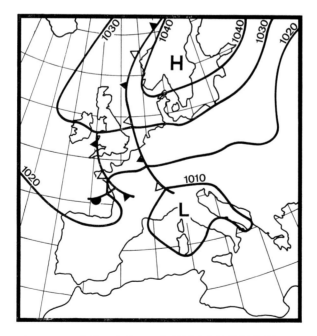

Fig.10. Easterly flow, February 9, 1956, 06h00 G.M.T. The strong high over Scandinavia steers extremely cold air from northern Russia into central Europe. Onset of an extended cold spell.

Alpine bow, in the gap between the Jura and the Alps, the "bise" may give rise to gale-force winds from the northeast, while outside of central Europe the "mistral" blows in the Rhône Valley and the "bora", a cold fall wind, sweeps Dalmatia.

However, winter is not predominantly characterized by strong cold advection from eastern Europe; it likewise brings gales from west and northwest when the vigorous Atlantic depressions cross the British Isles and the North Sea and propagate into Scandinavia (see Fig.9). Fresh developing secondary cyclones forming along the southern margin of a large primary cyclone may create extremely high peak gusts; e.g., SCHERHAG (1948) has analysed the hurricane-force wind occurrence of 27 October 1936: In the free atmosphere, only a few hundreds of metres above the ground, gales of about 190 km/h roared and in the North Sea high waves sank the lightship "Elbe I", while at the same time, there was spring tide to crown it all. This harmful coincidence of hurricane-force winds and spring tide is again and again a severe danger to the North Sea coast line because it is not possible to dike all sections adequately. SCHERHAG (1966) has also analysed the synoptic situation of a similar catastrophe which occurred on 16/17 February 1962 (see p. 29). Since northwest Germany often lies directly in the track of Atlantic storms, this type of severe weather may be expected repeatedly. On 13 November 1972 the atmospheric sea-level pressure at Hamburg dropped to the extremely low value of 955 mbar; although the fortunate coincidence of low tide prevented an extreme disaster, nevertheless, the storm caused not only very great damage but also the loss of 40 lives. SUSSEBACH (1968) has made a synoptic and statistical investigation of depressions accompanied by gale-force winds in central and eastern Europe; he found that each storm had individual features of its own whether it be an Atlantic storm crossing northern Germany or an Alpine föhn raging in the southern part of the area under consideration (see p. 22 and FREY, 1954).

In March, the weather generally settles down a little: the spell of bad weather, which frequently occurs in the first third of the month, is followed by the previously mentioned high-pressure situation about mid-month (early spring). This high does not have such strong marginal pressure gradients as the winterly anticyclones (Fig. 8), nevertheless, it strongly affects the daily weather. Since meanwhile the solar elevation has increased, the valleys in this situation have no longer periods of wide-spread ground fog or low stratus as earlier in winter anticyclonic weather. In the lowland, spring sets in with mild warm days; however, the nights are still cool. In the transitional months the diurnal temperature range almost reaches the same values as in summer. This is quite in contrast to the conditions in wide high valleys which in winter form strong surface inversions over the snowcover. There, winter may experience even higher diurnal temperature ranges than summer (see Table IV).

In April the weather intensifies again offering a variety of extended weather types such as the westerly flow followed by a cold air intrusion from the northwest with a rapid interchange of snow or rain showers on the one hand and sunny spells on the other hand (so-called "April weather"). For a change there may be short high-pressure spells or even an easterly flow and occasionally some early flat lows with spring thundershowers.

Furthermore, a characteristic feature is the southerly flow which occurs most frequently at this time of the year. Warm Mediterranean air is forced northward across the Alps resulting in the first fine warm spells; sometimes towards the end of March even a

TABLE IV

MEAN DIURNAL TEMPERATURE RANGE[1]

	Jan.	Feb.	Mar.	Apr.	May	June	July	Aug.	Sep.	Oct.	Nov.	Dec.
Bremen[2]	4.8	5.9	7.9	9.6	10.6	10.6	9.7	10.0	9.5	7.5	5.2	4.7
Munich[2]	7.0	8.5	10.2	10.7	11.4	11.3	11.1	11.3	11.2	9.6	6.7	6.1
Bever[3]	13.2	13.6	12.6	10.9	11.5	12.3	13.3	13.0	13.7	13.3	10.8	12.4

[1] Difference between mean max. temperature and mean min. temperature (°C).
[2] 1931–1960.
[3] Swiss high valley, 1,700 m above m.s.l., 1959–1968.

so-called summer day is experienced, i.e. a day with a temperature maximum of 25°C or more.

In individual years, the southerly flow may be so extremely strong that large amounts of dust are raised in north Africa and transported to central Europe. Over the northern lowlands the suspended dust particles merge with the natural and artificial air pollution; in the high mountains, however, the dustfall may be noticeable because the snow is given a reddish tint by the "blood rain". The firn fields melt more quickly in these years due to increased heat absorption. The northerly and the southerly flow greatly differ in skycolour: the pure Polar air during "April weather" gives the sky a deep blueness and an excellent visibility, whereas—except in the "föhn" valleys at the north of the Alps— the air flow from low latitudes enhances the turbidity and gives a whitish tint to the sky, especially when a considerable amount of dust is suspended in the air. The month that shows the greatest day-to-day temperature rise of the year is May. In this month, there is a great frequency of occurrence of slack pressure gradients, partly of thundery type, and partly of predominantly anticyclonic type. The high sun, one month before its climax, often causes such a rapid temperature rise over the land that the sea surface is considerably colder than the adjacent land (see Fig.6). This temperature difference due to the unequal heating of land and sea causes sudden invasions of cold air in western and central Europe (see Fig.12). Since there is almost no terrain obstacle that bars the cold air flowing from the North Sea into central Europe, destructive frosts have in individual years been observed in May south of the Alpine region. These cold spells certainly do not occur on a specific calendar date. It is an oversimplification for instance to associate the "ice saints" with May frosts, i.e. St. Pancras on 12 May, St. Gervais on 13 May, St. Boniface on 14 May and the "cold Sophia" on 15 May. The fact is that these spring frosts are widely scattered in individual years as shown in Fig.13.

Fig.1 (p.10) clearly shows a sudden discontinuity in the annual temperature rise: in early June, the temperature curve begins to flatten and, around the middle of the month, the mean temperature even drops. In contrast to the May frosts, this change of flow type occurs so frequently that it affects the seasonal trend of average temperature; this feature is the so-called "sheep cold" which, strangely to say, characterizes the onset of summer with the previously mentioned alternation of continental heating and cold maritime air intrusion. This is the period of the "European monsoon" the typical pressure pattern of which is shown in Fig.14.

The characteristic feature of the weather maps of mid-June is a low-pressure system over central Europe. These summer cyclones generally have weaker pressure gradients

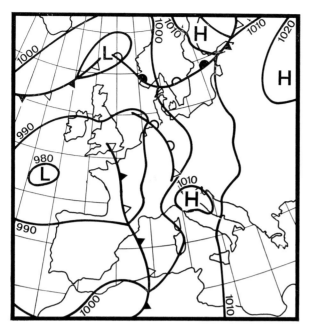

Fig.11. Southerly flow (föhn in Alpine valleys), April 10, 1963, 00h00 G.M.T. Warm Mediterranean air is forced across the Alps. The south slope of the Alps receives excessive orographic precipitation, whereas the northern lee side is characterized by the dry rain shadow.

Fig.12. Northerly flow, April 18, 1969, 00h00 G.M.T. Cold arctic air from the Polar area crosses the North Sea, advances to central Europe and penetrates into the Mediterranean resulting in a strong up-slope effect to the north of the Alpine barrier whereas the Alpine south slope is under the influence of a northerly föhn which mitigates the cold snap.

than the winterly depressions (see Fig.9), however, strong effects may be felt due to increased precipitation associated with the fronts, i.e. the transition zones between the warm and cold air masses, as well as due to convective motion of the air often accompanied by thunderstorm activity. In addition to this, the frontal zones of early summer

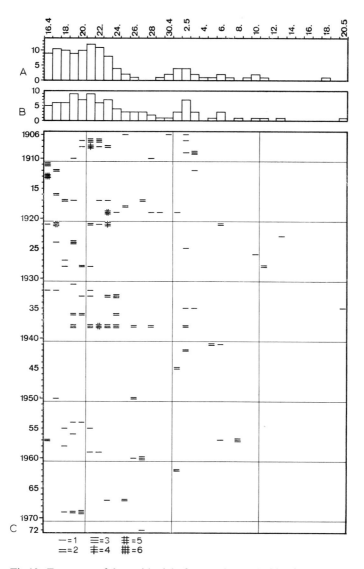

Fig.13. Frequency of days with night-frost at the north side of the Alps from April 16 through May 20: A. number of days with night-frost at Geneva (period of record: 1858–65 and 1874–1945); B. number of days with night-frost at Hallau near Schaffhausen, Switzerland (period of record: 1906–72); C. frequency of days with night-frost from year to year at Hallau (1906–72); the symbols indicate the various morning temperature minima: *1* = from −0.1°C through −0.9°C; *2* = from −1.0°C through −1.9°C; *3* = from −2.0°C through −2.9°C; *4* = from −3.0°C through −3.9°C; *5* = from −4.0°C through −4.9°C; *6* = from −5.0°C through −5.9°C. The frequency of night-frost undergoes a rapid decrease in the last third of April. However, there is a secondary maximum in the first days of May, whereas there is obviously no recurrence tendency during the so-called "ice saints" (12 through 14 May).

move more slowly than those of the depressions in winter, and they are associated not only with heavy precipitation on the south side of the Alps as for instance in connection with south föhn, but also with sporadic downpours over flat terrains without orographic rises as can be seen in Table V.

The monsoon influence gradually weakens, as the returns of the westerlies associated with the mean calendar dates 13 June, 27 June, 15 July, and 15 August, become less intense, because the difference between the air temperature and the surface-water temperature has decreased over the ocean.

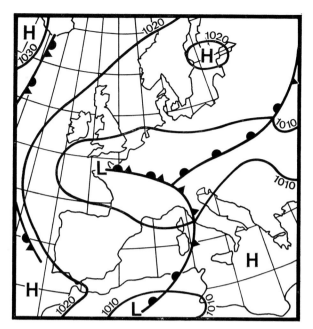

Fig.14. Low, June 10, 1971, 00h00 G.M.T. An extensive low has formed over western and central Europe due to the interaction of cold air advancing from the North Sea towards Iberia and warm summerly air of southeast Europe. The center of the low is characterized by dense cloudiness and heavy precipitation, in summer often thunderstorms. The strong gusts of wind of these summerly thunderstorms usually are of short duration in contrast to the winterly ones (Fig.9).

TABLE V

FREQUENCY OF DAILY AMOUNTS OF PRECIPITATION OF 60 MM OR MORE (Number of cases per centenary)

	I	II	III	IV	V	VI	VII	VIII	IX	X	XI	XII	Year
Hamburg[1]	—	—	—	—	—	2	1	—	1	1	—	—	5
Frankfurt[2]	—	—	—	—	—	2	2	3	—	—	—	—	7
Munich[3]	—	—	—	1	5	4	6	6	4	—	—	—	26
Zürich[4]	—	—	—	—	1	5	3	4	1	—	—	—	14
Lugano[5]	4	6	8	24	38	38	42	46	30	36	38	12	322
Vienna[6]	—	1.5	—	—	3	3	7	1.5	1.5	1.5	—	—	19
Salzburg[6]	3	—	—	—	5.5	5.5	14.5	11.5	4.5	4.5	3	—	49
Klagenfurt[6]	1.5	—	1.5	3	3	3	1.5	5.5	4.5	4.5	3	—	31

[1] German Weather Service archives (95 years).
[2] After Mollwo: Ber. des Deutschen Wetterdienstes, Nr. 43.
[3] German Weather Service archives.
[4], [5] Swiss Weather Service archives. Zürich: period 1864–1963; Lugano: period 1921–1970, reduced to 100 years.
[6] Austrian Weather Service archives, period 1901–1970, reduced to 100 years.

It can be inferred from long-year records that the temperature climax occurs around 20 July, i.e. about one month after the solstice (in analogy with the temperature minimum in winter). The absolute maximum of 40°C may be reached in isolated places only in the area under consideration during extremely hot summers as it can be seen from Table VI. The littoral seldom experiences temperature maxima above 32°C, whereas southeast Europe is much hotter.

Except for the coastline and the offshore islands, July generally does not only bring the annual temperature maximum but also the annual peak of thunderstorm activity. Then, the weather map shows the slack pressure gradients (Fig.15) typical for the warm season, resembling the pressure pattern in the tropics which hold the world's peak frequency of thunderstorm days. Thunderstorms not seldom are favourable to hail formation.

August does not bring any great change, yet the thunderstorm activity becomes less intense and less frequent. The stage of early autumn stabilization starts. Until about 10 September, the temperature continues to reach almost summer values (Fig.1, p.10) followed by a stronger drop at the end of high summer, however, without any particular date of transition. The transition from summer to autumn is gradual with the intermediate phase of the previously mentioned "Indian summer" which has features of both the foregoing as well as the forthcoming season.

TABLE VI

ABSOLUTE TEMPERATURE MAXIMA (°C)

Location	Date	Maximum	Period
Hamburg	July 2, 1952	37.7	1931–1970
Frankfurt	June 27, 1947	38.2	1881–1970
Munich	July 18, 1920	36.2	1881–1970
Vienna	July 14, 1832	38.7	1830–1960
Basel	July 2, 1952	39.0	1864–1973
Neustadt/Weinstrasse	July 2, 1952	39.6*	1925–1970

* Absolute temperature maximum of the Federal Republic of Germany until 1973.

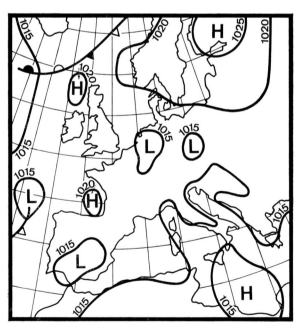

Fig.15. Flat low, July 12, 1955, 06h00 G.M.T. Summerly pressure pattern with low wind speed, cloudiness changing greatly with time and space and scattered thunderstorms with rain. Such slack pressure gradients are predominant in the warm season.

The pressure gradients in the early autumn sequence of anticyclones (Fig.16) are less in magnitude than those of the high-pressure situations found at the beginning of this round trip through the year. The autumn anticyclones are similar to the high pressure spells about mid-March although the frequency with which early morning fog occurs is greater, reaching a peak about mid-October. The lull of winds during this high-pressure period is followed immediately by a fairly sudden increase to high-wind speeds in winter, especially in the high mountains of the Alps and at the crest of the Black Forest (see Table VII).

Thus, this idyllic autumn spell normally comes to an abrupt end in the last third of October. The Atlantic westerlies push eastwards over Europe with new vigour, because the production of cold air in the Arctic Zone is resumed during the polar night. Frequently, the Atlantic depressions do not yet penetrate into central Europe, thus effecting southerly flow on their eastern flank in the central European area. This implies that the frequency with which the south föhn situations occur is increased; since the air temperature over the Mediterranean is still rather high, this rising air is laden with moisture resulting in excessive amounts of precipitation at the south slope of the Alps. Table VIII lists the maxima of daily precipitation amounts which have been measured in Germany, Austria and Switzerland. Local precipitation amounts, especially at the south slope of the Alps, can very well surpass these maxima. According to LAUSCHER (1971), 600–670 mm rain fell within 3 h in the Stifting Valley near Graz, Austria, on 16 July 1913, and 650 mm rain were reported to have fallen within 2 h at Schaueregg on 10 August 1916.

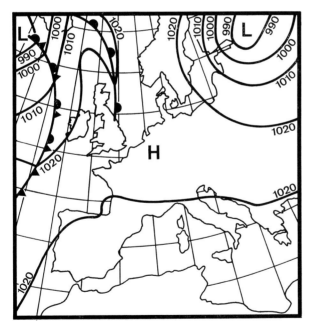

Fig.16. Early autumn high-pressure situation, September 24, 1970, 00h00 G.M.T. In contrast to the winterly high-pressure situation (Fig.8) the autumn pressure pattern is characterized by more pronounced zonal flow and weaker pressure gradients. This implies a settled weather period with early-morning fog in the valleys but predominantly sunny days and prolonged, rather high midday temperatures. The comparison of Fig.8 and Fig.16 clearly shows the seasonal variation of one of the 7 main weather types. Although each type always shows its typical circulation pattern, the resulting weather is decisively modified by the radiation balance of the season and the difference in the radiative properties of sea and land.

TABLE VII

MEAN ANNUAL VARIATION OF WIND SPEED (m/sec), 1951–1960

Station	Height (m above m.s.l.)	Jan.	Feb.	Mar.	Apr.	May	June	July	Aug.	Sep.	Oct.	Nov.	Dec.	Annual
Zugspitze	2,960	9.0	8.6	6.5	5.6	5.2	4.9	4.9	5.6	6.3	6.2	6.9	8.6	6.5
Wendelstein	1,735	6.2	5.7	5.2	4.4	4.4	4.0	4.1	4.0	4.0	4.2	4.4	5.5	4.7
Feldberg (Schwarzwald) (Black Forest)	1,486	10.8	9.9	8.1	7.7	6.9	6.1	6.6	7.0	7.2	7.8	8.5	9.8	8.0
Gr. Falkenstein	1,307	4.8	4.6	4.5	5.1	4.8	4.6	4.6	4.6	4.7	4.5	4.1	4.4	4.6
Hohenpeissenberg	977	5.2	4.9	4.6	4.3	4.1	4.0	4.2	4.1	4.3	4.1	3.9	4.8	4.4
Klippeneck	973	4.3	4.3	4.4	4.1	3.8	3.2	3.4	3.6	4.1	3.9	3.6	3.9	3.9
Wasserkuppe	921	6.8	6.1	5.7	5.1	4.3	4.2	4.6	4.7	5.1	5.5	5.5	5.9	5.3
Kl. Feldberg (Taunus)	805	7.0	6.4	6.0	5.8	5.3	5.0	5.1	4.7	5.7	5.9	6.3	6.3	5.8
Stötten	734	5.1	4.6	4.5	4.1	3.8	3.2	3.3	3.2	4.0	3.8	4.0	4.5	4.0
Gütsch*1 (Gotthard)	2,284	6.4	6.8	6.0	5.9	5.4	4.7	4.3	4.7	5.3	5.7	7.3	6.5	5.7
Säntis*2	2,500	7.2	7.0	6.1	5.8	5.0	5.5	6.2	6.6	5.9	6.3	6.6	7.1	6.3
Sonnblick*3	3,106	8.1	8.1	7.5	6.7	6.1	6.0	6.1	6.5	6.4	7.2	7.3	7.6	7.0
Plateau Rosà*1	3,488	8.6	8.4	7.6	7.0	6.5	5.1	5.0	5.4	5.3	5.4	7.9	8.1	6.7

*1 1961–1970.
*2 1901–1930.
*3 1901–1914 and 1927–1936.

TABLE VIII

MAXIMUM PRECIPITATION (mm/day)

Location	Date	Precip.
Bad Reichenhall (Germany)	Sept. 13, 1899*	222
Semmering (Lower Austria)	June 5, 1947	323
Dornbirn (Vorarlberg, Austria)	Aug. 31, 1910	336
Mosogno (Ticino, Switzerland)	Sept. 24, 1924	359

* The calendar date is somewhat doubtful.

The far-reaching penetration of the Atlantic depressions, which becomes more common in November, results in the wind over the continent changing direction from south to west. During these spells of gale-force winds in November the trees are stripped of their leaves. Later, central Europe is often affected by calm anticyclonic periods, bringing fog or low stratus and forming the transition to the starting point of this round trip through the year: the beginning of winter in December.

Climatic provinces

Details may be drawn from the References at the end of the chapter (see the subsections). For the area of central Europe under consideration, 4 main climatic provinces can be defined by their geographical latitude, their distance from the ocean and their altitude:

(*1*) the lowlands north of the central European highlands; (*2*) the German highlands and the Jura Mountains; (*3*) the northern Alpine foreland; (*4*) the Alps.

The lowlands north of the central-European highlands

It was thought unsuitable to deal with the lowland conditions in the territory of the Federal Republic of Germany and that of the Democratic Republic of Germany separately, because each of the climatic elements undergoes a gradual change from the predominant maritime influence in the lowland of the Federal Republic of Germany to the more continental features of the interior plain of the Democratic Republic of Germany.

The nearness of the North Sea and the Baltic attributes genuine maritime features to the northern and western parts of the lowlands, whereas the interior clearly is subject to increasing continentality as one proceeds eastwards. Though the northern lowlands hold numerous isolated hilly terrain features, the air, which invades the area from westerly, northerly or easterly directions, can almost freely penetrate towards the northern edge of the central-European highlands; this first major barrier shows up very well in the regional distribution of precipitation.

Since surface friction increases with surface roughness, the wind sweeps the lowlands more easily than the sheltered inner part of the highlands; in addition, the nearness of the northern cyclone tracks results in a much more frequent occurrence of gale-force winds, which reach their peak frequency in winter. Mention should be made here of the threat of tidal waves along the North Sea coast line which, in contrast to the Baltic coast line, is subject to the tides. A typical example is the flood disaster which occurred at Hamburg on 16/17 February 1962: maximum gusts of more than 40 m/sec coincided with a high tide which resulted in a flood-tide exceeding the mean high water by more than 4 m. It was the most severe storm surge Hamburg suffered since 1825.

The well marked predominance of westerly winds throughout the year is interrupted in March. During high-pressure spells over the continental part of Europe, which in early spring often are of rather long persistence, dry continental air flows rather quickly over the flat terrain from east to west. Then, the dry, sandy soil suffers from wind erosion, which may have detrimental effects on the winter grain. This may also lead to detrimental effects upon the economic activity. Persistent easterly surface flow results for instance in persistent low water in the German Bight so that only small ships can call at the great harbours of the estuaries, especially at Bremen and Hamburg.

During spells of fair weather, the influence of solar radiation by day and radiation to the sky at night causes land- and sea-breezes along the entire shore line of the North Sea and the Baltic. However, this cycle of diurnal local winds only affects the immediate coastal strip.

The oceanity of climate depends firstly on the thermal behaviour of the ocean. The heat capacity of sea water is almost double that of soil. Due to the turbulent water flow the heat is absorbed by a rather great mass of water resulting in a phase lag of the annual temperature extremes of surface water and air (see Fig.6 on p.16). Thus, the islands and the immediate coastal strip experience the mean maximum of air temperature as late as in August and the minimum in February. In addition, the retardating influence of the water masses levels off the mean diurnal range of air temperature in the coastal area. This unequal heating and cooling of adjacent land and water surfaces results in an excess of heat at the coast line in autumn and a deficit of heat in spring in comparison with the interior of the continent. This effect rapidly diminishes towards the inland and towards eastern Europe with decreasing distance from the sea. In winter, the coastal climate is modified by the ice conditions of the coastal waters; this is valid in particular for the Baltic Sea because it is not infrequently covered with ice. The formation and duration of the ice cover are not only functions of the meteorological parameters; they are also strongly influenced by the hydrographic and morphological conditions of the coastal area.

The increase of the annual range of air temperature towards the east, which is observed in the lowlands, is a measure of the increasing continentality. The main axis of increasing continentality runs from northwest towards southeast with an increase of the mean annual temperature range of 3.5 degrees, from 15.4°C at Jever (Federal Republic of Germany) to 18.9°C at Cottbus (Democratic Republic of Germany). Even along the coast line this increasing influence of continentality is felt: the mean annual temperature range is enhanced by 1.9 degrees from Jever to the Baltic sea-side place Wustrow (Democratic Republic of Germany) which has a mean range of 17.3°C.

However, the degree of continental heating in summer east of the Elbe River is not comparable with the degree of maritime mitigating in winter west of it; thus, the maritime influence in the western part of the lowlands is more powerful than the continental influence in the eastern part. The physical explanation of the regional temperature distribution (Figs.17, 18, pp.43, 44) must take into account the thermal conductivity of the soil; otherwise it would be hard to believe that by far the greatest risk of destructive frost in fall as well as in spring is a characteristic feature of some specific areas in the maritime western part, namely in the Lüneburger Heide and the peat-bogs at both sides of the Ems River.

The onshore sea air entering the continent is piled up as it is frictionally retarded by the land surface (coastal convergence) which results in the precipitation amount increasing in a coastal strip between about 10 km and 30 km inland (coastal lifting). This effect is not restricted to onshore flow; coastal convergence may occur in a surface flow which is parallel to the coast if, looking downstream, the land is to the right. Contrariwise, over the west coast of Schleswig-Holstein a surface flow from the north causes horizontal divergence of the airflow resulting in a lack of precipitation. Over the east coast of Schleswig-Holstein (Baltic Sea), however, a surface flow from the north causes horizontal convergence of the airflow resulting in increased precipitation over the coastal area. The predominance of westerly flow which most often brings precipitation to the lowlands, exerts a marked lee effect on the Baltic coast of Schleswig-Holstein as the surface flow is diverging when the air propagates from the land to the water surface of the Baltic due to the decrease of surface friction. This explains the generally low amounts of precipitation along the Baltic coast of Schleswig-Holstein. The relatively low hills in the lowlands, however, clearly show in the regional distribution of precipitation an increase in the precipitation amount: good examples for this orographic effect are the Lüneburger Heide (Federal Republic of Germany) and the Fläming upland (Democratic Republic of Germany) both of which have a mean annual precipitation surpassing that of their surroundings by 50 to 100 mm.

The seasonal variation of rainfall is characterized by a spring minimum in maritime western Europe. The probability for the occurrence of the smallest monthly precipitation amounts of the year is greatest in March (20% to 35% of all cases) and in February (10% to 25% of all cases); the March "dryness" is due to the increased frequency of intrusion of continental air in this period of the year. The probability for the occurrence of the highest monthly precipitation amounts of the year is greatest in August (15% to 35% of all cases) and in July (10% to 30% of all cases). There is also the remarkable feature of a fall maximum along the North Sea coast. The probability for the occurrence of the greatest monthly precipitation amount of the year is rather high in October (15% to 25% of all cases).

Snow cover generally is no major problem in the northern lowlands. Towards eastern Europe the duration of the snow cover gradually increases with the gradual lowering of winter temperatures, whereas in the west of this climatic province snow cover is an intermittent phenomenon due to the frequent occurrence of thaw.

The differences in the seasonal variation of temperature and moisture between land and sea affects the occurrence of fog: the maximum probability for fog over the continent is to be expected in autumn due to radiational cooling at night, whereas the coastal area has a main maximum of fog frequency in December and a secondary one in March; the latter is especially well marked over the open water of the North Sea and the Baltic, because in that month the temperature difference between the relatively warm water and the cold air flowing over it reaches a maximum.

The climate of the North Sea and Baltic coastal regions offers strong to moderate meteorological stimuli; the thermal effects are due to an increase in cooling power, mainly because of increased wind speed, while the radiation effects are due to the intensity of solar and sky radiation. Farther inland, there is a decline to mild, then weak and finally soothing meteorological stimuli due to the decrease of cooling power and solar as well as sky radiant intensity from the coast towards the interior. The lowlands north of the central-European highlands may very well experience spells of depressing human bioclimate due to an increase in heat load (frequent sultriness and high summer temperature) or due to the combination of damp and cold which occurs during winter in stagnant air, such as mist or fog over low ground; the latter implies a reduction of incoming radiation due to fog, and partly due to smog, particularly in the Ruhr Basin and the Bight of Cologne.

The German highlands and the Jura Mountains

General climatic features

The highlands south of the northern lowlands experience a decline of the maritime influence towards the south, while the influence of high-pressure conditions is felt more and more, especially in winter. Orographic effects are very well pronounced in this mountainous terrain, because the intrinsic climatic elements show a well marked relationship with height.

The entire region of this climatic province is subdivided by the Main River Valley generally running from east to west; however, it seems appropriate to begin with the discussion of those climatic features which are common to the entire climatic province concerned.

The general decrease of air temperature with height shows a seasonal variation: the vertical lapse rate is greatest in summer; therefore, during that time the temperature differences between the uplands and the valleys are most evident (see Figs. 17, 18, pp. 43, 44). In winter the vertical lapse rate on the average is at a minimum, because valleys, basins and troughs, due to radiational cooling, are often filled with stagnant cold air, the upper limit of which coincides with the top of the ground fog layer so that the uplands are considerably warmer due to bright solar radiation. These ground inversions in low valleys aggravate the problem of air pollution in industrial centres, because they prevent the exchange of air in these basins, which generally have low wind speeds anyway. Thus, the crests of the highlands are characterized by much smaller diurnal and annual temperature ranges; their situation in this respect resembles that of the ocean.

The individual mountain ranges are well marked by precipitation amounts which surpass those of their surroundings due to the upslope effect on their western slopes. This precipitation which falls from orographic clouds may extend for quite some distance windward of the base of the mountain barrier in accordance with the orientation of the mountain range. Because of the predominance of southwesterly flow those mountain ranges which are oriented from southwest to northeast are less subject to this upwind effect and do therefore not receive much surplus rain. The eastern slopes of the mountains are the lee sides which are characterized by dry rain shadow and little cloudiness. At the same level, there is a general decline of precipitation from north towards south as well as from west to east.

The increase of precipitation amounts with increasing height is greatest in winter, because then the upslope effects are most intense due to the lowering of the level at which lifting condensation occurs. In summer, convective precipitation, often accompanied by thunderstorms and hail, reduces these differences between valleys and crests, in individual cases even resulting in an inversion of precipitation amount. Of course, the mean precipitation charts nevertheless show the general increase of precipitation with height. The seasonal variation is characterized by a summer maximum of precipitation amount in low levels and a winter maximum on the mountain ridges. The general increase of snow-cover depth with increasing height renders winter sports possible, though the persistence of the snow cover does not bear comparison with that of the snow cover in the Alps.

In case of high-pressure situations, the highlands are favourable for the development of local winds such as the mountain and valley winds. Unfortunately, only a few of these systems of diurnal winds are well known, since they may play an important role in the air purification especially in densely populated areas (fresh air channelization).

The vertical distribution of fog of different types is a measure for the three-dimensional structure of the climate in the highlands. Low ground (valleys, basins and troughs) lies in the realm of radiation fog (ground fog); this fog type fills the valleys up to a certain height which is below the top of the ground inversion and such situations are characterized by a lack of air movement. Above this ground fog layer, there is the warm slope zone which seldom suffers from fog, thus being a therapeutically very favourable climatic zone of variable thickness. It reaches up to the base of the low stratus layer which, especially in winter, forms crest clouds over the highlands and, in case of supercooling in combination with strong winds, may cause detrimental rime or clear icing (e.g., damage to overhead lines, radio towers or edges of forests). The base of the crest clouds of some of the ridges represents the base of the precipitating clouds connected with bad weather in winter, whereas the upslope fog results in fog drip. Thus, low elevations have a relatively high frequency of fog occurrence (ground fog); there is a decrease of fog frequency with height due to the clearance in the slope zone, whereas at the base of the upslope fog the fog frequency jumps discontinuously (crest clouds). The climatic province under consideration is characterized by a wide range of bioclimatic types. The strain imposed by the atmospheric environment on man in the various stressful zones is due to: (*1*) heat load (sultriness, or heat in summer); (*2*) damp cold in stagnating air (ground fog areas); (*3*) reduced radiation received at the surface of the earth (same as (*2*)); and (*4*) smog in low elevations, especially in densely populated areas and basin locations.

With increasing height the soothing climate of the low valleys changes to a mountain climate with mild meteorological stimuli due to weak to moderate thermal effects which are a function of the exposure to the prevailing winds. Both solar and sky radiation in the highlands are less intense than on the coastal region and in the high mountains. Local valleys may have spells of moderate to strong meteorological stimuli due to a great diurnal range of temperature especially in case of a noteworthy temperature drop in the evening. The moderate to strong meteorological stimuli in the crest region of the highlands are due to: (*1*) increase of thermal effects due to an increase in cooling power (mainly caused by low values of air temperature); (*2*) increased intensity of solar and sky radiation; and (*3*) decreased atmospheric pressure as well as partial oxygen pressure.

Special regional features

There are two main factors which cause special regional features in the German highlands: the differences in the orientation of the mountain barriers effect a great variety of windward and lee effects with regard to the prevailing westerly winds, and the differences in terrain features create a variety of local climatic peculiarities.

The highlands north of 50°N

The influence of the highlands on the climatic regime is most obvious from the regional distribution of precipitation (Fig.22, p.50); the characteristic phenomena are the upslope effects on the west slopes and the lee effects on the east slopes of the mountain ranges. The upslope effect in this area reaches a maximum on the windward side of the Harz Mountains which extend farthest north into the west-wind belt and are notorious for their high rates of ice accretion. The "Bergisches Land", and upland to the west of the Rhenian Slate Mountains, in spite of its relatively modest height, receives abundant rain due to a very intense upwind effect to these windward slopes bordering the Bight of Cologne ("western Germany's water tower").

The westward bulge of the isohyets on rainfall maps far into the lowlands is exclusively due to this upwind effect of the uplands. The Thuringian Forest is also well known for its copious precipitation because, due to its orientation from southeast to northwest, it represents an effective barrier facing the steady westerly winds.

The orographic precipitation on the west slopes of the Thuringian Forest and the Harz Mountains annually yields amounts surpassing 1,000 mm; the greatest annual mean is reached on top of the Brocken (Democratic Republic of Germany) amounting to about 1,680 mm. Mention should be made of a noteworthy diversity in the precipitation regime of the Thuringian Forest and the Ore Mountains: the steady moist westerly winds act as crosswinds on the Thuringian Forest, whereas the moist southwest winds are in range with the course of the Ore Mountains, which in addition to this lie in the dry rain shadow behind the Fichtel Mountains and the Frankish Forest. The well marked rain shadow of the Thuringian Forest extends beyond the Thuringian Basin into the valleys of the Saale River and the Unstrut River, thus creating the largest precipitation shadow of the entire area of the German highlands with mean annual precipitation amounts of less than 500 mm.

In low elevations, the months which bring the greatest monthly rainfall amounts of the year are August with 30% of all cases and July with 25% of all cases, whereas the months which bring the smallest monthly precipitation amounts of the year are March with 30% of all cases and February with 20% of all cases; dry summer months are very rare occurrences. The bioclimate with moderate meteorological stimuli in the uplands begins at a height of about 500 m above m.s.l.; at about 600 m the meteorological stimuli become strong.

The highlands south of 50°N

The Upper Rhine Valley is famous for its favourable thermal conditions which are especially marked in the warm slope zones of the "Bergstrasse" (western edge of the "Odenwald" between Darmstadt and Heidelberg) and the "Weinstrasse" (eastern slope, i.e. lee side, of the Palatine uplands). On the other hand, the Upper Rhine Valley is prone to the formation of temperature inversions and air stagnation resulting in detrimental pollution effects due to the trapping of noxious fumes, particularly in the densely populated industrial centres of Mannheim and Ludwigshafen as well as of the Rhine–Main Basin.

Abnormally cool uplands are the "Baar" plateau (around Donaueschingen) as well as the Swabian Alb, where the karst valleys and basin locations are notorious for severe frost. In the course of the year the smallest monthly precipitation amounts are recorded in March with up to 25% of all cases and February with up to 20%, whereas the greatest monthly rainfall amounts occur in July with up to 40% of all cases and August with up to 20% of all cases. Since the crest heights of the highlands south of the Main River are generally greater than those north of it, the snow rate in the precipitation as well as the duration of snow cover south of the Main supersede those north of it, though generally both elements gradually increase from west to east.

The bioclimatic zone with moderate meteorological stimuli in these highlands is found at a little higher elevation than north of the Main River, namely starting at 600–700 m above m.s.l. Above an approximate height of 800 m the meteorological stimuli turn into strong ones. The altitude differentiation of the bioclimatic zones thus implies an increase from north to south.

The Jura Mountains

The Jura Mountain Range is oriented from northeast to southwest having a gradual increase in height from northeast to southwest from 800 to 1,000 m up to a peak height of 1,700 m above m.s.l. The Jura Mountains consist partly of extended ridges with longitudinal valleys the rivers of which force their way through these ridges in narrow canyons, partly of undulatory plateaus which have table heights of about 1,000 m above m.s.l., with embedded hollows. Both these terrain features are prone to the formation of temperature inversions; in winter this tendency is intensified by the snow which usually covers the plateaus; the extreme temperature minima of down to −39°C in the Swiss Confederation are found in La Brévine, the "Swiss Siberia".

The predominance of westerly winds results in an abundant orographic precipitation which on the crests in the west amounts to about 1,800–2,000 mm per year, whereas the low levels of the lee side between the Lake Leman and the Lake of Neuchâtel receive only half of this amount, i.e. about 950 mm per year.

The great thunderstorm activity of this mountain range (most of the meteorological stations have an average of 35 to 40 thunderstorm days per year) causes the precipitation maximum to occur during the summer season though the annual variation of precipitation is small. Calm wind conditions prevail in the valleys; however, the crests and peaks are notorious for gale-force winds. On top of the Chasseron (1,600 m above m.s.l.) the anemograph was destroyed after a gust of 50 m/sec had been recorded. But the extreme gust value of 57 m/sec which was observed on the exposed top of the Feldberg in the Black Forest is unlikely to be reached in the Jura Mountains due to the barrage effect of the Alps. A special phenomenon of the eastern Jura slope is the "joran", a katabatic wind from the northwest; however, its föhn effect is less intense than that of the Alps. The Alpine föhn often dissipates almost the entire deck of grey clouds, whereas the joran clears only the low clouds away.

The northern Alpine foreland

The climate of the northern Alpine foreland is mainly controlled by the influence of the Alps. This is especially the case during northerly flow due to the upwind effects which are not so much a function of the relative difference in height but rather a function of the distance windward of the base of the Alps.

For the climatic discussion, it is appropriate to divide the northern Alpine foreland into Swiss, German and Austrian sections.

The Swiss Midland

This densely populated land between Lake Leman and Lake of Constance is bordered by the Alps on the one side and the Jura Mountains on the other. The bottoms of its valleys are at an average height of 400 to 600 m above m.s.l. Numerous chains of hills run through it, most are oriented from northwest to southeast. The lowest parts of the plain are located at the southern foot of the Jura, so that the cold air flowing down the Alpine valleys at night is trapped there. Thus, ground fog or low stratus with tops between 600 and 700 m above m.s.l. builds up in the valleys in late autumn and winter, mainly before the surface is covered with snow. The onset of northeasterly winds may raise the cloud top to 900–1,200 m above m.s.l. or even higher. This explains the high frequency of days with fog in the Midland as well as the high mean value of cloudiness which is the highest one of the whole of Switzerland, not only in January (see Fig.25, p.54) but also in the annual mean. This winter cloud regime resembles that of the German highlands and is quite in contrast to the sunny inner Alpine valleys and the southern foothills of the Alps. In summer, the Midland is slightly less cloudy than the high mountains the peaks and crests of which are often covered by towering cumulus clouds (see Fig.26, p.55).

The prevailing surface flow in the eastern part of the Midland is the west wind quite in contrast to the situation in the western part where after the passage of a low-pressure system the surface air is drained through the gap between the Jura and the Alps into the lower Rhône Valley as soon as the upper wind veers to north. Then, gale-force northeasterly winds roar over the Mont Blanc bridge in Geneva; this wind, called the "bise", usually brings bright weather to the surroundings of Geneva, whereas farther to the east, heavy orographic clouds accumulate upon the Alpine north slope.

Lee effects cause a decline of the rainfall amount from the area east of Bern towards the western part of the Midland (see Fig.22). Quite in the north, to the west of the Lake of Constance, the rain shadow of the Black Forest is felt. The entire area of the Midland has the maximum precipitation in summer, mainly in the form of pre- or postfrontal showers or thundershowers, whereas in winter frontal precipitation predominates, of course modified by upslope or lee effects. Warm fronts near the Alps are dissipated mainly by the föhn effect, whereas cold fronts are strengthened due to the orographic lifting of the Polar air intruding from northwest. Most of the precipitation, especially in summer, is associated with cold fronts; thus, the upwind effect produces an increase in precipitation towards the base of the Alps even at the same level height, because the air flow is forced upward before the barrier slope is actually reached.

The German Alpine foreland

Numerous spots of marshy land are embedded in the German Alpine foreland mainly in the valleys of the Danube and its southern tributaries; this marshland is not only in winter prone to low temperatures; this also occurs in the transitional seasons. The lake effect of the numerous embedded lakes, such as the Lake of Constance, is felt mainly in winter resulting in a plus anomaly of temperature. In summer, the entire area is favoured by relatively warm temperatures in spite of the height above m.s.l. (tableland). The isohyets in the precipitation maps of Bavaria run from west to east, because the upwind effect of the Alps extends down to the Danube Valley. The mean annual precipitation amounts increase from 650 to 700 mm in the Danube Valley to more than 1,000 mm at the base of the Alps. The greatest mean monthly precipitation amounts can be expected in summer, because the warm season has the greatest frequency of occurrence of northwesterly flow ("European monsoon") connected with convective precipitation, not infrequently accompanied by thunderstorm activity or even hail. In the mean maps of the regional distribution of precipitation amounts, the upwind effect, i.e. the orographic precipitation on the northern slope of the Alps, overcompensates the effect of the southern föhn on the same slope.

The probability of the maximum monthly precipitation of the year occurring is 25% to 40% in July and only 5% to 25% in August. In the Bavarian Alpine foreland the probability of minimum monthly precipitation of the year is 15% to 35% in November; in February, the probability is 15% to 25%, and in March 10% of 20%.

The bioclimatic range covers weak meteorological stimuli in the north of the area which increase to moderate meteorological stimuli at the base of the Alps. The urban climate of Munich represents a singularity. Frequently, the entire area is a stressful zone either due to the discomfort during föhn spells or due to upslope conditions with orographic precipitation.

The Austrian Alpine foreland

The increasing continental influence is clearly felt in the Austrian section of the Alpine foreland, since the winter temperatures are lower and the summer temperatures higher than those in the corresponding sites of the western part of the Swiss Midland.

There is a significant difference in the precipitation amounts between upper and lower Austria. The western part receives considerably greater amounts obviously due to the upslope effect exerted by the southeasterly foothills of the Bohemian Forest on the one hand and of the Alps on the other hand; the orographic rainfall raises the annual amount of precipitation to between 800 mm and 1,000 mm, while in the lowland in Lower Austria the annual precipitation amounts are between 600 mm and 800 mm. A marked summer maximum of rainfall is typical in both parts. In Upper Austria, there are many locations where in 35% to 45% of all years July is the wettest month of the year.

Westerly surface winds are predominating; in the Vienna Basin, these winds are strengthened by the jet-effect due to the gap between the Alps and the Carpathians and channelled to the northwest. Thus, the highest surface wind gusts in all of Austria have been observed in this area where peak gusts in the lowland can surpass 45 m/sec; such high values are reached only in the peak region of the Alps and they are not exceeded even by the föhn storms of the Alpine valleys. Moreover, the föhn seldom reaches the ground in the Alpine foreland, most often a cold air layer a few hundred metres thick lifts the föhn above the valley floor. However, this surface layer of cold air does not mitigate but rather tends to aggravate the debilitating föhn stress which causes sensitive people to suffer from headaches, fatigue, etc.

The Alps

The Alps reach their greatest peak height in the west with the Mont Blanc (4,810 m above m.s.l.), however, the chain's greatest width is attained in Tyrol. Also located there is the lowest mountain pass across the Alps, namely the Brenner Pass (1,370 m above m.s.l.). Elsewhere in the surroundings the crest height is greater than 2,000 m above m.s.l. FLIRI (1962) has prepared detailed climatic charts to prove that it is not admissible to divide the Alpine area into only a north slope and a south slope. In between them, there is an inner Alpine zone with climatic features of its own. The numerous longitudinal valleys are situated in the rain shadow of storms from the north as well as from the south due to the double effect of the bordering mountain ranges. The only exception is the St. Gotthard area in Switzerland. There, the central dry zone does not exist due to the lack of a longitudinal valley. The best examples of well developed inner dry valleys are the Valais with as low an annual precipitation amount as 53 cm and next to it the Vintschgau (upper Adige Valley) with a mean annual precipitation of only 60 cm.

The inner Alpine valleys have also a temperature and fog regime of their own which essentially differs from that of the same levels in the outer regions. Fog is a rare phenomenon here, from which the numerous health resorts benefit. On the other hand, winter brings very strong temperature inversions to the snow-covered broad valleys so that temperature minima of less than −30°C have been observed. The record minimum temperature of −37°C having been recorded on top of the Sonnblick has been almost tied by the absolute temperature minima falling in between −33°C and −36°C in the valleys. Even in the lower Valais, freezing temperatures of −21°C have occurred; inversions may form at night even late in spring, so that the orchards which are not frost-resistant, must be protected from the destructive frost.

These temperature records have been broken by the extreme temperature minima in the deep holes of the vast karst area of the eastern Alps: in 1932 a temperature minimum of −53°C was recorded at an Alpine pasture called "Gstettner Alm". Above the warm slope zone, i.e. in the layer from a few hundred metres above the bottom of the valley up to the summit region, the observations yielded an average lapse rate of 0.4°–0.6°C per 100 m; this corresponds to the common value of vertical lapse rate in the free atmosphere at all latitudes.

The mountain stations of the Alps yield slightly higher means of wind speed than those of the highlands (Säntis, 2,500 m, 6.3 m/sec; Zugspitze, 2,960 m, 6.5 m/sec; Sonnblick, 3,106 m, 7.0 m/sec in comparison with the mean value of 6.4 m/sec obtained from mountain stations like Kleiner Feldberg at 805 m, Wasserkuppe at 921 m, Grosser Feldberg at 1,486 m). However, the peak gusts at the high-altitude stations of the Alps rarely exceed those of the Highlands at an approximate height of 1,500 m above m.s.l. The ice accretion also reaches a maximum in the layer between 1,500 m and 2,500 m above m.s.l., because during winter storms the temperatue of this layer frequently ranges from 0° to −5°C which corresponds to the maximum content of supercooled water.

The Swiss Alps

The great climatic differences between the north and the south side of the Alps are best experienced when crossing the Alps via the St. Gotthard Pass. Though there may be rather great variations in wind direction, it can be seen from the Tables II and III that westerly to northwesterly winds prevail so that the north side of the Alps is mostly subject to upslope effects while the south side of the Alps is subject to lee effects; only in spring and autumn, during spells of the south föhn, the reverse is true. Polar outbreaks of cold air hit the north side of the Alps with full vigour; however, after having been forced across the Alpine barrier they undergo adiabatic warming during their descent, blowing down the south slope of the Alps as the north föhn.

Therefore, at the height of 500 m above m.s.l. the mean annual temperature of the south slope is about 2°C higher than that of the north slope; this results in a horizontal temperature gradient which is almost twice the normal amount. The differences in cloudiness are even more conspicuous. The mean annual cloud cover is as low as about 50% over the south slope of the Alps, but it amounts to about 65% over the north slope; the total annual sunshine duration at Locarno is as high as 2,286 h, while at Altdorf it is only 1,485 h. Of course, it should be remembered that Altdorf has the disadvantage of a considerable shielding of the horizon due to the deep valley in which it is situated. Altdorf represents one of the typical föhn islands in the Alps with an average of 500 föhn hours per year occurring on about 64 days per year. The föhn is characterized by gusty south to southeast winds, adiabatic heating and lowered humidity, extremely good visibility and clarity of the atmosphere. The peak gusts amount to 39 m/sec. The föhn reaches its maximum frequency of occurrence in spring and autumn, but this differs greatly from one year to the other. Months with practically no föhn at all may occur, or on the contrary, there are spells of extremely high frequency of föhn. In 1962 for example, extremely strong föhn winds occurred resulting in havoc windfall in the forests in April as well as in November.

The precipitation regime, too, is different on both sides of the Alps. The mean annual precipitation amount of about 1,800 mm on the south side surpasses that of the north side where the average amount is only 1,200 mm. However, the reverse trend is found in the mean annual number of days with precipitation of at least 1 mm; a mean total of 105 days on the south side as against 140 days on the north side. These different totals reflect the difference in the nature of the precipitation. The south side experiences heavy showers falling from cumulus-type clouds, often accompanied by thunderstorm activity and hail in the Alpine foothills, with a well-marked winter minimum. On the contrary, the major part of the precipitation falling on the northern slopes of the Alps consists of frontal and orographic rain of only light to moderate intensity, though also often combined with thunderstorm activity and hail in the Alpine foothills. With increasing height, the annual totals of precipitation in the high Alps increase up to about 3,500 mm. The extreme total of 4,000 mm which has been recorded by the totalizer at the Mönchsgrat in the Bernese Alps appears somewhat doubtful. Measuring the precipitation in the high mountains is a very arduous task and the great difficulties have not yet been overcome as can be seen from the field test measurements with equipment that has been expressly designed for mountain sites (see SEVRUK, 1973).

The inner Alpine regions, however, namely Grisons (Graubünden) and Valais, are dry and have sunny valleys; in the valleys of the latter the scanty precipitation is rather uniformly distributed over the year.

The inner Alps and the south Alpine foothills are suitable sites for sanatoria and health resorts due to their favourable biometeorological and therapeutic features. The entire scale of meteorological stimuli is available.

(*1*) Soothing bioclimatic stimuli at the waterside of the lakes in Ticino and the east bank of the Lake Leman as well as at the waterside of the Vierwaldstättersee south of the Rigi Mountains.

(*2*) Moderate bioclimatic stimuli at heights between 700 and 1,000 m above m.s.l.

(*3*) Strong bioclimatic stimuli above 1,200 m above m.s.l., especially in high valleys, e.g., in the upper Engadine, which are subject to well developed mountain and valley winds (maloja wind).

The German Alps

The formation of layers of cold air in the Alpine valleys in winter also has an impact on the long record means of air temperature; this shows up for instance in the vertical distribution of monthly mean temperatures of the individual stations. Isothermal conditions are quite frequent in the layer between 700 m and 1,000 m above m.s.l. during the winter months giving evidence of the high frequency of occurrence of surface inversions in this season.

The Alpine valleys are affected by the föhn in quite a different way: it is mainly in the meridional transverse valleys that the föhn reaches the ground. As an example: Oberstdorf has an annual mean of about 20 days with föhn with two frequency maxima in November and April.

The highest peaks of the Bavarian Alps have annual precipitation amounts of more than 2,500 mm on the average. Valleys which are oriented from west to east, receive considerably less precipitation, whereas the mouths of meridional valleys may receive abundant rainfall due to orographic effects with northwesterly flow (e.g., in the Prien Valley near Hohenaschau at 600–700 m above m.s.l. the mean annual precipitation amounts to more than 2,000 mm).

With increasing height above m.s.l. the meteorological stimuli grow from moderate to strong, mainly due to: (*1*) thermal effects due to increased cooling power caused by low air temperature and wind exposure; (*2*) intense solar and sky radiation; (*3*) decreased atmospheric pressure and decreased partial oxygen pressure.

Down in the valleys, however, the bioclimatic stimuli are partly mild or weak.

The Austrian Alps

The Austrian section of the Alps is characterized by the great number of longitudinal valleys (Inn, Salzach, Enns, Mur–Mürz, Drava and Gail), resulting in a great number of individual partitions with different climatic features. The details are shown in comprehensive climatic monographs (LAUSCHER, 1932; STEINHAUSER et al., 1958, 1960; FRIEDRICH, 1970) as well as in individual maps (e.g., STEINHÄUSSER, 1955; STEINHAUSER, 1956, 1959, 1965). A description together with a physical explanation of the precipitation regime is given in STEINHAUSER (1955).

The vertical temperature distribution shows the same peculiarity as in the German section of the Alps. The normal lapse rate of 5°–6°C per 1,000 m is in winter interrupted in the layer between 700 m and 1,300 m above m.s.l., due to the frequent occurrence of ground inversions in the valleys during anticyclonic situations. The wide basins of the province Kärnten as well as the broad valley of Graz are particularly prone to extremely strong surface temperature inversions characterized by values of vertical temperature increase with height which resemble those of the Alpine foreland before the föhn has replaced the colder surface air. For example, according to FRIEDRICH (1970) on December 28, 1949 a morning temperature of −5.8°C was measured at the airport of Graz, whereas 100 m above ground the temperature was +7.5°C. Thus, at the bottom of these valleys destructive frost may be expected as late as May, in extreme cases even in the first week of June, and the first destructive frost may occur already by the end of September. The frequent occurrence of inversions also increases the number of fog days, thus the peak values of the annual total of fog days are found at Graz with 140 and at Klagenfurt with 89 fog days per year. In the southeast of the Austrian Alps, the sheltering effect of the northern mountain chain favours the cloudiness regime so that the Alpine north slope has a little higher mean annual cloudiness than has the southeast of the area.

However, the differences here are smaller than along the Swiss Alpine cross-section. There are considerable regional differences in sunshine duration in winter between the gloomy Danube Valley downstream of Linz and the sunny sides of the upper Inn Valley and of eastern Tyrol around Lienz. In December, the ratio of the average effective possible durations of sunshine at Linz is 18% to 42%, whereas in summer this ratio is 40% to 60%.

In the upper Drava Valley autumn is the wettest season in 30% to 40% of all years; anywhere else, in almost 70% of all years, summer is the season of maximum rainfall. In Steiermark the minimum precipitation can be expected in winter in 60% to 80% of all years. The influence of the rain bringing westerly and northwesterly winds is seen from the regional distribution of precipitation. Well marked precipitation minima are observed on the leeward side of the Arlberg mountain chain where the mean annual rainfall is only 600–800 mm in the upper Inn Valley. On the leeward side of the Tauern mountain chain the mean annual rainfall in large parts of the Mur Valley is only 700–1,000 mm. The maximum precipitation amounts of the eastern Alps are still doubtful. At any rate, they seem to exceed a total of 2,600 mm per year, because this threshold value is reached both in the eastern part of Vorarlberg, on top of the Feuerkogel to the east of Salzburg and in the central crest of the Alps at the highest meteorological mountain observatory on top of the Sonnblick which is 3,100 m above m.s.l. The extreme maxima of the western Alps are not reached due to the sheltering effect of the many mountain chains in the west. In addition, the orographic rainfall resulting from the south föhn is less abundant, because the mountain chains gradually decrease in height eastwards. Certainly, the frequency of föhn situations is high in Tyrol, but the orographic rainfall mainly occurs on the Italian slope, i.e., in South Tyrol.

In analogy with the western Alps thunderstorms and hail are common phenomena in the northern and southern Alpine foothills. The basins of Graz and Klagenfurt in the southeast of Austria are the ones mainly affected by hailstorms.

The map of potential therapeutic climates of Austria, published in FRIEDRICH (1970), clearly shows the favourable conditions of the Alps. This is particularly true for the southern parts which relatively seldom have fog or heavy precipitation and it applies mainly to the inner Alpine valleys. The interaction of exposure to the prevailing winds and the height above m.s.l. creates a great variety of therapeutic regions ranging from soothing bioclimate to strong meteorological stimuli. In addition to that, many health resorts have a scenic landscape which appeals to visitors.

The regional distribution of the climatic elements

For the reason of compatibility, the Figs.17–27 present almost the same climatic elements for the four climatic provinces of Central Europe under consideration as those presented for the neighbouring countries of western Europe in Vol.5 of this series (pp.152–157). The small scale of the maps inevitably resulted in emphasizing the general character rather than specific details of the climate particularly so for the highlands and the Alps. There is another inadequacy. The meteorological observational network in the high Alps is too sparse for yielding completely valid and comprehensive data. The maps are based upon the recent international standard period for normals, 1931–1960.

However, since some wartime observations are missing, it was inevitable to reduce some shorter series to this normal period by comparison with neighbouring stations. It is thought that methodological errors which might arise from this processing will be compensated by the small scale and the generalization of the maps. Local details or peculiarities cannot be read from these maps; they have to be taken from the climate charts of the individual countries which are listed in the References (pp.57–60).

Attention should be paid to the fact that the individual isolines have been labelled by interrupting the isolines and inserting values in these gaps; this is in contrast to the common practice in climatology of writing the corresponding values on that side of the isolines which is directed toward increasing values, thereby enhancing the lucidity of the maps.

Air temperature

The regional distribution of air temperature is discussed by considering the mean values of the two opposite months January and July. As has been mentioned before, along the North Sea coast the months showing the maximum and minimum temperatures of the year are shifted to February and August, respectively.

The months of January and July show characteristic differences in both the regional and the vertical distribution of temperature. The high frequency of surface temperature inversions in winter lowers the monthly mean of the vertical lapse rate to between 0.4°C and 0.5°C per 100 m. As mentioned before, the German Alps show an isothermal layer between 700 and 1,100 m above m.s.l. according to the 1931–1960 period of record, and the normal decrease of temperature with height does not begin below this layer. The same trend is valid for the remainder of the Alps as can be seen from the graph in STEINHÄUSSER (1955, p.150). It is shown that in the Austrian Alps the top of this isothermal layer is raised to about 1,300 m above m.s.l. Detailed maps of the regional temperature distribution indicated the low temperature means down in the valleys which trap the cold air, whereas the slope sides between 700 and 1,100 m or 1,300 m above m.s.l. have the benefit of rather mild temperatures. Thus, in the winter months the isotherms in the maps of the regional temperature distribution are more relaxed than in the summer months.

In January, the regional distribution of the surface temperature (Fig.17) is in analogy with that of western Europe (Vol.5). The isotherms run almost parallel to the shore line; of course, the area of the Baltic coast is markedly colder than that of the North Sea which is more under the influence of the Atlantic and, particularly, the Gulf Stream. Moreover, the regional temperature distribution is greatly influenced by the icing conditions of the Baltic Sea which undergo considerable variations from one year to another. Summing up, in winter the regional temperature distribution is controlled by the interaction between ocean and continent rather than by latitudinal effects.

The latter control applies, however, to summer so that in July (Fig.18)—as well as in the remainder of the summer months—the surface temperature gradually increases from north to south, partly counterbalanced by the influence of the increase in height above m.s.l., because the terrain gradually rises southward toward the Alps.

The mean vertical lapse rate is about 0.6°–0.7°C per 100 m resulting in tighter isotherms than in winter. For instance the north–south cross-section across the St. Gotthard

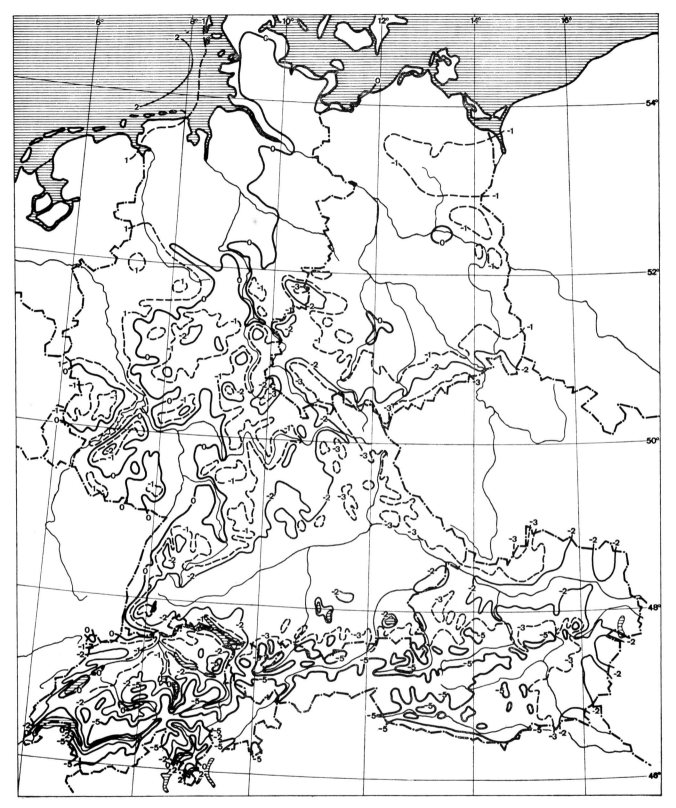

Fig.17. Mean temperature (°C) in January, 1931–1960.

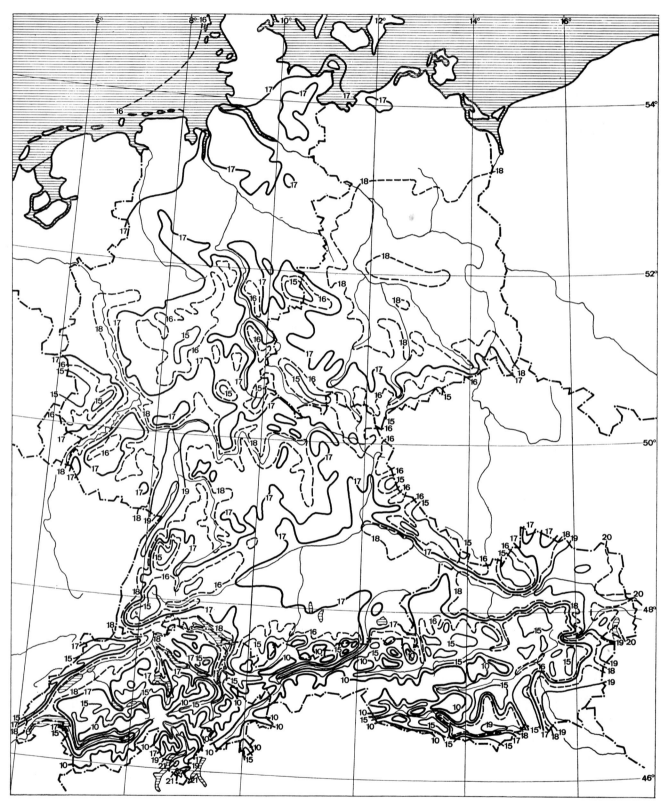

Fig.18. Mean temperature (°C) in July, 1931–1960.

at the same height above m.s.l. shows a temperature difference between the south and the north side of the Alps amounting to about 2°C per 1.5 degree of latitude in winter and about 2.5°C in summer. In this case, the latitudinal effect is intensified by the orographic effect; this is obvious from a comparison with the Atlantic area, where the meridional gradient for the same latitudinal distance is only about 0.5°–1°C.

The two maps of the regional temperature distribution (Figs.17, 18) do not give any idea of the diurnal variation of temperature. Thus, the mean diurnal variation of temperature computed from the difference between the daily maximum and minimum of air temperature (i.e., the aperiodic temperature variation) has also been mapped for the two key months January and July (Figs.19, 20). The daily maxima and minima of temperature normally have systematic local-time variations. The maximum temperature occurs between 14h00 and 15h00 local time, the minimum about one hour after sunrise. Advective processes may shift these times considerably. In high-pressure situations, the range of the diurnal variation of temperature mainly depends on the duration of incoming and outgoing radiation. Due to the difference in daylight hours during the course of the year, the radiation balance at the earth's surface is positive in summer and negative in winter. Therefore, in summer, the influence of continental heating prevails (Basel as well as Vienna Neustadt 11.5°C, Hamburg 9.9°C) whereas in winter the isolines of diurnal temperature variation tend to be zonal due to the effect of geographical latitude. There also are considerable local differences. High valleys of the Alps have the largest diurnal temperature variation, whereas on free peaks or slopes at the same height above m.s.l. diurnal temperature variation is rather small, as can be easily seen from the two station couples in Austria and Switzerland shown in Table IX.

The mean duration of the frost-free season (Fig.21) is defined as the interval between the last and first occurrences of below-freezing temperatures in spring and fall. These data are often unsuitable for showing the macroclimatic feature due to their relationship to the local microclimate as well as to the great variations of the first and last dates of frost from year to year. The grass temperature records (5 cm above ground) show a considerably shorter frost-free season, since the grass minima are much lower than the temperature minima measured in the instrument shelter at a height of 2 m above ground. Thus, it would be more appropriate to base any agricultural planning on the observations of the grass minimum. However, it is deemed recommendable not to use any of these data which are of little scientific value and to determine instead the probability of a temperature drop below specific freezing-temperature thresholds in individual

TABLE IX

MEAN VALUES OF MAXIMUM, MINIMUM AND DIURNAL VARIATION OF AIR TEMPERATURE (°C)

Station	Height above m.s.l. (m)	January			July		
		max.	min.	diurn. var.	max.	min.	diurn. var.
Sillian (valley)	1,080	−1.0	−12.0	11.0	22.6	8.0	14.6
Badgastein (slope)	973	−1.2	− 7.5	6.3	22.3	10.1	12.2
Bever (valley)	1,712	−3.1	−16.3	13.2	17.8	4.5	13.3
Arosa (slope)	1,818	−2.8	− 8.7	5.9	14.8	6.4	8.4

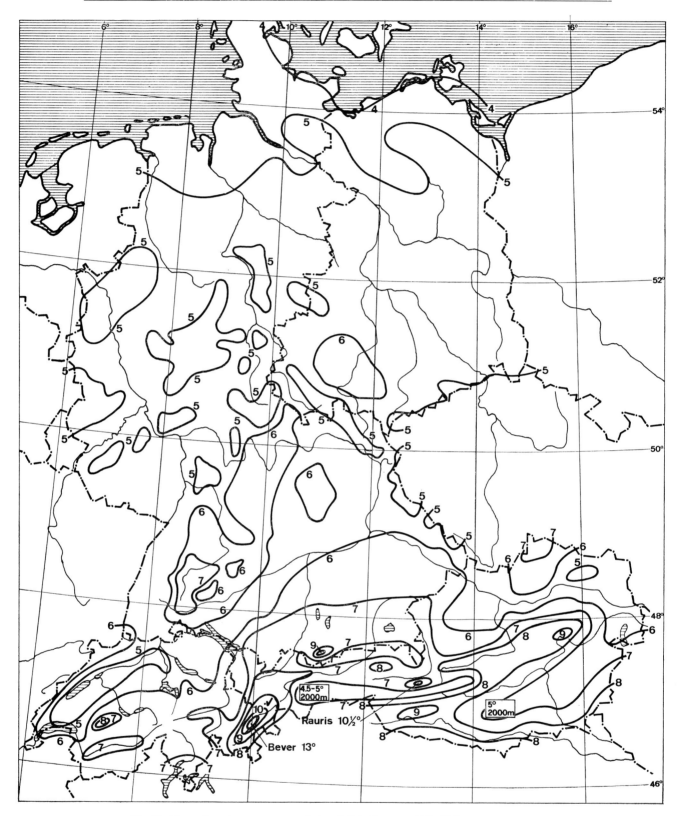

Fig.19. Mean diurnal temperature variation (°C) in January, 1931–1960.

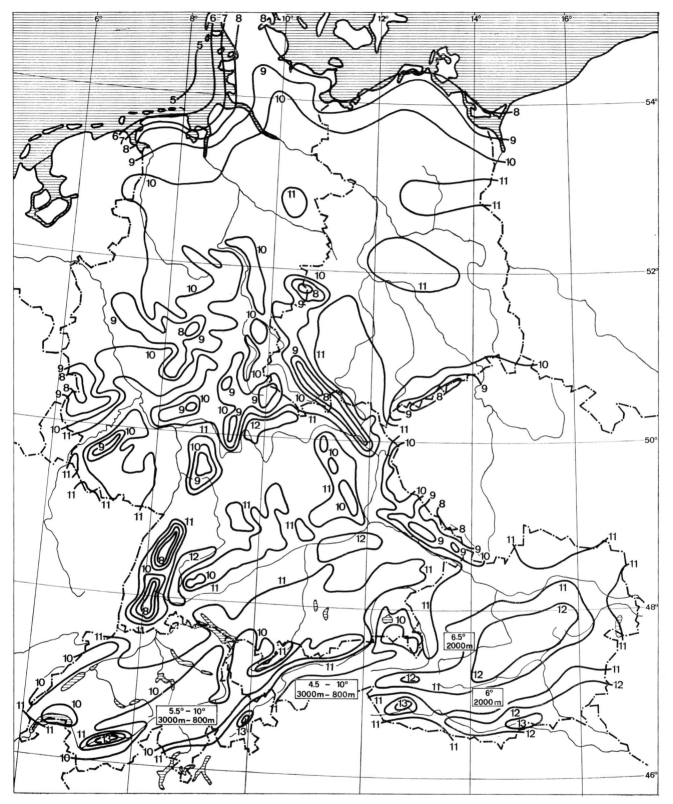

Fig.20. Mean diurnal temperature variation (°C) in July, 1931–1960.

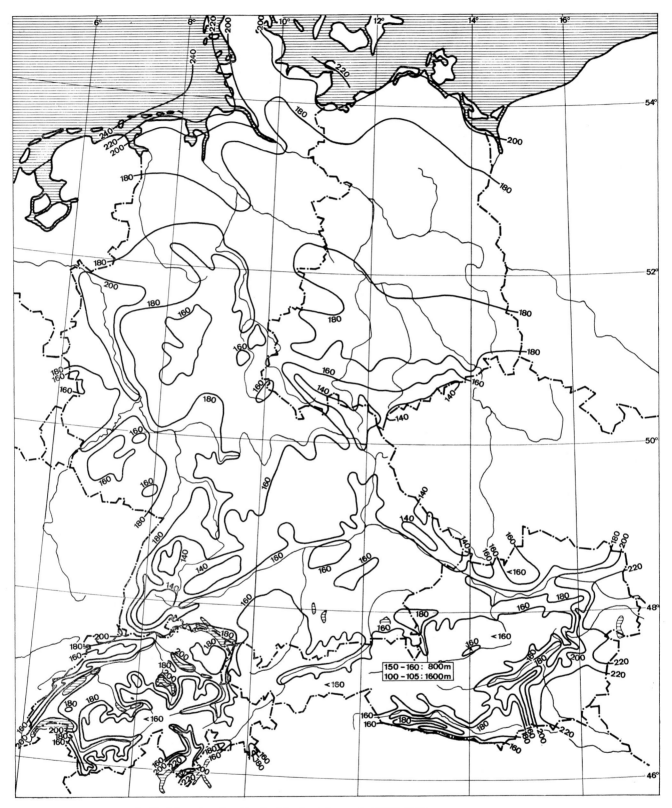

Fig.21. Mean duration (days) of the frost-free period 1931–1960.

months, as has been common practice for a long time already in the field of technical climatology.

In the Alps, the mapping of the duration of the frost-free season had to be restricted to the layer lower than 800 m above m.s.l., because above this level the isolines become too tight for being mapped. In Austria, the frost-free season at 800 m above m.s.l. lasts about 155 days per year. At a height of 1,600 m above m.s.l., however, it lasts only 104 days per year. In the high mountains above 2,500 m above m.s.l., spells of below freezing temperature may also be expected in summer.

Precipitation

In spite of the necessary generalization, the regional distribution of the mean annual precipitation amounts (Fig.22) shows very well the increase with height and the dominating influence of orography becomes obvious. The highest amount of precipitation falls on western and northern sides of mountains, whereas the rain-shadow areas are found on the eastern sides. This results in certain areas in a bulging of the isohyets far to the western or northern windward side, whereas on the eastern slopes of the mountains (after the moist air has been lifted over the orographic barrier) the prevailing subsidence results in a tightening of the isohyets. Unfortunately, due to the restriction in the spacing of the isohyets, this feature is not always clear.

Though the records of the totalizers are somewhat doubtful, it can be estimated that the maximum annual precipitation amounts to roughly 4,000 mm. Existence of a zone of maximum precipitation in the Alps below the crest height is still open to question. Anyway, the frequent redistribution of the snow load during snow-storms due to blowing or falling snow renders this problem meaningless for application purposes. The hydrologist in fact is mainly interested in the result of the metamorphosis of the snow, i.e. in the finite firn deposit in situ.

Homogeneity in annual precipitation amounts does not imply the same thing for the mean seasonal variation of precipitation. Since, however, in the climatic provinces under discussion the latter seldom materializes in individual years, it has been preferred to state the percentage frequency of occurrence of the maximum or minimum monthly precipitation amounts in the course of a year (see Chapter 5). The monthly precipitation amounts yield two maxima over the Alpine south side: the one in spring (May), the other one in autumn (October) due to the windward effect during the south föhn when the north side of the Alps experiences a lee effect. The remainder of the area concerned generally has a distinct summer maximum of rainfall (June through August) with the exception of the German highlands where a secondary precipitation maximum occurs in winter. With this exception, the main minimum of precipitation occurs anywhere else in winter (from north of the Alps to the North Sea: December or March; south of the Alps: January).

Sunshine

The regional distribution of the mean sunshine duration is affected by the variation in shielding of the horizon which is more than only a local feature. This problem may be avoided by mapping the effective possible duration of sunshine, but this type of data

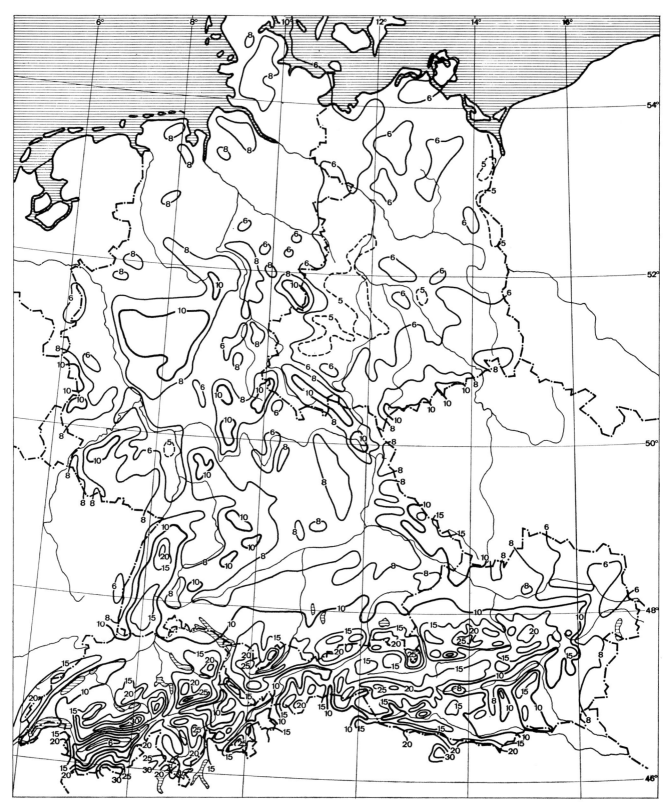

Fig.22. Mean annual precipitation. Isohyets shown in decimetres (1931–1960).

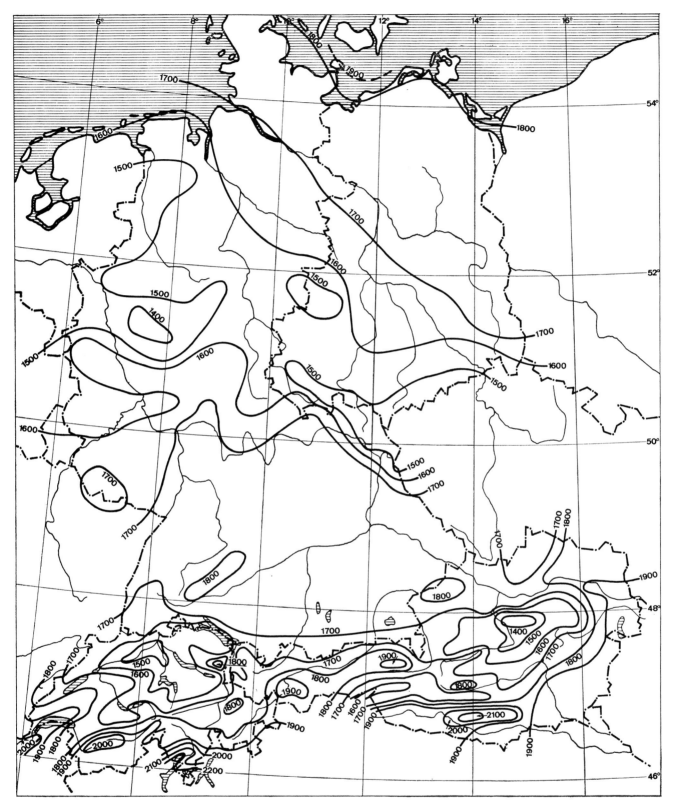

Fig.23. Mean annual duration of sunshine (h), 1931–1960.

was not available for the entire region under consideration. The maps of mean sunshine duration per year (Fig.23) and during the summer months from April through September (Fig.24) therefore cannot show anything but the general regional distribution, which is enhanced by the fact that the network of observations is considerably less dense than that of air temperature.

The latitudinal effect on the mean sunshine duration yields an increase from north to south; the differences in the annual mean (Fig.23) are more conspicuous than those of the summer season (Fig.24), because in the northern part of the map the cloudiness increases to very high values in winter. Remarkable local differences in sunshine duration are found in the Alps due both to the shielding effect of nearby mountain chains and to upslope and lee effects on the cloudiness. Mean values of annual sunshine duration of no more than 1,500 h are to be found in localities which, due to the shielding mountains south of them, do not receive any direct solar radiation in mid-winter for weeks. Nevertheless there are places in the central Alps with a mean annual sunshine duration of 2,000 h and some sites of the foothills of the southern Alps have more than 2,200 h (Fig.23). In summer, the regional differences are less outstanding.

Cloudiness

The mapping of mean cloudiness is somewhat debatable since this meteorological element is not obtained by measurements but by estimate. Only the synoptic stations are staffed by personnel who have been specially trained for this type of eye observation. The climatological stations on the other hand are operated by cooperative observers whose estimates may suffer from systematic errors. The maps therefore can emphasize only general features. Though nowadays the okta-unit is used in specifying cloud amount, the sky cover is given in tenths according to the former weather code which was applied during the long-record of cloud observations on which these maps are based. It was also deemed more intelligible to present the cloudiness in percentage amounts of sky covered. In accordance with Figs.17–20 the mean cloudiness has been mapped for January (Fig.25) and July (Fig.26).

In summer, the valleys are rather free of clouds, while the high mountain chains are favourable for the cloud formation and often produce cumulonimbus and thunderstorms. Winter is characterized by the formation of grey stratiform clouds due to either the temperature ground inversions in the valleys or to the orographic lifting on the high windward slopes. In between, there is the favoured slope zone beginning at a height of about 200–300 m above the bottom of the valley, which unfortunately does not show up because of the small scale of the maps (see SCHIRMER, 1970), while the "sunny islands" in the central Alpine valleys are quite obvious (Fig.25).

Winds

Surface-wind roses of several representative stations are presented in Fig.27. Westerly winds are prevailing along the North Sea coast and in the Austrian Danube Valley. Over the German highlands the effect of enhanced surface friction results in the predominance of southwesterly winds, which according to the exposure of the observation site may be deflected partly to the west or partly to the south. The Alpine area and the

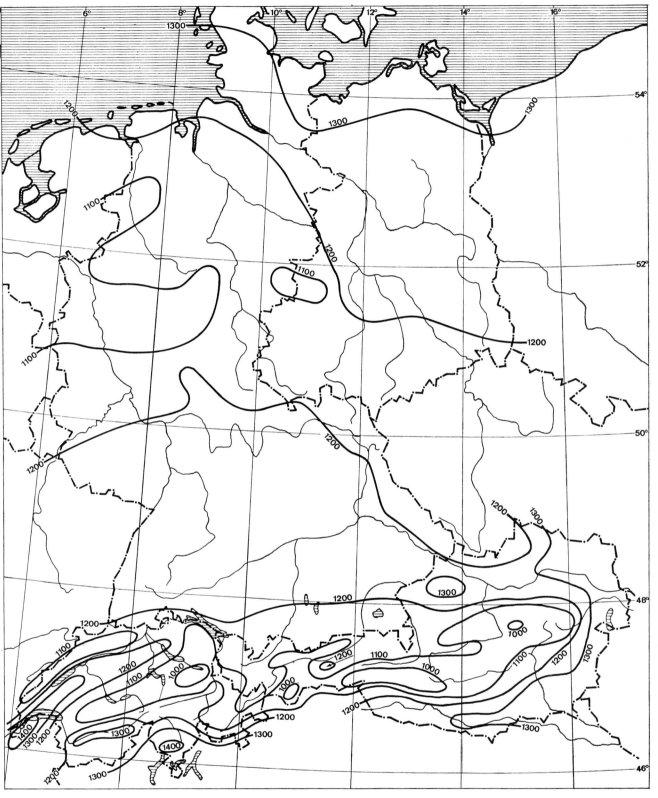

Fig.24. Mean duration of sunshine (h) during the 6 months April to September, 1931–1960.

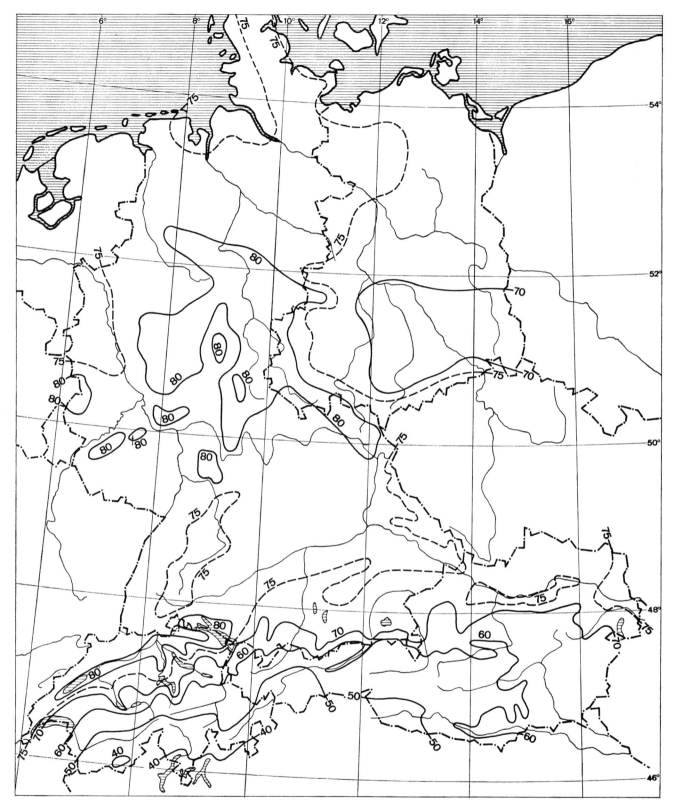

Fig.25. Mean cloudiness (%) in January, 1931–1960.

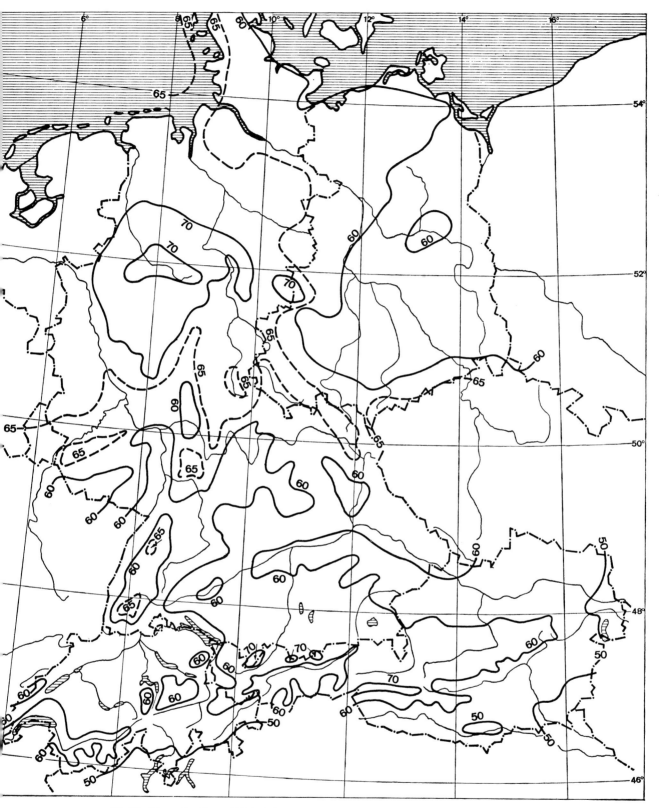

Fig.26. Mean cloudiness (%) in July, 1931–1960.

55

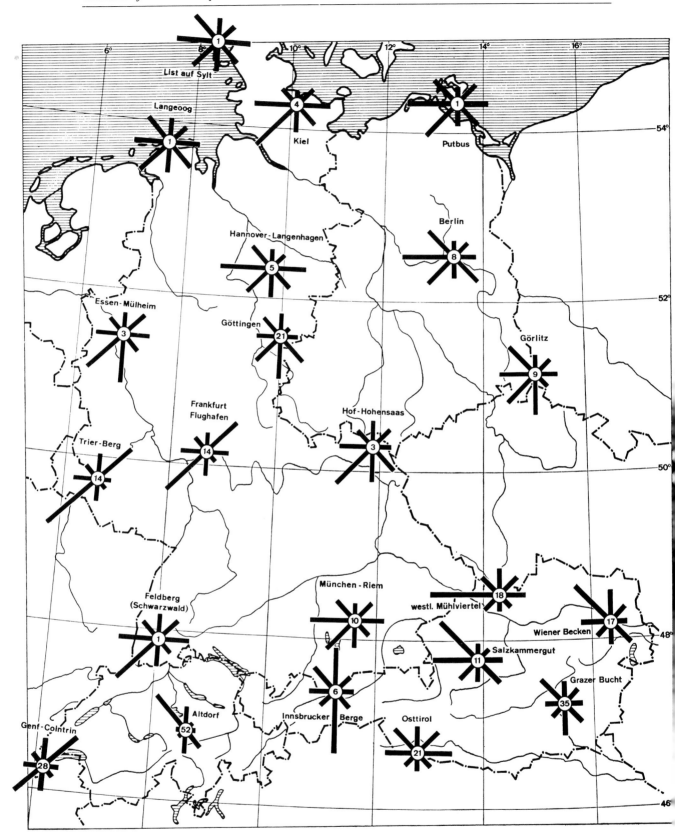

Fig.27. Annual mean wind distribution; figures in the circles are calms (%; 1 mm is 1.9%).

Vienna Basin have a prevailing northwesterly flow. The meridional transverse valleys in northern Tyrol channel the surface flow in such a way that it becomes parallel to the axis of the valleys (föhn). In the western part of the Swiss midland between the Jura and the Alps, the most frequent wind direction is northeast, namely the "bise". The great variety of local features of wind direction and speed in the highlands and the Alpine valleys is due to the channeling effect of deep valleys as well as to the system of diurnal winds, i.e., the mountain and valley winds. The land- and sea-breeze system is less pronounced. The physical explanation for the rather unusual great frequency of easterly winds in March has been given previously (see p.29).

The maps presented here can only touch upon the diversity of the weather processes and the climatic features. Large-scale maps—in some cases presenting additional climatic elements—have been published by the national weather services of the countries involved (see the References) which are recommended for more detailed information.

Climatic tables (pp. 61–73)

The following tables list the monthly and annual means of the essential climatic elements for selected stations. Unless otherwise specified, the mean values refer to the observational period 1931–1960; in individual cases, war gaps necessitated the reduction of climatological series to this normal period as did the relocation of stations. In the latter cases, the records have been reduced to the recent observation site (1973) for consistency. The absolute extremes of air temperature and daily precipitation amount have been evaluated from the longest period of record available because these two elements strongly depend on the length of record.

The authors wish to express their sincere thanks to all members of the climatic departments in the national Weather Services of the countries concerned for the climatic information obtained and the extra work load supplied by processing of special observations.

References

General works

BAUR, F., 1947. *Musterbeispiele europäischer Grosswetterlagen*. Dietrich, Wiesbaden, 35 pp.

BLÜTHGEN, J., 1966. *Allgemeine Klimageographie*. De Gruyter, Berlin, 720 pp.

BÜRGER, K., 1958. Zur Klimatologie der Grosswetterlagen. *Ber. Dtsch. Wetterd.*, 6 (45): 79 pp.

ENVIRONMENTAL SCIENCE SERVICES ADMINISTRATION (ESSA), 1966. *World Weather Records 1951–1960*, Vol. 2. Europe. U.S. Dept. of Commerce, Wash., D.C., 547 pp.

FLOHN, H., 1954. *Witterung und Klima in Mitteleuropa*. Hirzel, Stuttgart, 214 pp.

HESS, P. and BREZOWSKY, H., 1969. Katalog der Grosswetterlagen Europas. *Ber. Dtsch. Wetterd.*, 15 (113): 14 + 56 pp.

KÖPPEN, W. and GEIGER, R., 1932. *Handbuch der Klimatologie* (especially: Abt. E, Klimakunde von Mittel- und Nordeuropa). Bornträger, Berlin, Band 3, Teil M, pp.114–127.

LAUSCHER, F., 1971. Globale und alpine Klimatologie der Starkregen. *Hochwasser und Raumplanung. Schriftenr. Raumforsch. Raumplanung*, 11: 37–39.

SCHERHAG, R., 1948. *Neue Methoden der Wetteranalyse und Wetterprognose*. Springer, Berlin, 424 pp.

SCHNELLE, F., 1965, 1970. Beiträge zur Phänologie Europas, I und II. *Ber. Dtsch. Wetterd.*, 101 (Bd.14); 118 (Bd.16).

SCHÜEPP, M., 1950. *Wolken, Wind und Wetter* Büchergilde Gutenberg, Zürich, 263 pp.

STEINHAUSER, F., 1970. *Climatic Atlas of Europe.* (Maps of mean temperature and precipitation.) W.M.O., Geneva (VI) + 27 maps.

U.S. WEATHER BUREAU, 1948ff. *Monthly Climatic Data for the World.* U.S. Dept. of Commerce, Wash., D.C.

VON RUDLOFF H., 1967. *Die Schwankungen und Pendelungen des Klimas in Europa seit dem Beginn der regelmässigen Instrumenten-Beobachtungen (1670).* Vieweg, Braunschweig, 370 pp.

Federal Republic of Germany

BECKER, F., 1972. Bioklimatische Reizstufen für eine Raumbeurteilung zur Erholung. (With a map "Die bioklimatischen Zonen in der Bundesrepublik Deutschland") *Veröff. Akad. Raumforsch. Landesplanung. Forsch. Sitzungsber.*, 76: 45–61.

DEUTSCHER WETTERDIENST, 1950–1967. *Die Klimaatlanten der deutschen Bundesländer: Hessen, Bayern, Baden-Württemberg, Rheinland-Pfalz, Nordrhein-Westfalen, Niedersachsen; Schleswig-Holstein, Hamburg und Bremen.* Deutscher Wetterdienst, Offenbach a. Main.

FREITAG, E., 1967. Phänologische Gebietsmittel für die Bundesrepublik Deutschland. *Ber. Dtsch. Wetterd.*, 108 (Bd.15): 80 pp. + Karte.

MANIG, M. and SCHIRMER, H., 1961. Das Klima des südlichen Schwarzwaldes. Einflüsse der Änderung des natürlichen Abflusses auf das Klima. *Ber. Dtsch. Wetterd.*, 11 (77): 44 pp.

OTREMBA, E., 1971. Atlas der Deutschen Agrarlandschaft, 4. Lieferung (Teil 1, Blatt 7 und 8 von F. Schnelle und F. Witterstein. Steiner, Wiesbaden).

SCHERHAG, R., 1966. Aerologische und Synoptische Wettervorhersage. *Arch. Meteorol. Geophys. Bioklimatol., Suppl.*, 1: 24–53.

SCHIRMER, H., 1969. Langjährige Monats- und Jahresmittel der Lufttemperatur und des Niederschlags in der Bundesrepublik Deutschland für die Periode 1931–1960. *Ber. Dtsch. Wetterd.*, 115 (Bd.15).

SCHIRMER, H., 1970. Beitrag zur Methodik der Erfassung der regionalen Nebelstruktur. *Abhandlung des 1. Geographischen Instituts an der Freien Universität Berlin, Berlin*, 1970, pp. 135–146.

SUSSEBACH, I., 1968. Synoptisch-statistische Untersuchungen zu Sturmwetterlagen in Mittel- und Osteuropa. *Inst. Meteorol. Geophys. Freie Univ. Berlin, Meteorol. Abh.*, 86 (2) 105 pp.

Democratic Republic of Germany

BÖER, W., 1966. Vorschlag einer Einteilung des Territoriums der Deutschen Demokratischen Republik in Gebiete mit einheitlichem Grossklima. *Z. Meteorol.*, 17 (9–12): 267–275.

HENDL, M., 1966. *Grundriss einer Klimakunde der deutschen Landschaften.* Teubner, Leipzig, 95 pp.

METEOROLOGISCHER UND HYDROLOGISCHER DIENST der DEUTSCHEN DEMOKRATISCHEN REPUBLIK, 1953. *Klimaatlas für das Gebiet der Deutschen Demokratischen Republik.* Akademie-Verlag, Berlin (mit Ergänzungslieferungen 1959ff.).

MAEDE, A., 1958. *Methodische Voruntersuchungen für eine Agrarklimatologie der D.D.R.* Akademie-Verlag, Berlin, 39 pp. + 16 maps.

SEYFERT, F., 1972. *Phänologische Gebietsmittelwerte der Jahre 1957–1966 und des Gesamtzeitraumes 1947–1966 aus dem Höhenbereich 0 bis 300 m NN in der Deutschen Demokratischen Republik auf der Grundlage naturbedingter Landschaften.* Akademie-Verlag, Berlin, Band 13, Nr.102, 11 pp. and 90 tables.

Austria

ANONYMOUS, 1961–1972. *Atlas der Republik Österreich.* 5 Lieferungen hergestellt von der Kommission für Raumplanung Wien der österreichischen Akademischen Wissenschaften, Gesamtleitung: Hans Bobek. Freytagberndt, Wien.

FLIRI, F., 1962. Wetterlagenkunde von Tirol. Grundzüge der dynamischen Klimatologie eines alpinen Querprofils. *Tiroler Wirtschaftsstudien*, 13. Wagner, Innsbruck, 436 pp.

FRIEDRICH, F., 1970. *Klimatische Gegebenheiten. Strukturanalyse des österreichischen Bundesgebietes*, Springer, Wien, I/1, 21–34 + 13 maps.

HADER, F., 1969. Durchschnittliche extreme Tagesniederschlagshöhen in Österreich 1901–1950. *Arb. Zentralanst. Meteorol. Geodyn., Wien*, H5 (II) + 19 pp. + 1 map.

LAUSCHER, F., 1932. Neue klimatische Normalwerte für Österreich, 2. Teil. Bewölkung, Tage mit verschiedenen meteorologischen Ereignissen, Starkregen, Schneeverhältnisse. *Beih. Jahrb. Zentralanst. Meteorol. Geodyn., Wien*, 148: 28 pp.

STEINHAUSER, F., 1955. Die Verteilung der Besonnung in Österreich im Frühling, Sommer, Herbst und Winter. *Statist. Nachr.*, 10 (10): 2 pp. + 4 maps.

STEINHAUSER, F., 1955. Die neue Niederschlagskarte von Österreich für das Normaljahr 1901-1950. *Wetter Leben*, 7 (5-6): 95-100, + map.

STEINHAUSER, F., 1956. *Karte der Andauer der Schneedecke in Österreich 1901-1950*. Zentralanst. Meteorol. Geodyn., Wien.

STEINHAUSER, F., 1959. Karte der Windverteilung in Österreich, Windrosen für den Jahresdurchschnitt Zentralanst. Meteorol. Geodyn., Wien.

STEINHAUSER, F., 1965. Neue Karten der Schneeverhältnisse in Österreich. *Carinthia 2, Sonderh.*, 24: 241-250.

STEINHAUSER, F., ECKEL, O., and LAUSCHER, F., 1958, 1960. *Klimatographie von Österreich*, Lfg. 1-2. *Denkschr. Gesamtakad.*, 3. Österreich. Akad. d. Wiss.

STEINHÄUSSER, H., 1955. Hydrometeorologische Untersuchungen in den österreichischen Südalpen. *Österr. Wasserwirtsch.*, 7 (7, 10 + 12).

Switzerland

AMBROSETTI, F., 1971. Il Clima del sud delle Alpi. Boll. della Soc. Ticinese di Sci. Nat. 62/1971: 12-66.

BOUËT, M., 1972. Le foehn du Valais. *Veröff. Schweiz. Meteorol. Zentralanst.*, 26, 12 pp.

BOUËT, M., 1972. *Climat et météorologie de la Suisse romande*. Payot, Lausanne, 171 pp.,

FREY, K., 1954. Der Übergang einer antizyklonalen in eine zyklonale Föhnlage und weitere Beiträge zur Kenntnis der räumlichen Temperatur-, Druck- und Windverteilung bei Süd- und bei Nordföhn. *Inst. Meteorol. Geophys. Freie Univ. Berlin, Meteorol. Abh.*, II/3: 173-188.

ROSHARDT, A., 1946. Der Winter in der Innerschweiz. Eine vergleichende Studie auf Grund zwanzigjähriger Beobachtung. *Mitt. Naturforsch. Ges. Luzern*, 15, 143 pp.

SCHÜEPP, M., URFER, CH. and UTTINGER, H., 1960-1973. Klimatologie der Schweiz. *Beih. Ann. Schweiz. Meteorol. Zentralanst.*, 1959 ff.

SCHÜEPP, M., UTTINGER, H., BOUËT, M. and PRIMAULT, B., 1965-1974. *Atlas der Schweiz*. Eidg. Landestopogr., Wabern-Bern, Karten 11-13A.

Regional and local climates

ANONYMOUS, 1971. Das Klima von Berlin (II). *Abh. Meteorol. Dienstes DDR*, 13 (103): 187 pp.

ANTONIK, B., 1961. Das Klima von Potsdam (III). *Abh. Meteorol. Dienstes DDR*, 8 (61): 64 pp.

BAHR, R.-M., 1966. Das Klima von Berlin (I). *Abh. Meteorol. Dienstes DDR*, 10 (78): 48 pp.

BIDER, M., 1948. Vom Basler Klima. *Wirtschaft Verwaltung*, 4: 145-188.

BIDER. M., 1956. Klimatische Daten Basels für das praktische Leben. *Wirtschaft Verwaltung*, 1: 7-54.

BÖER, W., 1960. Das Klima von Potsdam (II). *Abh. Meteorol. Hydrol. Dienstes DDR*, 7 (53): 83 pp.

BRANICKI, O., 1963. Das Klima von Potsdam. Ergebnisse 60-jähriger Beobachtungen am Meteorologischen Observatorium 1893-1952. *Inst. Meteorol. Geophys. Freie Univ. Berlin, Meteorol. Abh.*, 32 (1): 121 pp.

FLIRI, F., 1973. Statistische Untersuchungen über den Zusammenhang von Südföhn und Gesamtklima in Innsbruck 1906-1972. *Beitr. Klimatol. Meteorol. Klimamorphol. Festschr. H. Tollner*, pp.45-57.

GOETZ, F. W. P., 1954. *Klima und Wetter in Arosa*. Huber, Frauenfeld, 148 pp.

GOLDSCHMIDT, J., 1950. Das Klima von Sachsen. *Abh. Meteorol. Hydrol. Dienstes DDR*, 1 (3): 33 pp.

HAUER, H., 1950. Klima und Wetter der Zugspitze. *Ber. Dtsch. Wetterd., US-Zone*, 16: 200 pp.

HENTSCHELL, G., 1953. Das Föhngebiet des Harzes. *Abh. Meteorol. Hydrol. Dienstes DDR*, 3 (23): 57 pp.

HEYER, E., 1962. Das Klima des Landes Brandenburg. *Abh. Meteorol. Hydrol. Dienstes DDR*, 9 (64): 62 pp.

KOCH, H.-G., 1953. *Wetterheimatkunde von Thüringen*. Fischer, Jena, 190 pp.

LAUSCHER, F. et al., 1959. Witterung und Klima von Linz. *Wetter Leben, Sonderh.*, VI: 235 pp.

PLEISS, H., 1951. Die Windverhältnisse in Sachsen. *Abh. Meteorol. Hydrol. Dienstes DDR*, 1 (6): 127 pp.

PLEISS, H., 1961. Wetter und Klima des Fichtelberges. *Abh. Meteorol. Hydrol. Dienstes DDR*, 8 (62): 323 pp.

Ševruk, B., 1973. Erfahrungen mit Totalisatoren mit schiefen, geneigten und bodenebenen Auffang-flächen im Einzugsgebiet der Baye de Montreux. *Veröff. Schweiz. Meteorol. Zentralanst.*, 30: 21 pp.

Steinhauser, F., 1938. *Die Meteorologie des Sonnblicks.* Springer, Wien, 1. Teil, 180 pp.

Steinhauser, F., Eckel, O. and Sauberer, F., 1955–1959. *Klima und Bioklima von Wien.* Österr. Ges. Meteorol., Wien, Teil I–III, 130 + 136 + 136 pp.

Tollner, H., 1952. Wetter und Klima im Gebiete des Grossglockners. *Carinthia 2. 14. Sonderh.*, 136 pp. + maps.

Various authors, 1973. *Klima und Witterung im Erzgebirge.* Edited by Hydrometeorologický ústav der ČSSR, und vom Meteorologischen Dienst der DDR, Akademieverlag, Berlin, 166 pp. and 25 maps.

Wustelt, J., 1962. Die Temperaturschwankungen in der Klimareihe von Jena und ihre Beziehungen zur Witterungskunde und zur Klimageschichte Europas. *Abh. Meteorol. Hydrol. Dienstes DDR*, 9 (66): 71 pp.

Zenker, H., 1964. Untersuchungen über Klima und Bioklima des Tales von Bad Berka. *Abh. Meteorol. Hydrol. Dienstes DDR*, 10 (73): 60 pp.

TABLE X

CLIMATIC TABLE FOR LIST AUF SYLT

Latitude 55°01′N, longitude 08°25′E, elevation 16 m

Month	Mean sta. press. (mbar)	Temperature (°C)				Mean vapour press. (mbar)	Precipitation (mm)	
		daily mean	daily range	extremes[1]			mean	max. in 24 h[2]
				max.	min.			
Jan.		0.8	3.7	8.6	−21.2	6.1	48	20
Feb.		0.4	3.9	10.4	−21.0	5.6	35	21
Mar.		2.3	4.5	13.9	−15.0	6.1	31	24
Apr.		6.4	5.3	21.4	− 4.0	7.7	34	20
May		10.8	6.2	25.1	1.3	10.0	40	32
June		14.2	5.7	29.6	4.8	12.5	42	60
July		16.4	5.3	31.6	9.1	14.8	65	104
Aug.		16.6	5.3	28.6	9.5	15.2	88	101
Sept.		14.2	4.9	24.5	4.9	13.2	79	38
Oct.		10.0	4.3	19.7	− 1.2	10.8	76	38
Nov.		5.9	3.6	13.1	− 4.8	8.4	60	40
Dec.		3.0	3.6	10.4	− 8.0	7.3	53	27
Annual		8.4	4.7	31.6	−21.2	9.8	651	104

Month	Number of days with			Mean cloudiness (%)	Mean sunshine (h)	Wind	
	precip. ⩾1 mm	thunderstorm	fog			preval. direct.	mean speed (m/sec)
Jan.	10.8	0.4	7.5	75	55	W	7.4
Feb.	8.3	0.1	7.7	72	75	W	6.5
Mar.	6.9	0.3	8.7	67	136	E	6.8
Apr.	7.9	0.4	4.5	61	203	NW	6.1
May	6.5	2.1	2.5	61	251	NW	6.4
June	6.5	1.6	1.0	64	256	W	6.2
July	10.0	3.7	0.7	68	241	W	6.7
Aug.	10.9	5.4	1.0	69	207	W	6.3
Sept.	11.8	2.9	1.7	60	181	W	6.6
Oct.	12.9	1.3	4.3	68	107	W	6.7
Nov.	12.4	0.8	6.5	78	48	SE	6.9
Dec.	12.0	0.8	7.3	80	31	SE	7.6
Annual	116.9	19.8	53.4	69	1,791	W	6.7

[1] 1937–1960; [2] 1891–1960.

TABLE XI

CLIMATIC TABLE FOR GREIFSWALD (City)

Latitude 54°06′N, longitude 13°27′E, elevation 2 m

Month	Mean sta. press. (mbar)	Temperature (°C)				Mean vapour press. (mbar)	Precipitation (mm)	
		daily mean	daily range	extremes			mean	max. in 24 h
				max.	min.			
Jan.	1013.8	− 1.0	4.8	11.8	−22.8	5.5	40	17.7
Feb.	1013.4	− 0.6	5.6	15.2	−27.2	5.5	33	21.5
Mar.	1014.5	2.4	6.7	21.4	−18.5	6.2	30	24.5
Apr.	1013.0	7.1	7.9	25.9	−10.6	8.0	39	26.6
May	1014.1	12.3	8.8	30.3	− 3.2	10.7	45	34.0
June	1014.2	16.1	9.1	33.7	2.5	13.9	55	39.0
July	1013.2	18.1	8.8	33.8	5.0	16.1	69	67.6
Aug.	1013.4	17.7	8.5	32.5	5.2	15.9	55	51.9
Sept.	1015.2	14.4	8.4	30.3	1.2	13.9	59	42.9
Oct.	1014.8	9.2	6.7	22.8	− 3.6	10.0	51	30.6
Nov.	1013.9	4.5	4.7	15.8	− 8.1	7.5	36	23.4
Dec.	1013.9	1.0	4.0	13.5	−17.4	6.4	41	19.7
Annual	1014.0	8.3	7.0	33.8	−27.2	9.9	553	67.6

Month	Number of days with			Mean cloudiness (%)	Mean sunshine (h)	Wind	
	precip. ⩾1 mm	thunderstorm	fog			preval. direct.	mean speed (m/sec)
Jan.	10.4	0.0	4.7	75	50	SW	5.8
Feb.	8.5	0.0	4.5	74	66	SW	5.5
Mar.	6.8	0.3	5.3	62	136	W	6.0
Apr.	8.4	0.9	1.5	62	181	W	5.8
May	7.0	2.6	1.9	57	268	NE	5.3
June	8.4	3.6	1.1	57	270	W	4.9
July	10.0	4.8	1.3	61	252	W	4.4
Aug.	9.1	4.4	2.0	59	221	SW	4.4
Sept.	9.0	1.7	3.0	55	181	SW	4.6
Oct.	9.6	0.4	4.5	65	112	SW	4.9
Nov.	9.0	−	6.3	77	51	SW	5.1
Dec.	10.0	0.0	6.7	77	46	SW	5.8
Annual	106.2	18.7	42.8	65	1,834	SW	5.3

TABLE XII

CLIMATIC TABLE FOR LÜBECK

Latitude 53°54'N, longitude 10°42'E,

Month	Mean sta. press. (mbar)	Temperature (°C)				Mean vapour press. (mbar)	Precipitation (mm)	
		daily mean	daily range	extremes[1]			mean	max. in 24 h[2]
				max.	min.			
Jan.	1012.6	0.1	5.0	13.5	−24.3	5.7	54	32
Feb.	1014.0	0.5	5.6	16.8	−27.2	5.3	45	21
Mar.	1016.1	3.3	7.5	21.3	−16.4	6.0	39	23
Apr.	1016.2	7.7	8.8	28.4	− 7.0	7.6	48	23
May	1016.5	12.2	9.8	34.0	− 3.0	10.0	56	39
June	1015.5	15.8	9.9	34.2	0.9	12.8	62	42
July	1013.6	17.7	9.4	35.6	4.0	14.8	85	55
Aug.	1013.0	17.2	9.2	34.5	2.9	15.1	85	78
Sept.	1015.1	14.0	8.9	32.4	− 0.1	12.4	60	51
Oct.	1015.5	9.5	7.1	24.4	− 7.5	10.0	59	48
Nov.	1015.4	5.2	4.9	17.6	−14.3	7.7	54	39
Dec.	1011.6	2.0	4.6	15.2	−16.5	6.8	51	24
Annual	1014.6	8.8	7.6	35.6	−27.2	9.5	698	78

Month	Number of days with			Mean cloud-iness (%)	Mean sun-shine (h)	Wind	
	precip. ≥1 mm	thunder-storm	fog			preval. direct.	mean speed (m/sec)
Jan.	10.3	0.2	6.0	74	55	SW	4.3
Feb.	9.2	−	6.5	74	68	W	3.9
Mar.	7.9	0.3	3.6	63	138	E	4.0
Apr.	9.3	0.6	2.1	57	196	NE	3.6
May	8.7	2.8	0.9	56	250	NE	3.6
June	9.1	3.0	1.1	58	239	W	3.5
July	10.5	3.9	1.0	60	246	W	3.6
Aug.	11.0	5.5	2.4	64	202	SW	3.3
Sept.	10.1	0.9	2.9	54	189	SW	3.3
Oct.	10.2	0.1	5.6	69	110	SW	3.3
Nov.	10.1	0.1	7.2	78	45	SE	3.7
Dec.	9.9	0.3	7.0	80	31	SW	4.2
Annual	116.3	17.7	46.3	66	1,769	SW	3.7

[1] 1881−1960; [2] 1931−1960.

TABLE XIII

CLIMATIC TABLE FOR HAMBURG

Latitude 53°38'N, longitude 10°00'E, elevation 14 m

Month	Mean sta. press. (mbar)	Temperature (°C)				Mean vapour press. (mbar)	Precipitation (mm)	
		daily mean	daily range	extremes[1]			mean	max. in 24 h[2]
				max.	min.			
Jan.	1012.7	0.0	4.9	14.4	−22.8	5.7	57	30
Feb.	1014.0	0.3	5.5	17.2	−29.1	5.3	47	27
Mar.	1016.0	3.3	7.6	21.1	−14.3	6.1	38	35
Apr.	1016.3	7.5	9.4	27.6	− 7.1	7.6	52	35
May	1016.7	12.0	10.9	32.1	− 5.0	10.1	55	38
June	1015.7	15.3	10.6	34.5	1.3	12.8	64	47
July	1014.0	17.0	9.9	35.1	3.4	14.9	82	68
Aug.	1013.2	16.6	10.0	35.7	2.4	15.1	84	42
Sept.	1015.4	13.5	9.5	32.3	− 1.2	12.5	61	65
Oct.	1015.8	9.1	7.4	25.1	− 5.9	10.1	59	61
Nov.	1015.5	4.9	4.9	17.3	−13.5	7.9	57	44
Dec.	1011.7	1.8	4.4	13.1	−16.4	6.9	58	25
Annual	1014.8	8.4	7.9	35.7	−29.1	9.6	714	68

Month	Number of days with			Mean cloud-iness (%)	Mean sun-shine (h)	Wind	
	precip. ≥1 mm	thunder-storm	fog			preval. direct.	mean speed (m/sec)
Jan.	11.5	0.3	6.9	79	51	SW	5.0
Feb.	10.2	0.2	6.6	77	64	SW	4.3
Mar.	7.9	0.5	4.2	68	131	E	4.6
Apr.	10.1	0.7	2.3	65	186	NW	4.1
May	9.7	3.7	1.1	66	230	W	4.0
June	10.3	3.2	0.7	67	222	W	3.9
July	11.9	5.9	1.6	70	220	W	4.0
Aug.	11.5	6.7	2.3	71	183	SW	3.8
Sept.	10.3	1.8	2.9	61	171	SW	3.8
Oct.	10.5	0.5	6.4	71	100	SW	4.0
Nov.	10.6	0.4	7.4	81	44	SE	4.3
Dec.	10.9	0.3	6.2	84	28	SW	5.0
Annual	125.4	24.2	48.6	72	1,630	SW + W	4.2

[1] 1881−1960; [2] 1912−1960.

TABLE XIV

CLIMATIC TABLE FOR POTSDAM

Latitude 52°23'N, longitude 13°04'E, elevation 81 m

Month	Mean sta. press. (mbar)	Temperature (°C)				Mean vapour press. (mbar)	Precipitation (mm)	
		daily mean	daily range	extremes			mean	max. in 24 h
				max.	min.			
Jan.	1004.9	− 1.1	5.4	12.6	−22.5	5.2	44	34.0
Feb.	1004.5	− 0.3	6.1	16.9	−23.0	5.3	39	23.8
Mar.	1005.6	3.3	8.5	21.7	−17.3	5.9	32	20.8
Apr.	1004.1	8.3	10.1	27.7	− 7.0	7.6	42	25.5
May	1005.2	13.4	11.2	32.6	− 3.6	10.1	47	33.4
June	1005.3	16.8	11.0	35.6	3.6	12.7	66	69.2
July	1004.3	18.4	10.5	38.4	6.4	14.9	71	34.7
Aug.	1004.5	17.7	10.5	37.4	5.7	14.7	71	68.0
Sept.	1006.3	14.2	10.0	33.7	0.1	12.4	45	54.2
Oct.	1005.9	8.9	7.9	25.0	− 6.4	9.6	47	37.3
Nov.	1005.0	4.2	5.2	17.8	−10.1	7.5	46	27.9
Dec.	1005.0	0.7	4.7	15.4	−19.5	6.0	40	17.1
Annual	1005.1	8.7	8.4	38.4	−23.0	9.3	590	69.2

Month	Number of days with			Mean cloudiness (%)	Mean sunshine (h)	Wind		Mean daily global radiation (ly/day)
	precip. ⩾1 mm	thunderstorm	fog			preval. direct.	mean speed (m/sec)	
Jan.	9.6	0.2	6.6	76	57	W	5.3	57
Feb.	9.1	0.1	4.8	76	68	W	5.3	105
Mar.	7.8	0.3	5.3	64	147	E	5.3	215
Apr.	9.1	2.0	2.3	64	182	W	4.9	325
May	8.3	4.2	1.4	62	235	W	4.2	437
June	9.1	5.6	0.8	62	246	W	4.2	478
July	10.3	6.8	1.6	65	238	W	4.2	442
Aug.	9.2	5.8	2.4	61	220	W	4.2	378
Sept.	7.9	2.3	3.5	57	188	W	4.4	279
Oct.	8.4	0.5	7.0	67	118	W	4.9	151
Nov.	9.3	0.1	9.3	79	52	W	4.9	63
Dec.	8.6	0.0	9.8	77	46	SW	5.1	41
Annual	106.7	27.9	54.8	68	1,797	W	4.6	248

TABLE XV

CLIMATIC TABLE FOR BROCKEN

Latitude 51°48'N, longitude 10°37'E, elevation 1,142 m

Month	Mean sta. press. (mbar)	Temperature (°C)				Mean vapour press. (mbar)	Precipitation (mm)	
		daily mean	daily range	extremes			mean	max. in 24 h
				max.	min.			
Jan.	881.8	− 4.6	4.5	14.0	−25.9	3.9	158	153.7
Feb.	881.8	− 4.7	4.6	12.0	−28.4	4.1	126	52.7
Mar.	882.4	− 2.0	5.3	12.9	−17.0	4.7	94	48.9
Apr.	881.2	1.2	5.6	20.3	−11.3	5.7	105	82.2
May	882.5	5.7	6.2	22.6	−10.5	7.9	96	29.8
June	882.5	9.1	6.2	24.4	− 2.8	9.7	115	124.2
July	881.4	10.8	5.8	27.3	− 0.1	11.6	143	75.1
Aug.	881.8	10.7	5.6	27.8	0.0	11.2	117	75.1
Sept.	883.8	7.9	5.2	24.4	− 4.5	9.6	105	63.3
Oct.	882.8	3.6	4.5	19.6	−10.3	7.2	122	60.3
Nov.	881.8	− 0.3	4.3	15.1	−15.2	5.6	115	65.8
Dec.	882.0	− 3.0	4.0	12.9	−24.8	4.4	126	68.9
Annual	883.4	2.9	5.1	27.8	−28.4	7.2	1,422	153.7

Month	Number of days with			Mean cloudiness (%)	Mean sunshine (h)	Wind	
	precip. ⩾1 mm	thunderstorm	fog			preval. direct.	mean speed (m/sec)
Jan.	18.4	0.2	27.0	83	56	SW	14.2
Feb.	15.2	0.2	24.7	81	63	W	14.2
Mar.	13.9	0.4	24.1	72	112	W	11.3
Apr.	14.7	1.9	22.8	75	122	W	11.3
May	13.7	4.0	22.0	72	170	SW	8.8
June	13.3	4.9	21.7	72	181	W	8.8
July	14.7	5.8	24.0	78	160	W	9.0
Aug.	14.3	5.0	23.7	75	156	W	9.0
Sept.	13.9	1.5	23.4	72	128	SW	10.4
Oct.	14.6	0.2	26.7	79	83	SW	12.0
Nov.	15.3	0.2	26.6	84	45	SW	12.9
Dec.	16.5	0.3	27.2	83	48	SW	13.3
Annual	178.5	24.6	293.9	77	1,324	SW, W	11.3

TABLE XVI

CLIMATIC TABLE FOR DRESDEN-WAHNSDORF

Latitude 51°07'N, longitude 13°41'E, elevation 246 m

Month	Mean sta. press. (mbar)	Temperature (°C)				Mean vapour press. (mbar)	Precipitation (mm)	
		daily mean	daily range	extremes			mean	max. in 24 h
				max.	min.			
Jan.	984.6	− 1.2	4.7	14.1	−23.8	4.9	38	25.5
Feb.	984.2	− 0.7	5.6	16.6	−27.0	5.1	36	46.7
Mar.	984.8	3.2	7.1	20.8	−14.7	6.0	37	30.0
Apr.	983.8	8.2	8.7	28.7	− 5.8	7.6	46	41.0
May	984.8	13.0	9.6	31.7	− 3.3	10.4	63	68.9
June	985.5	16.5	9.7	34.0	5.3	12.7	68	47.6
July	985.0	18.1	9.3	36.6	6.8	14.7	109	114.1
Aug.	985.0	17.8	9.3	36.8	5.3	14.3	72	62.2
Sept.	986.6	14.4	8.7	33.0	1.0	12.3	48	37.0
Oct.	985.9	9.1	6.9	25.6	− 6.7	9.3	52	33.1
Nov.	985.0	4.3	4.6	19.1	− 9.4	7.1	42	37.2
Dec.	984.7	0.4	4.3	13.2	−20.3	5.6	37	30.9
Annual	985.0	8.6	7.4	36.8	−27.0	9.2	648	114.1

Month	Number of days with			Mean cloud-iness (%)	Mean sun-shine (h)	Wind		Mean daily global radiation (ly/day)
	precip. ≥1 mm	thunder-storm	fog			preval. direct.	mean speed (m/sec)	
Jan.	9.3	0.2	6.6	75	61	W, SE	5.8	68
Feb.	8.4	0.4	6.0	75	71	W	5.5	121
Mar.	8.1	0.5	6.5	66	129	W	5.1	218
Apr.	8.8	2.2	4.1	66	166	W	5.1	323
May	9.5	5.8	3.1	65	225	W	4.2	421
June	9.5	6.2	2.2	65	230	W	3.8	441
July	11.2	7.5	2.8	65	227	W	4.2	408
Aug.	8.9	5.9	3.1	61	214	W	4.0	365
Sept.	8.1	2.3	4.4	59	167	W	4.4	271
Oct.	8.7	0.7	5.9	66	117	W, SE	4.9	168
Nov.	8.2	−	7.8	76	60	W	5.1	74
Dec.	8.6	0.1	9.0	75	54	W, SE	5.3	48
Annual	107.3	31.8	61.5	68	1,721	W	4.9	244

TABLE XVII

CLIMATIC TABLE FOR ERFURT

Latitude 50°59'N, longitude 10°58'E, elevation 315 m

Month	Mean sta. press. (mbar)	Temperature (°C)				Mean vapour press. (mbar)	Precipitation (mm)	
		daily mean	daily range	extremes			mean	max. in 24 h
				max.	min.			
Jan.	976.8	− 1.6	5.4	13.5	−24.4	5.1	33	18.4
Feb.	976.4	− 0.8	6.5	17.6	−22.7	5.2	31	28.5
Mar.	977.1	2.8	8.8	21.5	−19.4	6.1	28	25.9
Apr.	976.0	7.5	9.9	30.2	− 8.1	7.5	34	29.2
May	977.1	12.1	10.7	31.5	− 3.8	10.1	58	39.1
June	977.8	15.5	10.5	33.1	2.0	12.7	67	48.8
July	977.2	17.3	10.3	36.5	5.2	14.4	71	47.1
Aug.	977.4	16.5	10.4	37.0	5.0	14.1	55	34.3
Sept.	978.8	13.1	10.0	33.1	− 0.3	12.1	46	58.5
Oct.	978.2	8.0	8.1	26.8	− 8.0	9.2	45	35.0
Nov.	977.2	3.8	5.4	20.1	− 9.3	7.2	34	20.5
Dec.	977.0	− 0.1	5.2	17.4	−23.8	5.7	30	17.9
Annual	977.2	7.8	8.4	37.0	−24.4	9.1	532	58.5

Month	Number of days with			Mean cloud-iness (%)	Mean sun-shine (h)	Wind	
	precip. ≥1 mm	thunder-storm	fog			preval. direct.	mean speed (m/sec)
Jan.	8.4	0.1	5.5	76	50	SW	3.6
Feb.	7.7	0.1	4.1	73	70	SW	3.2
Mar.	7.1	0.3	4.6	62	123	SW	2.8
Apr.	8.4	1.3	1.5	62	157	SW	3.0
May	8.7	4.6	2.0	61	200	NE	2.4
June	8.5	5.2	0.5	62	215	W	2.6
July	9.5	5.1	0.6	65	211	SW	2.6
Aug.	8.6	5.0	0.9	62	198	SW	2.4
Sept.	7.4	1.6	3.0	58	158	SW	2.6
Oct.	7.8	0.3	5.2	65	105	SW	2.8
Nov.	7.9	−	7.2	78	52	SW	2.8
Dec.	8.0	0.1	6.2	75	42	SW	3.4
Annual	98.0	23.7	41.3	67	1,581	SW	2.8

TABLE XVIII

CLIMATIC TABLE FOR MARBURG/LAHN

Latitude 50°49'N, longitude 08°46'E, elevation 243 m

Month	Mean sta. press. (mbar)	Temperature (°C)				Mean vapour press. (mbar)	Precipitation (mm)	
		daily mean	daily range	extremes*1			mean	max. in 24 h*2
				max.	min.			
Jan.	1015.4	− 0.6	5.1	12.5	−28.5	5.3	55	29
Feb.	1015.0	0.3	6.3	16.5	−23.4	5.3	49	30
Mar.	1015.4	4.1	9.1	21.6	−17.3	6.3	35	23
Apr.	1016.3	8.2	10.3	28.8	− 7.4	7.5	43	30
May	1016.3	12.6	11.5	33.1	− 4.2	10.1	52	42
June	1015.9	15.6	11.7	36.5	− 0.8	12.7	66	53
July	1015.1	17.1	11.2	37.0	3.6	14.5	67	54
Aug.	1014.6	16.4	11.1	34.4	3.3	14.4	71	61
Sept.	1016.7	13.2	10.4	32.0	− 1.3	12.1	50	34
Oct.	1017.2	8.4	8.0	25.9	− 7.9	9.6	48	28
Nov.	1017.1	4.2	5.0	17.7	−16.7	7.3	51	38
Dec.	1014.6	0.7	4.2	14.8	−22.1	6.4	51	33
Annual	1015.8	8.4	8.7	37.0	−28.5	9.3	638	61

Month	Number of days with			Mean cloud- iness (%)	Mean sun- shine (h)
	precip. ≥1 mm	thunder- storm	fog		
Jan.	11.3	0.1	5.7	76	32
Feb.	9.9	−	6.2	68	67
Mar.	7.8	0.4	3.5	57	139
Apr.	8.2	1.2	1.7	50	192
May	8.4	3.8	1.8	52	227
June	9.6	3.9	2.2	57	212
July	10.5	6.1	2.6	56	214
Aug.	10.8	5.0	5.5	56	188
Sept.	8.9	1.5	8.7	57	155
Oct.	8.8	0.1	12.6	64	96
Nov.	10.3	0.2	8.9	78	30
Dec.	10.0	0.1	9.6	81	14
Annual	114.5	22.4	69.0	63	1,566

*1 1881−1960; *2 1891−1960.

TABLE XIX

CLIMATIC TABLE FOR AACHEN

Latitude 50°47'N, longitude 06°06'E, elevation 202 m

Month	Mean sta. press. (mbar)	Temperature (°C)				Mean vapour press. (mbar)	Precipitation (mm)	
		daily mean	daily range	extremes*1			mean	max. in 24 h*2
				max.	min.			
Jan.		1.8	5.1	17.1	−19.2	6.1	72	39
Feb.		2.2	5.8	20.5	−20.3	6.0	59	45
Mar.		5.6	7.9	23.2	−11.9	7.1	49	32
Apr.		8.9	8.7	29.3	− 5.4	8.0	63	47
May		12.9	9.7	34.8	− 1.3	10.5	67	62
June		16.0	9.4	36.6	2.2	13.1	77	68
July		17.6	9.0	37.0	5.9	14.9	75	74
Aug.		17.2	8.9	37.2	3.4	14.8	82	61
Sept.		14.5	8.6	34.3	0.3	12.8	68	67
Oct.		10.1	7.1	30.2	− 5.7	10.3	64	51
Nov.		6.0	5.0	20.6	−13.2	8.0	67	43
Dec.		3.1	4.5	17.6	−16.5	7.2	62	38
Annual		9.7	7.5	37.2	−20.3	9.9	805	74

Month	Number of days with			Mean cloud- iness (%)	Mean sun- shine (h)	Wind	
	precip. ≥1 mm	thunder- storm	fog			preval. direct.	mean speed (m/sec)
Jan.	12.9	0.2	6.3	76	51	SW	4.1
Feb.	11.1	0.3	5.7	70	74	SW	3.3
Mar.	9.2	0.5	4.3	65	125	SW	3.1
Apr.	11.1	1.2	3.2	60	178	SW	2.9
May	10.2	4.3	2.7	62	205	SW	2.3
June	10.7	4.1	2.0	66	200	SW	2.4
July	11.4	5.5	1.5	69	190	SW	2.6
Aug.	11.8	4.5	1.9	64	188	SW	2.7
Sept.	10.3	2.5	2.5	59	160	SW	2.6
Oct.	10.5	0.5	5.1	62	123	SW	2.9
Nov.	11.6	0.3	5.6	74	62	S	3.2
Dec.	11.5	0.2	5.5	76	49	SW	3.8
Annual	132.3	24.1	46.3	67	1,605	SW	3.0

*1 1881−1960; *2 1891−1960.

TABLE XX

CLIMATIC TABLE FOR BAMBERG

Latitude 49°53'N, longitude 10°53'E, elevation 282 m

Month	Mean sta. press. (mbar)	Temperature (°C)				Mean vapour press. (mbar)	Precipitation (mm)	
		daily mean	daily range	extremes*[1]			mean	max. in 24 h*[2]
				max.	min.			
Jan.		− 1.2	5.5	13.6	−29.7	5.2	50	44
Feb.		− 0.2	6.9	19.9	−30.1	5.2	42	19
Mar.		3.8	9.4	24.4	−17.0	6.3	35	24
Apr.		8.4	10.6	29.2	− 9.7	7.6	40	35
May		13.2	11.2	33.3	− 4.1	10.5	60	52
June		16.6	11.2	34.5	1.1	13.3	73	75
July		18.2	10.9	38.0	3.0	14.9	82	60
Aug.		17.1	10.9	38.3	0.7	14.5	68	70
Sept.		13.7	10.6	33.3	− 2.5	12.1	50	54
Oct.		8.3	8.8	26.5	− 7.5	9.3	51	45
Nov.		3.8	5.6	21.9	−18.0	7.2	45	40
Dec.		0.2	4.6	16.0	−27.3	6.1	47	54
Annual		8.5	8.8	38.3	−30.1	9.4	643	75

Month	Number of days with			Mean cloud-iness (%)	Mean sun-shine (h)	Wind	
	precip. ⩾1 mm	thunder-storm	fog			preval. direct.	mean speed (m/sec)
Jan.	10.8	0.1	2.3	78	44	S	2.6
Feb.	9.0	−	3.1	71	64	S	2.6
Mar.	7.7	0.2	3.0	61	135	NW	2.8
Apr.	8.5	1.0	1.0	58	170	NW	2.7
May	9.6	4.7	2.1	58	217	NW	2.6
June	10.4	6.1	1.7	62	202	W	2.3
July	11.0	6.3	1.7	60	217	W	2.0
Aug.	9.8	5.9	4.0	62	190	S	2.0
Sept.	8.8	1.6	6.2	56	157	W	2.0
Oct.	8.4	0.7	9.0	63	105	S	2.1
Nov.	8.9	0.1	3.7	78	45	S	2.1
Dec.	9.3	−	4.6	84	28	S	2.1
Annual	112.2	26.7	42.4	66	1,574	S	2.3

*[1] 1881−1960; *[2] 1891−1960.

TABLE XXI

CLIMATIC TABLE FOR KARLSRUHE

Latitude 49°01'N, longitude 08°23'E, elevation 115 m

Month	Mean sta. press. (mbar)	Temperature (°C)				Mean vapour press. (mbar)	Precipitation (mm)	
		daily mean	daily range	extremes*[1]			mean	max. in 24 h*[2]
				max.	min.			
Jan.	1016.8	0.8	5.9	16.0	−25.4	5.7	66	51
Feb.	1015.8	1.8	7.2	19.3	−23.1	5.7	56	30
Mar.	1015.4	6.0	10.0	24.0	−15.0	6.9	43	35
Apr.	1016.5	10.1	10.8	29.6	− 6.0	8.1	59	44
May	1016.3	14.4	11.8	34.4	− 3.4	10.9	66	52
June	1016.3	17.7	11.4	37.2	2.3	14.3	84	47
July	1016.1	19.5	11.1	38.8	5.8	15.7	76	38
Aug.	1015.4	18.6	11.2	37.0	4.0	15.5	80	68
Sept.	1017.2	15.2	10.5	33.9	− 1.2	13.3	66	43
Oct.	1018.0	9.8	8.8	27.0	− 6.8	10.3	56	79
Nov.	1017.7	5.3	6.1	20.1	−12.5	7.7	57	38
Dec.	1016.1	1.7	5.3	17.8	−21.5	6.7	52	57
Annual	1016.5	10.1	9.3	38.8	−25.4	10.1	761	79

Month	Number of days with			Mean cloud-iness (%)	Mean sun-shine (h)	Wind	
	precip. ⩾1 mm	thunder-storm	fog			preval. direct.	mean speed (m/sec)
Jan.	10.9	−	5.5	79	52	SW	2.8
Feb.	9.3	0.2	5.7	74	72	SW	2.3
Mar.	8.6	0.6	3.1	62	152	SW	2.1
Apr.	9.5	1.8	0.7	61	188	SW	2.3
May	9.7	4.3	1.0	60	236	SW	2.0
June	11.0	6.5	0.8	65	214	SW	2.2
July	10.7	6.2	0.3	60	239	SW	2.3
Aug.	10.1	5.4	1.7	61	218	SW	2.0
Sept.	9.6	2.6	5.1	58	181	SW	2.0
Oct.	8.7	0.5	10.0	64	128	SW	1.7
Nov.	10.0	−	8.9	81	48	SW	1.7
Dec.	9.5	0.3	8.1	81	43	SW	2.3
Annual	117.6	28.4	50.9	67	1,771	SW	2.3

*[1] 1881−1960; *[2] 1891−1960.

TABLE XXII

CLIMATIC TABLE FOR AUGSBURG

Latitude 48°22'N, longitude 10°54'E, elevation 490 m

Month	Mean sta. press. (mbar)	Temperature (°C)				Mean vapour press. (mbar)	Precipitation (mm)	
		daily mean	daily range	extremes*[1]			mean	max. in 24 h*[2]
				max.	min.			
Jan.		− 1.7	5.6	16.1	−26.5	4.8	50	46
Feb.		− 0.5	6.8	20.0	−28.2	5.1	47	28
Mar.		3.8	8.9	24.4	−15.2	6.4	42	36
Apr.		8.2	10.0	29.0	− 7.5	7.7	47	51
May		12.7	10.8	31.5	− 2.5	10.4	82	87
June		16.1	10.6	34.8	3.3	13.3	107	66
July		17.8	10.7	37.5	3.6	14.8	111	58
Aug.		17.2	10.6	37.0	2.5	14.4	80	54
Sept.		13.8	10.1	33.2	− 1.3	12.3	71	71
Oct.		8.2	8.3	27.5	− 7.8	9.3	57	36
Nov.		3.3	5.5	21.7	−17.7	6.9	48	60
Dec.		− 0.5	4.8	17.4	−20.9	5.9	46	33
Annual		8.2	8.6	37.5	−28.2	9.3	788	87

Month	Number of days with			Mean cloudiness (%)	Mean sunshine (h)	Wind	
	precip. ⩾1 mm	thunderstorm	fog			preval. direct.	mean speed (m/sec)
Jan.	11.4	0.2	7.4	72	67	W	2.8
Feb.	10.1	0.1	7.3	69	75	W	2.7
Mar.	8.3	0.6	5.3	56	147	E	2.7
Apr.	9.5	0.8	1.5	54	179	W	3.1
May	11.9	3.9	1.8	53	220	E	2.9
June	14.0	6.9	2.0	58	209	W	2.8
July	13.3	6.7	0.9	54	238	W	2.7
Aug.	11.5	4.9	1.7	51	223	W	2.4
Sept.	10.4	1.8	6.0	48	181	W	2.4
Oct.	9.0	0.1	11.0	58	129	W	2.3
Nov.	8.6	0.1	10.0	76	55	E	2.3
Dec.	9.5	0.1	10.1	76	48	W	2.4
Annual	127.5	26.2	65.0	60	1,771	W	2.6

*[1] 1881–1960; *[2] 1891–1960.

TABLE XXIII

CLIMATIC TABLE FOR VIENNA (Hohe Warte)

Latitude 48°15'N, longitude 16°22'E, elevation 203 m

Month	Mean sta. press. (mbar)	Temperature (°C)				Mean vapour press. (mbar)	Precipitation (mm)	
		daily mean	daily range	extremes			mean	max. in 24 h
				max.	min.			
Jan.	993.4	− 1.4	4.9	16.7	−21.9	4.5	40	21
Feb.	992.0	0.4	5.5	18.5	−22.6	4.8	43	44
Mar.	992.2	4.7	7.9	24.0	−11.2	5.9	45	41
Apr.	990.7	10.3	8.8	27.3	− 3.2	7.8	45	38
May	990.7	14.8	9.0	32.6	− 0.3	11.1	70	93
June	991.9	18.1	8.8	36.1	4.1	13.8	67	66
July	991.5	19.9	9.2	38.3	8.8	15.3	83	67
Aug.	991.8	19.3	9.0	34.2	8.0	15.2	72	76
Sept.*	993.8	15.6	8.7	31.6	− 0.1	12.8	41	85
Oct.	993.8	9.8	7.1	27.8	− 3.1	9.6	56	54
Nov.	993.5	4.8	4.6	19.6	− 8.8	7.1	53	45
Dec.	993.6	1.0	3.9	16.5	−15.3	5.3	45	50
Annual	992.4	9.8	7.2	38.3	−22.6	9.5	660	93

Month	Number of days with			Mean cloudiness (%)	Mean sunshine (h)	Wind	
	precip. ⩾1 mm	thunderstorm	fog			preval. direct.	mean speed (m/sec)
Jan.	8.1	0.0₃	7.9	75	57	W	3.2
Feb.	7.3	0.2	6.9	71	84	W	3.2
Mar.	7.7	0.3	5.4	61	138	W	3.2
Apr.	7.6	1.4	1.5	60	184	W	3.1
May	8.5	4.0	0.5	57	235	W	3.1
June	8.9	5.8	0.4	56	249	W	3.1
July	9.4	6.7	0.1	53	266	W	3.2
Aug.	8.7	5.8	0.8	49	250	W	2.8
Sept.	6.6	1.4	2.4	48	199	W	2.7
Oct.	7.5	0.2	7.1	59	129	W	2.6
Nov.	7.8	0.0₃	10.0	78	55	W	3.1
Dec.	7.9	0.1	8.9	79	45	W	2.9
Annual	96.0	25.7	51.9	62	1,891	W	3.0

TABLE XXIV

CLIMATIC TABLE FOR ST. PÖLTEN

Latitude 48°13′N, longitude 15°38′E, elevation 282 m

Month	Mean sta. press. (mbar)	Temperature (°C)				Mean vapour press. (mbar)	Precipitation (mm)	
		daily mean	daily range	extremes			mean	max. in 24 h
				max.	min.			
Jan.	984.6	− 2.2	5.5	15.9	−26.1	4.5	32	18
Feb.	983.0	− 0.7	7.2	18.4	−26.8	4.8	37	35
Mar.	983.0	3.7	8.8	24.3	−17.7	5.9	40	107
Apr.	981.8	9.0	10.7	28.3	− 6.2	8.0	49	71
May	981.9	13.7	11.5	32.1	− 2.4	11.6	89	78
June	983.7	17.0	11.7	36.2	2.2	14.2	102	87
July	983.8	18.8	11.5	39.0	7.3	15.7	108	78
Aug.	983.8	18.0	11.6	36.7	3.4	15.5	92	64
Sept.	985.6	14.5	11.2	32.3	− 2.1	13.0	58	53
Oct.	984.9	8.8	8.6	27.4	− 6.0	9.5	53	38
Nov.	984.1	3.9	5.5	20.0	−13.6	7.0	45	51
Dec.	984.8	0.0	5.0	15.4	−24.0	5.5	36	32
Annual	983.8	8.7	9.1	39.0	−26.8	9.6	741	107

Month	Number of days with			Mean cloud-iness (%)	Mean sun-shine (h)	Wind	
	precip. ≥1 mm	thunder-storm	fog			preval. direct.	mean speed (m/sec)
Jan.	8.8	0.0	4.4	74		W	2.8
Feb.	8.0	0.0	2.9	71		W	3.0
Mar.	7.1	0.1	3.7	62		W	3.0
Apr.	7.9	0.9	1.5	62		W	2.7
May	10.1	3.7	1.0	60		W	2.8
June	10.6	4.7	0.7	60		W	3.2
July	11.2	5.9	0.5	57		W	3.0
Aug.	11.6	4.3	1.6	52		W	2.6
Sept.	8.0	1.0	3.1	52		W	2.3
Oct.	7.9	0.2	7.4	63		W	2.6
Nov.	7.9	0.0	7.3	76		W	2.8
Dec.	7.8	0.0	6.8	79		W	2.7
Annual	106.9	20.8	40.9	64		W	2.8

TABLE XXV

CLIMATIC TABLE FOR FELDBERG

Latitude 47°52′N, longitude 08°00′E, elevation 1,486 m

Month	Mean sta. press. (mbar)	Temperature (°C)				Mean vapour press. (mbar)	Precipitation (mm)	
		daily mean	daily range	extremes[1]			mean	max. in 24 h[2]
				max.	min.			
Jan.		− 4.3	5.2	12.7	−22.8	3.9	163	101
Feb.		− 4.1	5.5	14.8	−30.7	3.7	154	100
Mar.		− 1.2	5.4	14.2	−17.8	4.7	116	79
Apr.		1.4	5.7	20.6	−14.2	5.5	111	81
May		5.8	5.6	24.2	−10.0	7.3	127	177
June		9.0	5.4	25.1	− 2.7	9.6	164	88
July		10.8	5.4	28.0	− 0.4	10.7	164	126
Aug.		10.7	5.0	27.2	− 0.8	10.3	170	67
Sept.		8.4	5.2	24.6	− 5.4	9.2	147	101
Oct.		4.0	5.0	19.4	−10.1	6.7	144	129
Nov.		0.3	4.6	16.8	−14.5	5.1	152	114
Dec.		− 2.8	4.8	12.9	−21.6	4.4	120	133
Annual		3.2	5.3	28.0	−30.7	6.8	1,732	177

Month	Number of days with			Mean cloud-iness (%)	Mean sun-shine (h)	Wind	
	precip. ≥1 mm	thunder-storm	fog			preval. direct.	mean speed (m/sec)
Jan.	16.3	0.5	25.0	77	75	SW	10.7
Feb.	14.3	0.3	21.2	74	92	SW	9.9
Mar.	12.7	0.8	21.9	70	140	SW	8.1
Apr.	13.9	1.9	20.3	72	154	SW	7.7
May	15.0	6.2	19.9	72	194	SW	6.9
June	15.2	9.1	22.6	75	168	SW	6.1
July	15.2	9.3	22.5	69	196	SW	6.6
Aug.	14.1	7.6	21.1	70	181	SW	7.0
Sept.	13.7	2.9	21.8	65	154	SW	7.2
Oct.	12.5	1.1	22.5	63	143	SW	7.8
Nov.	13.3	0.4	22.0	72	94	SW	8.5
Dec.	14.3	0.4	22.9	72	82	SW	9.8
Annual	170.5	40.5	263.7	71	1,673	SW	8.0

[1] 1921–1960; [2] 1891–1960.

TABLE XXVI

CLIMATIC TABLE FOR SALZBURG

Latitude 47°48′N, longitude 13°00′E, elevation 435 m

Month	Mean sta. press. (mbar)	Temperature (°C)				Mean vapour press. (mbar)	Precipitation (mm)	
		daily mean	daily range	extremes			mean	max. in 24 h
				max.	min.			
Jan.	965.3	− 2.5	7.3	15.2	−30.4	4.6	73	77
Feb.	964.0	− 1.1	8.5	19.4	−30.6	4.9	70	55
Mar.	964.2	3.7	10.2	24.2	−18.1	6.1	70	49
Apr.	963.6	8.3	10.6	30.0	− 9.2	8.0	89	64
May	963.9	13.2	11.3	31.2	− 3.4	10.9	127	104
June	966.1	16.0	10.9	35.0	0.2	13.5	167	103
July	966.3	17.8	11.0	36.2	5.2	15.2	191	135
Aug.	966.0	17.1	10.6	36.3	2.0	15.1	163	120
Sept.	967.0	14.0	10.7	32.6	− 3.0	12.8	111	62
Oct.	966.0	8.4	9.7	26.0	− 8.3	9.3	82	50
Nov.	965.3	3.3	7.4	21.5	−16.7	6.7	70	56
Dec.	965.4	− 0.9	6.3	16.3	−27.7	5.2	65	47
Annual	965.2	8.1	9.5	36.3	−30.6	9.4	1,278	135

Month	Number of days with			Mean cloud-iness (%)	Mean sun-shine (h)	Wind	
	precip. ≥1 mm	thunder-storm	fog			preval. direct.	mean speed (m/sec)
Jan.	12.8	0.2	7.5	72	74	SE	2.2
Feb.	12.1	0.3	6.4	68	97	SE	2.0
Mar.	11.5	0.6	3.5	62	141	NW	2.2
Apr.	13.5	1.4	1.5	66	158	SE	2.3
May	14.9	5.6	0.9	65	190	SE	2.1
June	16.2	6.8	0.8	64	208	SE	2.0
July	16.3	7.6	0.9	62	219	SE	2.1
Aug.	15.1	5.7	1.6	55	218	SE	2.0
Sept.	12.2	2.2	4.4	55	180	SE	1.7
Oct.	11.0	0.2	7.7	63	135	NW	1.6
Nov.	11.5	0.2	8.9	74	70	SE	1.8
Dec.	12.1	0.2	10.1	74	64	SE	1.8
Annual	159.2	31.0	54.2	65	1,754	SE	2.0

TABLE XXVII

CLIMATIC TABLE FOR GARMISCH-PARTENKIRCHEN

Latitude 47°30′N, longitude 11°06′E, elevation 704 m

Month	Mean sta. press. (mbar)	Temperature (°C)				Mean vapour press. (mbar)	Precipitation (mm)	
		daily mean	daily range	extremes[1]			mean	max. in 24 h[2]
				max.	min.			
Jan.	1018.3	− 4.3	8.0	17.3	−27.0	4.4	96	64
Feb.	1017.1	− 2.9	8.9	21.4	−29.3	4.4	87	83
Mar.	1016.2	1.9	10.0	24.2	−19.0	5.7	78	62
Apr.	1016.7	6.4	10.5	28.5	−12.1	7.1	86	66
May	1016.6	10.9	10.9	31.7	− 6.2	9.7	134	83
June	1016.7	14.2	10.6	32.7	0.0	12.8	174	95
July	1016.7	16.0	10.6	34.9	2.0	14.1	200	75
Aug.	1016.3	15.2	10.5	34.2	1.0	14.0	171	98
Sept.	1018.1	12.4	10.4	31.4	− 4.0	11.9	126	74
Oct.	1018.9	6.6	9.5	27.8	− 9.5	8.7	89	48
Nov.	1019.0	1.7	7.8	24.0	−16.4	6.3	78	52
Dec.	1018.0	− 2.7	7.2	17.1	−22.7	5.2	69	124
Annual	1017.4	6.3	9.6	34.9	−29.3	8.7	1,388	124

Month	Number of days with			Mean cloud-iness (%)	Mean sun-shine (h)	Wind	
	precip. ≥1 mm	thunder-storm	fog			preval. direct.	mean speed (m/sec)
Jan.	11.0	−	2.1	69	82	S	1.1
Feb.	11.2	−	2.1	69	99	N	1.8
Mar.	10.6	0.1	0.6	66	151	N	1.5
Apr.	12.2	1.6	0.4	69	154	N	1.7
May	14.3	5.0	−	69	173	N	1.5
June	16.5	8.2	0.1	73	154	N	1.3
July	15.9	7.5	−	68	180	N	1.3
Aug.	15.8	7.2	0.2	64	186	N	1.3
Sept.	12.2	2.1	0.8	59	172	N	1.5
Oct.	10.4	0.7	2.1	59	146	N	1.4
Nov.	10.0	0.1	3.4	70	91	N	1.2
Dec.	10.2	−	3.8	67	71	S	1.1
Annual	150.3	32.5	15.6	67	1,659	N	1.3

[1] 1889–1960; [2] 1893–1960.

TABLE XXVIII

CLIMATIC TABLE FOR ZÜRICH

Latitude 47°23'N, longitude 08°34'E, elevation 569 m

Month	Mean sta. press. (mbar)	Temperature (°C)				Mean vapour press.*3 (mbar)	Precipitation (mm)	
		daily mean	daily range*1	extremes*2			mean	max. in 24 h*3
				max.	min.			
Jan.	949.2	− 1.1	5.6	16.0	−18.6	4.9	74	54
Feb.	948.7	0.3	7.9	18.2	−24.8	5.0	70	46
Mar.	948.4	4.5	9.0	22.3	−11.5	6.0	66	45
Apr.	948.3	8.6	10.4	29.2	− 6.7	7.5	80	47
May	948.7	12.7	11.2	31.8	− 2.0	10.0	107	90
June	951.0	15.9	11.4	34.9	3.3	12.3	136	81
July	951.2	17.6	11.4	36.4	5.6	13.8	143	60
Aug.	951.0	17.0	10.7	34.7	4.1	13.8	131	65
Sept.	951.8	14.0	10.3	31.4	− 0.4	12.4	108	103
Oct.	950.7	8.6	8.3	25.7	− 6.1	9.3	80	38
Nov.	949.4	3.7	5.8	22.6	−11.8	6.7	76	47
Dec.	949.2	0.1	5.0	16.1	−19.3	5.4	65	50
Annual	949.8	8.5	8.9	36.4	−24.8	8.9	1,136	103

Month	Number of days with			Mean cloud-iness (%)	Mean sun-shine (h)	Wind		Mean evap.*4 (mm)
	precip. ⩾1 mm	thunder-storm*5	fog			preval. direct.*6	mean speed*6 (m/sec)	
Jan.	11.5	0.1	6.5	83	46	W/SW	2.6	17
Feb.	10.4	0.3	5.1	73	79	W	3.0	30
Mar.	9.4	0.5	2.6	62	149	W	3.1	49
Apr.	11.0	2.4	1.0	63	173	W	3.2	66
May	12.7	5.2	0.6	63	207	W	2.9	92
June	13.0	7.9	0.7	63	220	W	2.9	81
July	12.9	8.7	0.5	59	238	W	2.7	83
Aug.	12.8	7.3	1.0	58	219	W	2.6	72
Sept.	10.4	2.3	3.4	61	166	W	2.4	55
Oct.	10.0	0.3	6.6	71	108	W	2.3	31
Nov.	10.3	0.4	6.7	82	51	W	2.8	20
Dec.	10.2	–	7.1	86	37	W	3.2	18
Annual	134.6	35.4	41.8	69	1,693	W	2.8	614

*1 1959–1968; *2 1901–1964; *3 1901–1960; *4 1954–1963, Wild scales; *5 Zürich-Kloten, 1958–1969; *6 Zürich-Kloten, 1961–1970.

TABLE XXIX

CLIMATIC TABLE FOR INNSBRUCK

Latitude 47°16'N, longitude 11°24'E, elevation 582 m

Month	Mean sta. press. (mbar)	Temperature (°C)				Mean vapour press. (mbar)	Precipitation (mm)	
		daily mean	daily range	extremes			mean	max. in 24 h
				max.	min.			
Jan.	948.2	− 2.8	7.5	18.5	−26.6	4.1	57	58
Feb.	947.2	− 0.5	8.7	17.5	−26.9	4.4	52	34
Mar.	947.0	4.8	10.8	24.8	−16.9	5.3	43	39
Apr.	946.3	9.3	11.8	28.6	− 4.7	6.9	55	39
May	946.6	13.8	12.3	32.6	− 2.0	9.5	77	61
June	948.8	16.7	12.4	35.9	0.6	12.3	114	54
July	949.2	18.1	12.1	36.9	4.2	14.1	140	65
Aug.	949.2	17.4	11.4	34.5	3.3	14.0	113	76
Sept.	950.2	14.6	11.6	30.7	− 0.8	11.9	84	52
Oct.	949.0	9.0	10.2	24.8	− 4.2	8.5	71	49
Nov.	948.0	3.4	7.4	23.0	−15.2	6.1	57	39
Dec.	948.6	− 1.1	6.5	17.9	−24.8	4.7	48	44
Annual	948.2	8.6	10.2	36.9	−26.9	8.5	911	76

Month	Number of days with			Mean cloud-iness (%)	Mean sun-shine (h)	Wind	
	precip. ⩾1 mm	thunder-storm	fog			preval. direct.	mean speed (m/sec)
Jan.	8.5	0.1	3.1	63	73	W	1.2
Feb.	7.7	0.0	2.2	62	105	W	1.3
Mar.	7.0	0.1	0.6	58	158	E	1.4
Apr.	9.3	0.5	0.3	63	166	E	1.6
May	10.7	2.7	0.1	64	189	E	1.5
June	14.3	6.4	0.1	64	190	E	1.4
July	13.9	8.4	0.1	61	210	E	1.2
Aug.	12.7	5.6	0.1	60	199	E	1.2
Sept.	10.0	2.2	0.6	55	177	E	1.2
Oct.	8.4	0.3	2.1	58	145	E	1.2
Nov.	7.9	0.0	3.5	66	86	W	1.2
Dec.	7.8	0.0	3.5	64	68	W	1.2
Annual	118.2	26.3	16.3	62	1,766	E	1.3

TABLE XXX

CLIMATIC TABLE FOR SANTIS

Latitude 47°15'N, longitude 09°20'E, elevation 2,500 m

Month	Mean sta. press. (mbar)	Temperature (°C) daily mean	daily range*1	extremes*2 max.	min.	Mean vapour press.*3 (mbar)	Precipitation (mm) mean	max. in 24 h*2
Jan.	743.0	−9.0	5.6	4.3	−32.0	2.5	202	125
Feb.	742.7	−9.0	5.5	5.7	−30.4	2.5	180	88
Mar.	744.5	−6.6	5.4	7.2	−23.7	3.0	164	128
Apr.	746.3	−4.1	5.2	11.6	−19.9	3.6	166	129
May	749.5	0.4	5.2	17.8	−15.3	5.3	197	172
June	753.4	3.6	5.6	17.0	−9.4	6.7	249	183
July	754.8	5.6	5.5	20.5	−5.6	7.6	302	110
Aug.	754.8	5.5	5.6	18.7	−5.7	7.0	278	112
Sept.	754.1	3.5	5.3	16.0	−13.0	6.3	208	96
Oct.	750.2	−0.6	4.8	12.5	−16.8	4.5	183	91
Nov.	746.4	−4.5	4.9	10.4	−21.4	3.4	190	168
Dec.	744.1	−7.6	5.1	6.1	−30.1	2.7	168	115
Annual	748.7	−1.9	5.3	20.5	−32.0	4.6	2,487	183

Month	Number of days with precip. ≥1 mm	thunder- storm*4	fog	Mean cloud- iness (%)	Mean sun- shine (h)	Wind preval. direct.*2	mean speed*2 (m/sec)
Jan.	14.3	0.1	18.1	65	112	WSW	8.0
Feb.	13.3	0.1	16.8	65	123	WSW	7.5
Mar.	12.9	0.5	17.5	63	166	WSW	6.5
Apr.	14.7	1.7	20.3	70	160	WSW	5.9
May	14.5	5.1	21.8	72	184	WSW	5.4
June	17.1	7.0	23.4	74	174	WSW	5.7
July	16.6	7.7	23.8	71	196	WSW	6.3
Aug.	16.2	6.1	23.1	69	186	WSW	6.6
Sept.	13.1	2.6	19.1	63	170	WSW	6.2
Oct.	11.8	0.3	15.6	61	163	WSW	6.6
Nov.	11.7	0.1	15.3	62	129	WSW	7.0
Dec.	13.4	0.1	16.6	61	117	WSW	7.7
Annual	169.6	31.4	231.4	66	1,880	WSW	6.6

*1 1959–1968; *2 1901–1960; *3 1901–1964; *4 1941–1960.

TABLE XXXI

CLIMATIC TABLE FOR SONNBLICK

Latitude 47°03'N, longitude 12°57'E, elevation 3,106 m

Month	Mean sta. press. (mbar)	Temperature (°C) daily mean	daily range	extremes max.	min.	Mean vapour press. (mbar)	Precipitation (mm) mean	max. in 24 h
Jan.	686.3	−13.2	4.7	1.0	−31.3	1.8	115	76
Feb.	685.9	−13.0	4.7	3.4	−36.6	1.8	108	65
Mar.	688.2	−11.2	4.6	1.7	−30.2	2.2	112	63
Apr.	690.6	−8.2	4.6	5.0	−23.2	3.0	153	50
May	694.6	−3.8	4.3	9.4	−19.8	4.3	136	108
June	698.7	−0.6	4.3	12.0	−10.6	5.5	142	51
July	700.6	1.6	4.4	12.8	−10.5	6.4	154	34
Aug.	700.3	1.4	4.3	12.0	−10.0	6.3	134	49
Sept.	699.6	−0.5	4.0	9.9	−15.5	5.2	104	37
Oct.	695.2	−4.3	4.0	8.6	−18.6	3.7	118	57
Nov.	690.9	−8.3	4.0	4.8	−23.2	2.8	108	49
Dec.	688.0	−11.4	4.4	1.4	−30.4	2.1	111	62
Annual	693.3	−6.0	4.4	14.2	−36.6	3.8	1,495	108

Month	Number of days with precip. ≥1 mm	thunder- storm	fog	Mean cloud- iness (%)	Mean sun- shine (h)	Wind preval. direct.	mean speed (m/sec)
Jan.	13.3	0.0	21.3	67	109	N	7.3
Feb.	12.0	0.0	19.6	67	118	N	7.1
Mar.	13.3	0.0	22.6	70	147	SW	6.7
Apr.	16.1	0.1	24.5	77	139	SW	5.8
May	15.3	1.4	26.7	81	148	SW	5.2
June	16.7	4.8	25.5	80	150	N	5.1
July	17.5	7.1	26.6	78	167	SW	5.1
Aug.	16.2	5.8	25.9	74	165	SW	5.2
Sept.	12.3	1.9	22.8	67	161	SW	5.5
Oct.	12.2	0.1	20.9	65	147	SW	6.2
Nov.	12.2	0.1	20.2	66	112	SW	6.6
Dec.	12.7	0.0	20.7	65	108	SW	6.8
Annual	169.8	21.3	277.3	71	1,671	SW	6.0

TABLE XXXII

CLIMATIC TABLE FOR GRAZ

Latitude 46°59'N, longitude 15°27'E, elevation 342 m

Month	Mean sta. press. (mbar)	Temperature (°C)				Mean vapour press. (mbar)	Precipitation (mm)	
		daily mean	daily range	extremes			mean	max. in 24 h
				max.	min.			
Jan.	976.4	− 3.8	8.6	13.6	−23.0	4.1	31	29
Feb.	975.2	− 1.5	10.8	18.5	−20.0	4.5	35	41
Mar.	975.3	3.4	11.0	23.7	−14.8	5.8	34	21
Apr.	974.1	9.0	13.1	27.5	− 3.8	7.9	52	40
May	974.6	13.7	12.0	32.5	− 1.4	11.1	93	55
June	975.9	17.1	12.1	36.0	3.2	14.0	126	90
July	975.7	19.0	12.1	37.1	4.9	15.7	114	105
Aug.	976.2	18.0	12.3	35.6	4.1	15.4	91	49
Sept.	977.8	14.3	11.9	32.9	− 0.3	13.0	80	70
Oct.	977.4	8.6	10.7	27.0	− 6.4	9.5	79	56
Nov.	977.1	3.3	7.4	20.0	− 9.0	6.8	57	44
Dec.	977.0	− 1.3	6.7	15.0	−19.0	5.1	48	28
Annual	976.1	8.3	10.6	37.1	−23.0	9.4	840	105

Month	Number of days with			Mean cloud- iness (%)	Mean sun- shine (h)	Wind	
	precip. ⩾1 mm	thunder- storm	fog			preval. direct.	mean speed (m/sec)
Jan.	5.9	0.0	14.4	68	75	S	1.2
Feb.	5.8	0.1	11.7	62	107	S	1.4
Mar.	5.9	0.3	6.8	60	151	N	1.6
Apr.	7.9	1.8	4.7	60	178	S	1.7
May	10.8	6.1	5.8	61	210	S	1.8
June	11.4	7.9	5.6	58	234	NW	1.6
July	11.3	8.2	6.3	53	259	NW	1.6
Aug.	10.5	6.8	9.6	51	240	NW	1.4
Sept.	8.4	2.7	12.1	53	188	NW	1.4
Oct.	8.5	0.4	13.6	61	133	S	1.2
Nov.	8.3	0.0₃	12.6	74	70	S	1.2
Dec.	6.8	0.0	17.1	74	61	N	1.1
Annual	101.5	34.3	120.3	61	1,906	S	1.4

TABLE XXXIII

CLIMATIC TABLE FOR KLAGENFURT

Latitude 46°39'N, longitude 14°20'E, elevation 448 m

Month	Mean sta. press. (mbar)	Temperature (°C)				Mean vapour press. (mbar)	Precipitation (mm)	
		daily mean	daily range	extremes			mean	max. in 24 h
				max.	min.			
Jan.	964.7	− 5.3	6.5	11.7	−27.0	4.0	39	45
Feb.	963.0	− 2.6	9.1	14.9	−24.5	4.6	42	53
Mar.	963.7	3.1	10.5	21.2	−18.7	5.9	39	62
Apr.	962.4	8.7	11.3	27.8	− 7.1	8.1	69	45
May	963.0	13.3	11.0	30.8	− 4.3	10.8	88	50
June	965.2	17.0	11.3	34.7	1.6	14.0	124	63
July	964.4	18.6	11.6	37.4	4.0	15.9	122	68
Aug.	963.7	17.7	11.2	34.3	3.1	15.6	102	57
Sept.	966.2	14.1	10.0	32.4	− 0.4	13.3	87	106
Oct.	965.6	8.1	8.6	23.6	− 8.0	9.7	87	91
Nov.	965.2	2.3	5.4	17.5	−11.7	6.8	73	58
Dec.	965.4	− 2.5	4.3	11.8	−23.0	5.0	54	53
Annual	964.3	7.7	9.2	37.4	−27.0	9.5	926	106

Month	Number of days with			Mean cloud- iness (%)	Mean sun- shine (h)	Wind	
	precip. ⩾1 mm	thunder- storm	fog			preval. direct.	mean speed (m/sec)
Jan.	5.6	0.0	13.4	69	74	NW	0.8
Feb.	6.0	0.1	8.1	59	116	NW	1.0
Mar.	5.7	0.3	4.0	57	165	E	1.4
Apr.	8.7	1.6	2.5	60	188	E	1.6
May	10.7	4.6	2.8	63	210	E	1.4
June	11.8	8.0	1.9	61	231	E	1.4
July	10.9	8.9	1.8	54	256	E	1.3
Aug.	9.9	7.0	4.7	53	241	E	1.2
Sept.	7.8	3.4	9.0	57	183	E	1.2
Oct.	8.4	1.0	14.3	66	118	E	1.0
Nov.	7.2	0.2	13.3	78	54	NW	1.0
Dec.	6.9	0.0	13.3	79	42	NW	0.8
Annual	99.6	35.1	89.1	63	1,878	E	1.2

TABLE XXXIV

CLIMATIC TABLE FOR GENEVA (Observatory)

Latitude 46°12'N, longitude 06°09'E, elevation 405 m

Month	Mean sta. press. (mbar)	Temperature (°C)				Mean vapour press.*3 (mbar)	Precipitation (mm)	
		daily mean	daily range*1	extremes*2			mean	max. in 24 h*2
				max.	min.			
Jan.	969.6	1.1	6.0	17.3	−14.9	5.6	63	46
Feb.	968.9	2.2	7.8	20.6	−18.3	5.7	56	42
Mar.	967.8	6.1	9.3	22.9	−11.4	6.8	55	49
Apr.	967.8	10.0	10.7	27.5	− 4.5	8.3	51	45
May	967.8	14.1	11.2	32.0	− 1.7	11.1	67	45
June	970.0	17.8	11.8	35.7	3.8	13.7	89	58
July	970.2	19.9	12.5	38.3	5.3	15.2	64	63
Aug.	969.8	19.1	11.5	36.5	4.8	15.2	94	71
Sept.	971.0	15.8	10.6	34.8	1.2	13.7	99	80
Oct.	970.2	10.3	8.7	26.0	− 3.7	10.3	72	61
Nov.	969.2	5.7	6.3	23.2	− 8.5	7.5	83	76
Dec.	969.4	2.1	5.3	20.8	−16.0	6.1	59	54
Annual	969.3	10.3	9.3	38.3	−18.3	9.9	852	80

Month	Number of days with			Mean cloud-iness (%)	Mean sun-shine (h)	Wind		Mean evapor.*4 (mm)
	precip. ≥1 mm	thunder-storm*5	fog			preval. direct.*6	mean speed*6 (m/sec)	
Jan.	9.2	–	3.5	79	54	SW	3.0	12
Feb.	7.6	0.6	1.9	66	98	SW	3.6	22
Mar.	7.5	0.3	0.6	55	169	NE	3.5	36
Apr.	8.0	1.1	0.3	53	206	NE	3.2	52
May	9.1	3.4	0.1	53	243	NE	2.9	62
June	9.7	5.9	0.2	48	269	NE	2.8	65
July	7.4	7.2	0.2	43	297	SW	2.7	81
Aug.	9.1	6.2	0.3	46	266	SW	2.7	62
Sept.	8.9	2.5	0.8	52	198	NE	2.3	47
Oct.	8.5	0.7	3.1	63	131	NE	2.2	18
Nov.	9.4	0.6	3.9	78	61	NE	2.7	19
Dec.	8.9	0.1	4.3	82	44	SW	3.5	15
Annual	103.3	28.6	19.2	60	2,036	NE	2.9	491

*1 1959–1968; *2 1901–1960; *3 1901–1964; *4 Changins/Nyon, 1964–1972; *5 Geneva-Cointrin, 1958–1969; *6 Geneva-Cointrin, 1961–70.

TABLE XXXV

CLIMATIC TABLE FOR LUGANO

Latitude 46°00'N, longitude 08°57'E, elevation 276 m

Month	Mean sta. press. (mbar)	Temperature (°C)				Mean vapour press.*3 (mbar)	Precipitation (mm)	
		daily mean	daily range*1	extremes*2			mean	max. in 24 h*2
				max.	min.			
Jan.	983.6	1.9	7.3	24.6	−12.5	5.0	62	70
Feb.	982.4	3.6	8.5	24.8	−14.0	5.1	67	80
Mar.	982.8	7.5	9.5	27.0	− 6.6	6.6	98	107
Apr.	981.8	11.7	10.3	31.4	− 2.7	8.5	148	92
May	982.0	15.4	10.4	32.6	0.5	11.8	214	125
June	983.2	19.3	10.7	34.4	3.3	14.5	198	134
July	983.0	21.4	11.0	38.0	8.0	16.2	185	129
Aug.	983.1	20.5	10.8	36.4	6.5	16.3	196	263
Sept.	985.1	17.4	9.8	36.0	2.0	14.1	159	139
Oct.	984.5	12.1	8.7	28.2	− 2.4	10.6	173	155
Nov.	983.9	6.9	6.8	22.8	− 6.0	7.4	147	100
Dec.	983.5	3.1	6.8	21.4	− 9.6	5.6	95	86
Annual	983.2	11.7	9.2	38.0	−14.0	10.1	1,742	263

Month	Number of days with			Mean cloud-iness (%)	Mean sun-shine (h)	Preval. wind direct.*3			Mean wind speed*6 (m/sec)	Mean evapor.*4 (mm)
	precip. ≥1 mm	thunder-storm*5	fog			7h30	13h30	21h30		
Jan.	6.0	–	1.5	50	117	N	S	N	2.0	47
Feb.	5.9	0.1	0.9	47	143	N	S	N	2.0	53
Mar.	7.2	0.8	0.5	51	171	N	S	N	2.1	80
Apr.	9.9	3.5	0.1	52	186	N	S	N	2.1	105
May	13.1	6.9	–	57	191	N	S	N	2.0	117
June	12.0	7.9	–	48	234	N	S	N	1.9	101
July	10.0	11.1	–	42	268	N	S	N	1.9	135
Aug.	10.4	9.4	–	43	243	N	S	N	1.8	111
Sept.	8.9	3.8	0.0	47	189	N	S	N	1.8	73
Oct.	9.0	1.7	0.2	54	147	N	S	N	1.8	57
Nov.	8.3	1.6	1.0	54	110	N	S	N	1.9	40
Dec.	7.3	–	1.2	51	102	N	S	N	1.9	43
Annual	108.0	46.8	5.4	50	2,101	N	S	N	1.9	962

*1 1959–1968; *2 1901–1960; *3 1901–1964; *4 1954–1963, Wild scales, Pregassona/Lugano; *5 1958–1969, Locarno–Aeroporto; *6 1951–1970.

Chapter 3

The Climate of Poland, Czechoslovakia and Hungary

WINCENTY OKOŁOWICZ

The history of meteorological records and climatological summaries

Czechoslovakia, Poland and Hungary have a meteorological and climatological research history of many years and even centuries.

At the Jagellonian University in Cracow a number of professors have carried out fairly regular weather records noted generally in the margins of the astrological calendars of the years 1487–1585. Very detailed records of this kind were made by the Rev. Marcin Biem of Olkusz, one of the teachers of Copernicus, for many years the Rector of the Cracow University. His observations of atmospheric phenomena refer to a period of more than 40 years, with the exception of some intervals. These chronicles of the weather belong to the oldest non-instrumental meteorological observations carried out for many years.

The first observations with the use of instruments were carried out for some time in Warsaw, on the break of the years 1654 and 1655, in the framework of the so-called Florentine meteorological network organised on the initiative of Ferdinand II, the great Toscanian prince.

In the 18th century further trials were made of meteorological observations, for longer or shorter periods of time, in several scores of places, particularly in countries of the eastern part of central Europe. For instance, in the Astronomical Observatory at the Collegium Clementinum in Prague, regular instrumental observations had already commenced in 1769 (fragmentary ones in 1752); in Warsaw they started in 1776 and have been carried out regularly since 1779; in Vilna sporadic observations commenced in 1753 and regular ones began in 1777; for Cracow the years were 1792 and 1824, respectively, etc.

On the basis of the observations begun in 1776 by J. F. Bończa-Bystrzycki (a royal physicist and astronomer in Warsaw), and continued from 1803 by A. Szeliga-Magier (professor in physics), W. Jastrzębowski published in 1828, *Carte météorographique de la capitale du Royaume de Pologne*. Its second edition under the title *Carte climatologique de Varsovie comme point central de l'Europe* was published in 1846. It was one of the two earliest monographs on the climate of a town in the world literature.

Josef Stepling and Anton Strnad, the first directors of the Astronomic Observatory in Prague, were the precursors of meteorological researches in Czechoslovakia. A. Strnad had initiated the establishment of a system of meteorological stations which was eventually founded in 1817. The results of observations and the reports on the influence of weather conditions on the vegetation of agricultural cultures were published in annuals,

Using these annuals Karel Kreil, the sixth director of the Astronomic Observatory in Prague, prepared his *Climatology of Bohemia* (1864); this was the first description of the climate of Czechoslovakia. K. Kreil organised the Austrian Central Meteorological Service (Meteorol. Zentralanstalt für Österreich) and became its first director in 1851. It is worth mentioning that in Czechoslovakia an ombrometrical network existed which comprised in 1880 over 700 stations. Later, a similar system was organised in Moravia (in the environs of Brno).

In Hungary, which was autonomous within the Austro-Hungarian empire, a state network of meteorological stations was first organised on the initiative of the Hungarian Academy of Sciences. The Central Institute of Meteorology and Earth Magnetism headed this network, which comprised over 40 stations and in the early 20th century over 140. This institute was formally approved in 1870.

In Poland, divided in the 19th century among the neighbouring powers (Russia, Germany and Austro-Hungary), "regional" networks of meteorological stations have been developed. The Vilna system was the oldest and it comprised nearly 60 stations equipped with uniform instruments and working according to a common instruction written by J. Mickiewicz, professor of physics at the Vilna University. This system was discontinued in 1831, after the insurrection against the Russian Tsar. Similar regional systems were also organised in Galicia (in 1865, initiated by the Physiographic Commission of the Cracow Scientific Society) and in central Poland, the "Warsaw network" with the Central Station and Meteorological Office in Warsaw, on the initiative of the Sugar Section of the Society for Industry and Commerce. The "Warsaw network" consisted of nearly 65 meteorological and 270 pluviometrical stations. Its directors were successively: Kwietniewski, Merecki, and from 1909 onwards Gorczyński.

Before the outbreak of World War I, the meteorological systems organised by the three partitioning powers (Austro-Hungary, Germany and Russia) were in operation on the territory of Poland. Of the formerly existing networks only the "Warsaw network" still exists. It has its central office in Warsaw and was transformed after the recovery of independence in 1919, into the State Meteorological Institute, headed by W. Gorczyński. At the same time, in Czechoslovakia an independent state meteorological service was established with the central office in Prague.

World War II brought great losses to meteorological services; in Poland they were completely destroyed. In Poland and Czechoslovakia newly established meteorological and hydrological services were joined together; in Hungary they operate independently (FICKER, 1951; HRGIAN, 1959; GREGOR, 1966; ROJECKI, 1966).

Geographical position

The area of eastern central Europe under consideration in this chapter extends from slightly below 46°N in its southern part to almost 55°N in its northern region. The width of the zone lying between those parallels is slightly greater than the shortest distance between the Adriatic and the Baltic Sea. The part of Czechoslovakia extending farthest west (almost 12°E) lies beyond the line connecting the two seas Venice Bay and Pomorze Bay, and the easternmost corner of Poland comes within reach of the meridian passing through Riga (ca.24°E). The point at the Drava River where the Hungarian territory extends farthest south (approximately to the parallel passing through Trieste), and the promontory Rozewie at the Baltic in Poland, advanced northernmost, lie at approximately the same longitude, between the meridians 18° and 19°E. Eastward from here the area broadens constantly while it merges with the great east European lowland (see Fig.1).

The fairly considerable meridional extent of the discussed area (about 9°) leads to the expectation of relatively large differences in solar conditions resulting from the possible differences in length of the day and altitude of the sun at the point of culmination. The location in a site where the relatively narrow and disintegrated body of western Europe merges into the much wider Eurasian land area suggests that oceanic as well as continental air masses may readily penetrate into this area. The lowlands of northern and central Poland which are not protected by any elevations on their west, north, and east sides, offer to both kinds of air masses very easy access. The western part of Czechoslovakia (Czech Massif) is, on the other hand, surrounded on all sides by mountains up to 1,000 m and more a.s.l. The greatest altitudes are found in the Sudetes (up to and over 1,600 m height) and, separated from them by the Moravian Gate, the Carpathian Mountains (reaching in the Tatra 2,655 m a.s.l.). Together they form the largest mountain complex of the discussed area. The east part of Czechoslovakia (i.e. Slovakia), beginning from the northern border and extending to the middle of the country, is covered by a number of mountain ranges, which lower in southward direction. A major plain appears here in the southwest part of the country on the Danube, to where the Small Hungarian Plain extends. The latter is separated from the Great Hungarian Lowland by a mountain range of 700 to about 1,000 m a.s.l. The major part of both lowlands lies on Hungarian territory. They are surrounded by mountains on all sides. The flow of fresh maritime air masses to low-lying terrains of this type is impeded. Moist air masses passing over mountains are always undergoing some transformation, e.g. by föhn processes. Thus, not only the greater distance from sea but also the location in basins, surrounded by mountains, contribute to the fact that the climate shows more distinctly continental features in some parts of Czechoslovakia and Hungary than in Poland.

Sunshine and radiation

Table I shows approximate length of the day during the summer and winter solstice for the northern, central, and southern parts of the area.

Another essential factor for the influx of radiation energy is the sun's altitude. The greater this altitude, the lesser the depletion of solar radiation in the atmosphere. The changes in altitude of the sun (at culmination) are shown in Table II.

With the solar altitudes (h) given in Table II, the relative (optical) thickness of the atmosphere (m) in summer varies from 1.2 in north Poland to 1.1 in south Hungary (for $h = 90°$, $m = 1.0$). With high solar altitude in the second half of June, the lessening of radiation due to the length of the radiation path in the atmosphere thus shows a difference of hardly 10% between the north and the south of the entire examined area. A similar difference in depletion of radiation is noted also in summer between south Hungary and the Tropic of Cancer areas. The differing amounts of radiation energy reaching the ground in summer in the areas between the Baltic and the Danubian plain

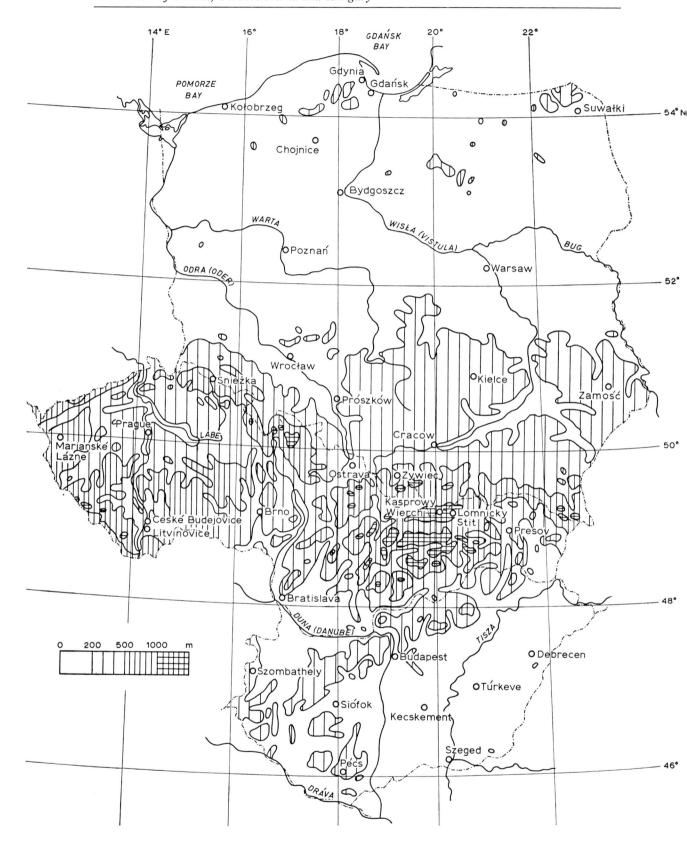

Fig.1. Localities mentioned in the tables and in the text.

TABLE I

EXTREME DAY LENGTHS IN SELECTED LOCALITIES[1]

	Gdańsk, Gdynia (ca. 54°30′N)	Prague, Cracow (50°05′N)	Szeged, Pécs (46°10′N)
Longest day	17 h 10 min	16 h 20 min	15 h 45 min
Shortest day	7 h 10 min	8 h 00 min	8 h 25 min
Difference	10 h 00 min	8 h 20 min	7 h 20 min

[1] Refraction accounted for with ca. 5 min accuracy.

TABLE II

ANGLE OF SOLAR ALTITUDE (h) AT CULMINATION DURING SUMMER (S.S.) AND WINTER SOLSTICES (W.S.)

	North Poland	Warsaw	Prague	Budapest	South Hungary
s.s.	59°	61°	63.5°	66°	68°
w.s.	12°	14°	16.5°	19°	21°

are probably more strongly affected by the length of the day than by the relative thickness of the atmospheric layer. In this surmise other possible causes are not taken into account. However, in the winter season the situation described above is different: if $h = 12°$, $14°$, $16°$, $19°$, $21°$, then $m = 4.7, 4.1, 3.6, 3.0, 2.8$. This implies that in north Poland (e.g. in Gdynia) the intensity of solar radiation in winter is depleted much more than in south Hungary. Furthermore, the length of the day in Gdańsk is in December over one hour shorter than, for example, at Szeged. This substantial difference in potential influx of solar radiation energy in winter is due exclusively to the difference in geographical latitude. Actually, though, also in summer the direct solar radiation is in Poland even smaller than that in areas located further to the south, although that might be expected because of the length of the day.

This situation is apparent from the data in the climatic tables (pp.115ff.) representing, among other things, mean monthly sunshine in selected localities.

In none of the Polish localities the mean sunshine duration in June exceeds 260 h. It is longest in Gdynia, which is located on Gdańsk Bay and is partly protected towards the west by the highest elevations of the Pomorze lake district (up to 329 m a.s.l.). Sunshine duration in that district (Chojnice) is in June about 30 h shorter than in Gdynia and approximately 25 h shorter than in the Mazury lake district (Suwałki).

In Czechoslovakia and Hungary the maximum sunshine duration which exceed the Polish maximum in June, occurs in July, and reaches over 300 h on the Hungarian lowland. Also in June, the sunshine duration is longer than in Poland, in spite of the shorter day. This fact indicates the occurrence of much greater cloudiness north of the Sudetes and the Carpathians than in the areas south of them, notably the Hungarian plains.

The foregoing analysis of geographical and insolation conditions facilitates the understanding of the distribution and annual course of solar radiation. Unfortunately, however, the data concerning that important climate-forming agent are scanty and inhomogeneous, and are presented here for the purpose of general information only (Table III).

TABLE III

MEAN TOTAL RADIATION (kcal./cm²) FOR VARIOUS PLACES DURING DIFFERENT PERIODS

	Jan.	Feb.	Mar.	Apr.	May	June	July	Aug.	Sept.	Oct.	Nov.	Dec.	Year
Gdynia[1]	1.4	3.1	6.2	10.1	14.0	15.5	14.1	11.7	7.8	4.3	1.9	0.9	91.0
Suwałki[2]	1.0	2.6	6.6	9.3	12.4	14.0	12.7	10.9	7.5	3.6	1.3	0.7	82.6
Chojnice[2]	1.1	2.8	6.4	8.9	12.1	13.5	11.8	10.2	7.7	3.7	1.4	0.7	80.3
Warsaw[2]	1.2	2.8	6.5	8.8	11.4	12.6	11.9	10.9	7.9	4.2	1.5	0.7	80.4
Cracow[2]	1.6	3.1	6.4	9.2	11.7	12.6	12.8	11.3	7.9	4.6	1.9	1.1	84.2
Debrecen[3]	2.7	4.0	8.2	11.7	15.8	16.2	17.3	15.9	12.3	7.3	2.8	2.1	116.3
Budapest[4]	2.1	3.6	7.4	10.6	13.6	14.8	15.4	13.2	9.8	5.8	2.1	1.5	99.9
Szeged[3]	2.9	4.4	8.0	11.0	14.6	14.9	15.2	14.0	10.3	6.8	2.7	2.4	107.2
Pécs[3]	2.9	4.1	7.9	11.2	13.3	13.9	14.5	13.5	9.4	5.9	2.3	2.1	101.0

[1] 1932–1937; after STASZEWSKI and UHORCZAK, 1966.
[2] 1951–1960; after PASZYŃSKI, 1966.
[3] 1958–1962; after ANONYMOUS, 1967.
[4] 1936–1960; after ANONYMOUS, 1967.

The selected places are located north of the Carpathians and in the Great Hungarian Plain.

At the end of spring and with summer approaching, the monthly radiation amounts are greater in the northern lake belt of Poland (Chojnice, Suwałki) than in central and south Poland (Warsaw, Cracow). But all of them are smaller than in Hungary, in spite of the fact that in this period of the year the days are shorter in Hungary than in the Baltic coast areas. In the winter months (with minimum radiation in December and January) radiation amounts in Hungary are 2 to 3 times greater than in Poland. Also the annual mean values indicate the markedly privileged position of the Hungarian plain compared with the Polish lowlands, which receive only 80% or even less of the total radiation energy obtained per horizontal unit in the Danubian lowlands.

The discussed areas are considerably warmer (particularly in winter) than those located at approximately the same geographical latitudes in Asia (e.g. on Lake Baikal or Sakhalin), or in North America (e.g. southward of Hudson Bay), in spite of the fact that the solar conditions are roughly similar. This fact is due to the circulation of the atmosphere, which will be discussed in the following section.

Atmospheric circulation and winds

General circulation

Central Europe is located in a zone of prevailing westerlies, as a result oceanic influences are acting on the whole area, though the various parts are not uniformly affected by them. Hungary borders almost directly with the zone of Mediterranean climates where the prevalence of the westerlies is distinctly marked only in the cold half of the year, while during the warm season the Mediterranean countries are under the influence of the Azores high which fairly often reaches into the Danubian plains and does so much more frequently than to the lowlands sited north of the Carpathian Mountains.

In the European mean-pressure field of January, the axis of a high-pressure belt runs from Madrid through Vienna and Bratislava, then eastward between Odessa and Kiev. The 1,022 mbar isobars, encompassing in this belt the areas of highest pressure in the west and the east, reach on the one side from the Azores high almost to the Pyrenees and on the other from the Asiatic high to the east part of the Don Basin. Central Europe is thus placed in a saddle, half-way between the eastern peripheries of the Azores high and the western peripheries of the Siberian high. With this distribution of mean pressure in January southwest winds predominate north of the Sudeten and the Carpathians, while winds of a northern direction prevail over the central Danubian areas (ANONYMOUS, 1964).

In older maps, designed on the basis of observations during the period 1851–1900, several additional second-order high-pressure centers are marked: one over the southern Carpathians (near Slovakian Rudava), two over the Alps (the larger is over the eastern part of the Alps), and one over the Iberian Peninsula. They are characterized by pressure above 1,021 mbar (isobar 766 mm Hg). In this way the Hungarian plain was overlain during the period 1851–1900 by a lower pressure separating the local Carpathian highs from the Alpine highs (GORCZYŃSKI, 1917).

A more detailed and accurate picture of the isobars is given in the maps of the *Klima-Atlas von Ungarn* (ANONYMOUS, 1960), based on a fairly large amount of material originating from the period 1901–1950. Although the distribution of the isobars differs in details from the 1851–1900 pattern, the general features are similar. In the mean-pressure field, for the period 1901–1950, in January a local baric low (1020.5 mbar) subsists over the Great Hungarian Plain with its center between Danube and Tisza. Thence the atmospheric pressure increases in western direction toward the Alps (1021.2 mbar) and in directions north and northeast toward the Carpathians (1020.8 mbar) within Hungarian territory. The isobars pass, in January, over the Polish lowland from west-southwest to east-northeast, which explains the preponderance of wind directions from southwest in the mean wind field.

In summer, when southwestern Europe and the Mediterranean countries are occupied by the Azores high and in Asia low pressure persists, the southern component of the surface winds in Poland abates and the westerlies become dominant. In Czechoslovakia during summer winds from the sector northwest to north (particularly over the western part of the country) are dominant while the northern winds prevail over a smaller area than they would in winter—namely in eastern Slovakia. The preponderance of the northeast–southeast and southeast winds which are typical in winter, subsides in the south and the west part of Slovakia (OTRUBA, 1964). The wind directions with an eastern component, occurring in Slovakia in winter, are consistent with the described pressure distribution, namely the tendency for a development of a local high in the mountains of central and southern Slovakia, and of a local low over the Great Hungarian Plain. Similar conditions often prevail in this region in spring as well and primarily in fall. In October, the lowest mean pressure (less than 1018.0 mbar) occurs over the area between Danube and Tisza, increasing westward in the direction of the Alps and north- and eastward toward the Carpathians (to 1018.4 mbar within the Hungarian frontier). In fall the Hungarian low-pressure area is not closed on the south side, facilitating the inflow of moist-air masses from the Mediterranean Sea to the basin of the Hungarian plains.

The foregoing general characteristics of mean baric pressure distribution and the associated winds require a supplementary note concerning the disturbances developing on the polar and the arctic front. They are caused by the formation and mainly the east- or northeastward migration (most often in winter) of whole families of cyclones.

Migratory baric systems and their part in weather formation

Migratory anticyclones and cyclones exercise a particular effect on weather conditions in Poland, Czechoslovakia and Hungary owing to the relative proximity of the source regions of strongly contrasting air masses. Depending on the season of the year and on the location and size of the baric systems as well as the duration of their persistence the weather over these countries (or parts of them) becomes very cool or distinctly warm, moist or dry. The variance in annual frequency of the different circulation types and their associated weather types is ultimately decisive for the climatic conditions and their regional differentiation.

In order to examine the regularities prevailing in weather and climate, it is first of all necessary to distinguish the circulation types from the associated atmospheric conditions typical for the respective area. Several attempts have been made in this direction. So, for instance, PÉCZELY (1957) distinguishes for Hungary 13 such types, including among them 7 anticyclonic (A) and 6 cyclonic (C) circulation types.

The anticyclonic types (A) are marked with supplementary symbols showing the location of the high-pressure center in respect to Hungary (according to the cardinal points n, e, s, w), or over selected geographical regions: over Great Britain—B; over Feno-Scandinavia—F; location over Hungary or the near-Carpathians is not marked with a special sign.

The cyclonic types carry the following supplementary indexes: m for meridional circulation; z for zonal circulation; w for cases in which Hungary is located in the forefield of the low-pressure system (with warm fronts); c for Hungary in the rear field (with cold fronts); M denoting lows of Mediterranean origin; C without additional symbol: lows with the center over Hungary (as for A). General information on the frequency occurrence of the particular circulation types over Hungary is given in Fig.2, plotted after table II-b of PÉCZELY's work (1957).

19 circulation types were distinguished for Czechoslovakia by Rein and Konček (REIN, 1959), of which 8 are anticyclones (a) and 11 cyclones (c). The authors use additional letter symbols to indicate the direction of movement: Wa denoting e.g. west anticyclonic; NWa northwest anticyclonic circulation, and so forth. They also used other symbols, e.g. for the highs (H) and for the lows (O) if their centers were situated over central Europe (REIN, 1959; OTRUBA, 1964, pp.29–39).

The greatest occurrence frequency shows the type H with an annual mean of more than 12% (almost 16% in fall), and the type Wc—12% annual mean (18% in winter) (KONČEK et al., 1966).

The circulation types over Poland were worked out by LITYŃSKI (1968), who used a three-class, equal-probability method and three parameters: a zonal and a meridional circulation index and types of circulation: cyclonic, "zero" or anticyclonic. In this manner he obtained 27 types (with three classes for every of the 3 parameters). The circulation indexes were computed for a pair of meridians (0°–35°E) and a pair of

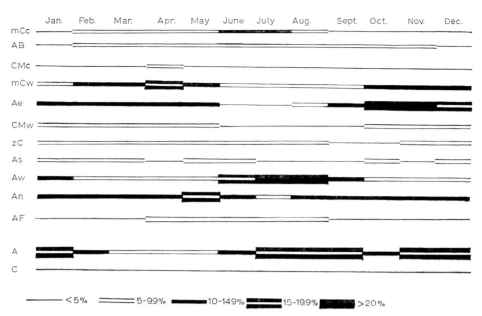

Fig.2. Circulation types after PÉCZELY (1957).

parallels (40°–65°N), enclosing a quadrant with central Europe in the centre. The circulation was defined according to the atmospheric pressure prevailing over central Poland (Warsaw). The pressure distribution indicates that, in agreement with the selected classification method, in January one should attribute to cyclonic circulation all those cases where the pressure over Warsaw was less than 1,011 mbar, while to anticyclonic circulation cases were attributed where pressure was above 1,022 mbar; the intermediate cases belong to a group called "zero" circulation. In July the analogous boundaries were 1,011 and 1,016 mbar, respectively.

It is worth mentioning that Péczely in his work for Hungary applied a uniform value of 1,015 mbar pressure (measured in Budapest) for the whole year to distinguish cyclonic from anticyclonic circulations.

The frequency occurrence of circulation types in Poland is shown graphically in Fig.3 after LITYŃSKI (1968). The character of circulation is indicated here by the direction of advection (similarly as in determination of wind direction) and an index. A zero used in connection with a direction symbol denotes a lack of relatively high velocity (advection), generally within the range of 0 to $\pm$ 3 m/sec. The symbol "–" indicates the directions N and E. The index "zero" is used when there was no distinct cyclonic or anticyclonic circulation (pressure in Warsaw 1,011–1,022 mbar in winter, 1,011–1,016 mbar in summer). The deviations from mean temperatures and mean precipitation amounts associated in general with different circulation types are given in Table IV. These baric data were collected by Lityński from source materials for a 67-year period (1900–1966). It is difficult to compare the three afore-mentioned modes used for typification of circulation systems over Poland, Czechoslovakia and Hungary. Each of them is based on a different method of type selection, which to some extent finds its expression also in the number of types chosen and amounts of symbols used. We also have to take into consideration the fact that the Rein–Konček scheme is based on a much shorter period, namely the period 1951–1960 (KONČEK et al., 1966). An attempt to compare the Polish

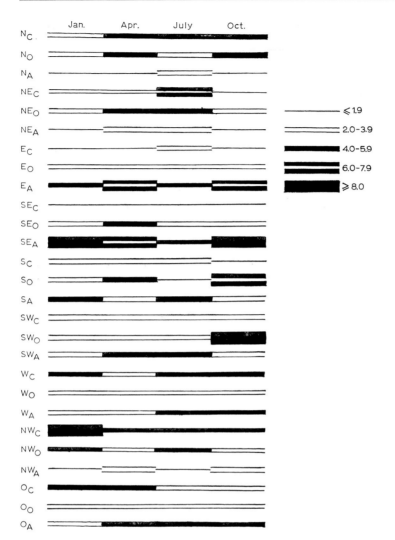

Fig.3. Circulation types after LITYŃSKI (1968).

TABLE IV

POSITIVE (+) AND NEGATIVE (−) DEVIATIONS[1] FROM MEAN TEMPERATURE/MEAN PRECIPITATION SUMS, ASSOCIATED WITH
DIFFERENT CIRCULATION TYPES

	Jan.	Apr.	July	Oct.		Jan.	Apr.	July	Oct.		Jan.	Apr.	July	Oct.
N_C	o/+	−/o	−/o	−/+	SE_C	o/o	o/o	o/o	o/o	W_C	+/o	o/+	o/o	o/+
N_0	o/o	−/−	−/o	−/o	SE_0	−/o	o/−	+/o	+/−	W_0	+/o	o/o	o/o	o/o
N_A	o/−	−/−	o/−	−/−	SE_A	−/−	+/−	+/−	−/−	W_A	o/−	o/−	+/−	+/−
NE_C	−/+	−/+	−/+	−/+	S_C	+/−	+/o	+/o	+/o	NW_C	+/+	−/+	−/+	o/+
NE_0	−/−	−/o	o/o	−/o	S_0	+/−	+/−	+/o	+/−	NW_0	+/o	−/o	−/o	−/o
NE_A	−/−	−/−	o/−	−/−	S_A	−/−	+/−	+/−	+/−	NW_A	o/o	−/−	o/−	o/−
E_C	−/o	−/+	−/+	−/+	SW_C	+/o	o/o	+/o	+/o	0_C	o/o	o/+	o/o	o/+
E_0	−/o	o/o	+/o	−/o	SW_0	+/−	+/o	+/o	+/−	0_0	o/o	o/−	o/o	o/o
E_A	−/−	−/−	+/−	−/−	SW_A	o/−	+/−	+/−	+/−	0_A	o/−	+/−	+/−	o/−

[1]o = no deviation.

and the Hungarian systems on the basis of a list of five winter and five summer months (January resp. July), selected at random from the first decade of the current century and from the years after 1950, yields approximately the following summary results:

(*1*) In January, with 35 cases of anticyclonic circulation over Poland (after LITYŃSKI, 1968), 29 anticyclonic and 6 cyclonic ones (of which 5 with lows of Mediterranean origin) were noted in Hungary (after PÉCZELY, 1957).

(*2*) In July, with 47 cases of anticyclonic circulation over Poland, 45 such cases were observed over Hungary, while only 2 cyclonic ones (mCc, while N_A and 0_A were noted in Poland).

(*3*) In January, with 76 cases of cyclonic circulation over Poland, in Hungary as many as 27 cases of anticyclonic type and only 49 cases of cyclonic type occurred; among the afore-mentioned 27 anticyclonic types as many as 24 occurred at a time when over Poland the circulations W_C and NW_C or 0_C were noted. In cases with W_C and NW_C in Poland, types A_w and A dominated over Hungary; with 0_C in Poland, type An dominated over Hungary.

(*4*) In July, with 64 occurrences of cyclonic types in Poland, Hungary had 26 days with anticyclones and 38 with cyclones.

Here, like the cases afore-mentioned, there is a general accordance regarding the direction of circulation although its types are sometimes defined differently for Poland and Hungary.

From the foregoing notes the following observations can be made:

(*1*) In winter, during high-pressure weather in Poland, Hungary may be under the influence of low pressure, mostly of Mediterranean origin in approximately 17% of cases (over the Mediterranean Sea occurs intense cyclogenesis at that time of the year).

(*2*) In summer, when Poland has high-pressure weather, this type of weather nearly always prevails in Hungary too.

(*3*) When Poland has low-pressure weather in winter, Hungary remains in 36% of all cases under the influence of a high, since at that period of the year Hungary lies often in the high-pressure belt extending from the Azores to Asia.

(*4*) In summer, if Poland is under cyclonic circulation, Hungary has in 41% of cases high-pressure weather, most frequently owing to the effect of the Azores high (type A_w). These observations have orientative nature only, since they were derived on basis of only 5 selected winter and 5 summer months, that is, based on situations from 155×2 days.

Examples of weather situations

For further illustration of possible formation of weather conditions in the eastern part of central Europe in individual cases some particular examples may be added here.

On January 22, 1962 (00h00 GMT) southern Europe was overlain by a high-pressure ridge (over 1,025 mbar) from which emanated a high-pressure wedge (up to 1,015 mbar) through the eastern Baltic countries toward the North Cap. North of Scotland a deep low (967 mbar) was present. In its reach a secondary low-pressure system developed over Skagerrak (980 mbar) on this day at 12h00 GMT.

The flow of heat advection was directed from southwest Europe towards Scandinavia (at the 500–1,000 mbar level). At 00h00 GMT of January 23, 1962 the low-pressure

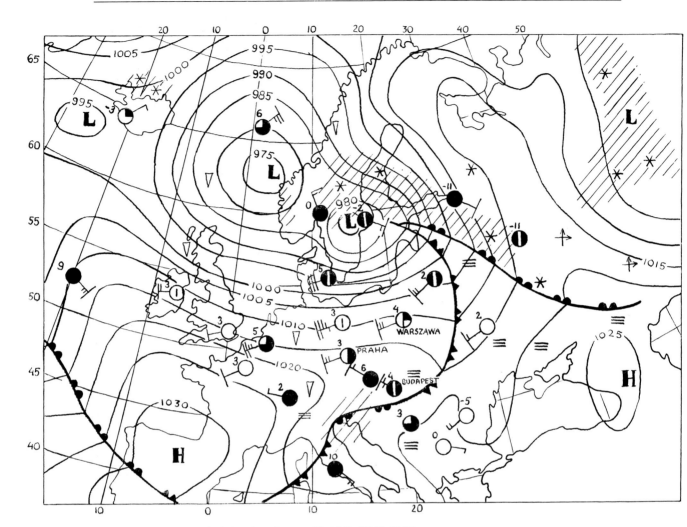

Fig.4. Synoptical situation, January 23, 1962, 00h00 GMT.

center over Skagerrak had already shifted to the environs of Stockholm, deepening to 974 mbar. The temperatures noted on this day in central Europe were positive (Fig.4). This case belonged to the type W₀ (after Lityński's classification), in which the temperatures are above normal, and the precipitation amount below normal (cf. Table IV). A situation with cold air advection (type NE₀) is shown in Fig.5. The advection does not yet reach Hungary. A stationary front separates it from Poland and Czechoslovakia. On the following days, a high-pressure wedge expanded eastward, forming there new anticyclonic centers. After the passing of several lows and advection of continental polar air at the end of January, negative temperatures prevailed in all central Europe, with lowest values in Budapest on January 29 (−8°C at midday) and next day in Warsaw (−21°C at 00h00 GMT and −16°C at midday; Fig.6). At that time part of Hungary was located in the zone of ample precipitation occurring at the margin of a low-pressure system of Mediterranean origin, it had remained in this position since January 8, 1963 (Fig.5). It is worth mentioning that the greatest rainfall and snow precipitation in fall and winter are brought to the Danube region by Adriatic cyclones, which are frequent in the period October to April.

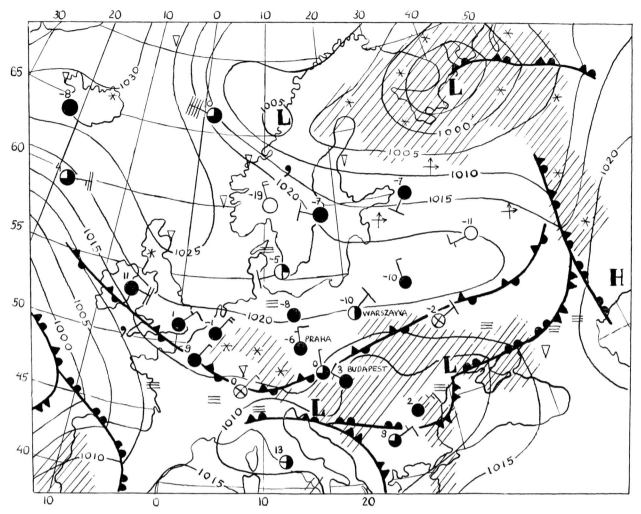

Fig.5. Synoptical situation, January 8, 1963, 12h00 GMT.

With the circulation occurring with those cyclones, it sometimes happens that warm moist air-streams pass over the mountains and reach the Hungarian plains, overriding the cold air masses stagnant over them during several days. This produces the development of advective inversion and a lengthy period of cloudy weather with precipitations (ANONYMOUS, 1960a, p.57).

An example of hot high-pressure weather was July 25, 1963 (Fig.7), when at 00h00 GMT 21°C in Warsaw, and 20°C both in Prague and Budapest was observed. This day was the last of a series of warm days with the highest temperatures of that month. On July 26, after the passing of a cold front, the temperatures decreased by 10°C and more. At that time, the area encompassed by heat advection at the 500–1,000 mbar level extended from the Iberian Peninsula through Europe to the northeast. That day belonged to the circulation type SW$_A$.

In the transition periods of the year, except in baric situations specific for spring or fall, weather characteristic for the warm, as well as for the cold periods may occur. Thus, for example, in Poland there are in the transition periods commonly more days with no or only small cloud cover than in summer or winter. Those days may be warm or cool

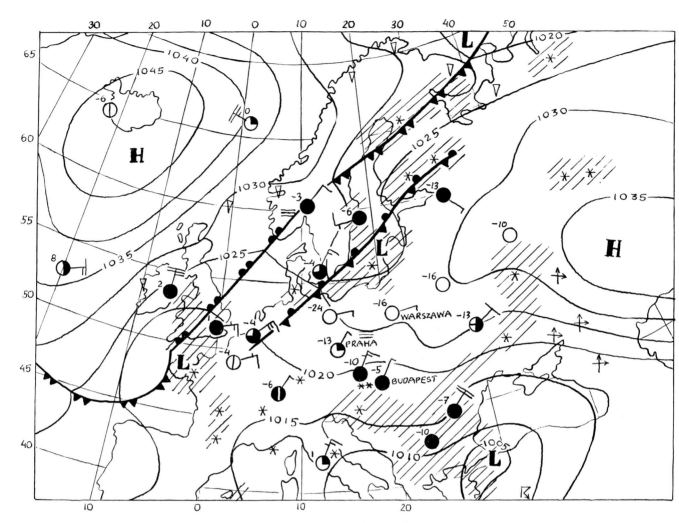

Fig.6. Synoptical situation, January 30, 1963, 12h00 GMT.

depending on the dominant directions of airmass advection. Warm days may occur in early spring or late fall, separated from the summer season by cool periods.

Cases with cold-air advection of arctic origin are shown in Fig.8 and 9. Fig.8 (April) shows that the minimum temperature in Poland in some localities fell to −5°C, late frosts also occurred in Czechoslovakia and Hungary. On April 12, 1968 the midday temperature rose to 5°C in Warsaw and Prague and to 9°C in Budapest. On the day preceding the situation shown in Fig.9, high-pressure centers formed over Scandinavia, then an anticyclone encompassed the whole Baltic Sea and the adjacent regions, the next day moving further eastward and expanding in both directions (from the Sudetes unto Novaya Zemlya). In this period late frosts developed (minimum −5°C at 2 m height) in Poland. Under similar conditions (when a weak high covered part of Sweden and the Baltic Sea from the Polish coast to the south part of the Gulf of Bothnia) late frosts appeared again in north Poland on May 31 of that year.

The situations of May 3 and 31 (Fig.9) are interesting in several respects. Thermal factors play a certain part in their genesis. In Scandinavia (notably in the mountainous regions) temperature increase in spring is impeded by the expenditure of heat for snow-

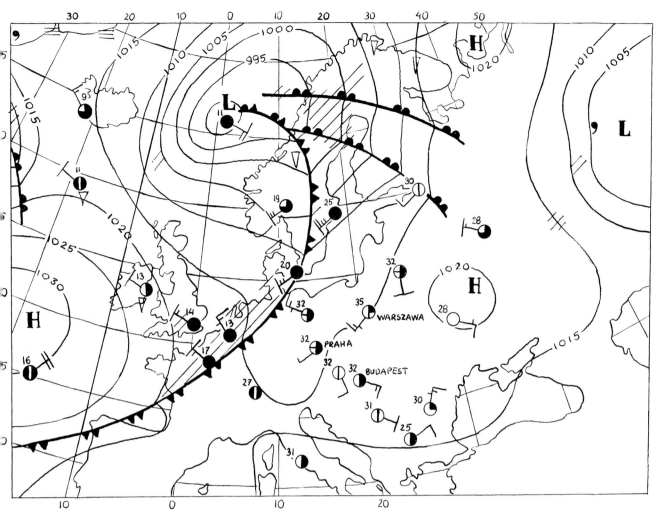

Fig.7. Synoptical situation, July 25, 1963, 12h00 GMT.

melting. The waters of the Baltic are still relatively cold. In consequence, in spring a tendency prevails to formation of local high-pressure systems over those areas. Advection from the north of distinctly cold-air masses and also frequent occurrence of late spring frosts, harmful to fruit and horticulture, are associated with that situation.

In spring the tendency for development of local high-pressure systems near the Baltic contributes to the phenomenon that, on the Baltic coast in this season and the first half of summer the greatest number of days with sunshine occur. The higher occurrence frequency of north winds in northern Poland (Table V) may perhaps be connected with this too.

Since the onset of spring starts in eastern Europe later than in central and southern Europe, the eastern circulation may also contribute to the occurrence of spring frosts in central Europe, for instance in situations similar to that shown in Fig.6.

The cooling of the land that begins in fall sometimes causes the formation of local high-pressure centres, notably in mountainous regions and the east of Europe (Fig.10). The appearance of several small highs in the mountains of central Europe would seem to indicate their thermally conditioned genesis. In fall, however, more frequently an

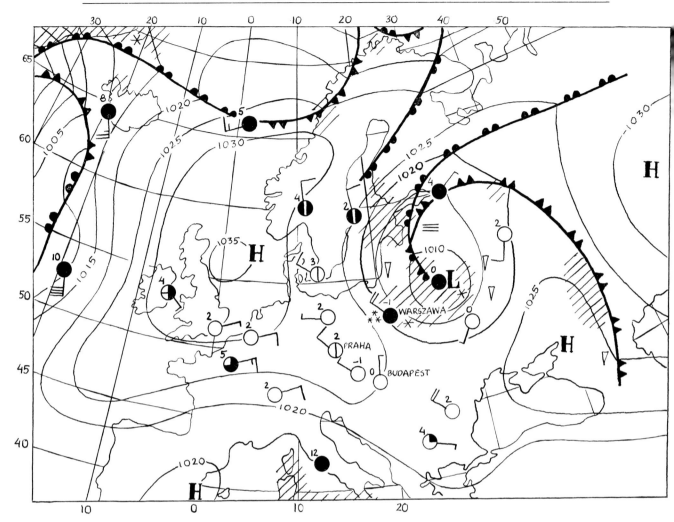

Fig.8. Synoptical situation, April 12, 1968, 00h00 GMT.

TABLE V

FREQUENCIES OF WIND DIRECTIONS (%)

	N	NE	E	SE	S	SW	W	NW	C
Gdynia (1951–60):									
Jan.	3.1	3.5	3.4	6.7	19.1	20.0	24.6	14.3	5.0
Apr.	14.4	8.9	4.3	9.6	10.8	7.8	14.0	18.8	11.6
July	8.8	9.1	5.7	6.4	7.9	10.6	22.8	19.2	9.5
Oct.	3.5	1.8	4.9	10.2	18.1	19.9	21.5	12.9	7.2
Year	7.7	5.5	5.0	9.4	14.3	13.3	20.6	16.4	7.8
Suwałki (1951–60):									
Jan.	4.9	7.3	6.0	14.7	10.4	24.9	16.6	9.0	6.2
Apr.	11.6	10.5	10.4	14.7	7.8	10.3	12.2	12.4	10.1
July	8.1	12.0	8.7	5.9	6.0	16.0	19.2	15.7	8.4
Oct.	3.5	6.0	7.1	14.1	11.2	21.6	17.3	8.4	10.8
Year	8.0	9.0	8.3	13.8	7.7	16.3	16.4	11.8	8.7
Chojnice (1951–60)									
Jan.	2.9	8.0	8.4	6.7	9.0	21.9	22.1	12.1	8.9
Apr.	13.1	9.8	8.8	10.1	6.7	9.0	12.4	16.3	13.8
July	9.7	10.5	4.9	5.0	5.4	14.8	22.1	16.9	10.7
Oct.	2.0	6.6	12.6	9.0	6.1	21.9	17.0	9.6	15.2
Year	5.9	10.1	10.2	8.5	6.5	15.5	17.0	13.6	12.7

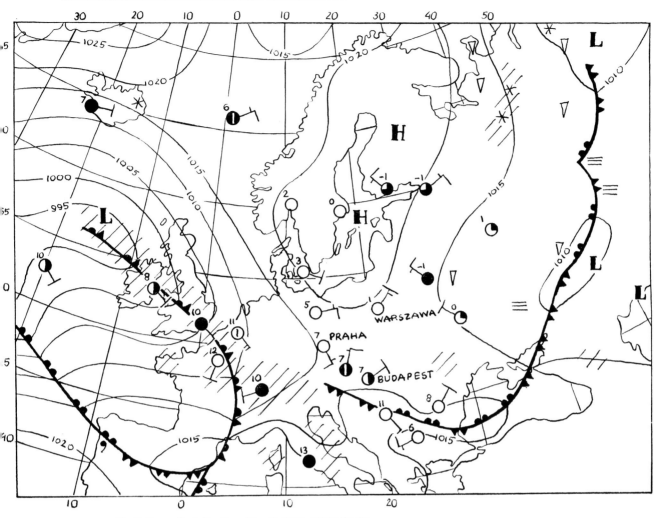

Fig.9. Synoptical situation, May 3, 1965, 00h00 GMT.

TABLE V (*continued*)

	N	NE	E	SE	S	SW	W	NW	C
Poznań (1951–60):									
Jan.	3.9	6.4	8.7	9.2	11.9	23.4	19.5	9.6	7.4
Apr.	12.0	8.7	10.9	10.2	6.9	10.0	16.7	14.2	10.4
July	9.7	7.0	6.5	5.6	8.1	18.6	23.0	12.7	8.8
Oct.	2.0	5.5	12.3	12.0	12.0	21.7	15.5	6.8	12.2
Year	6.2	6.8	12.4	9.9	9.0	17.2	17.6	10.5	10.4
Warsaw (1951–60):									
Jan.	3.2	6.2	7.3	14.2	12.7	18.3	22.4	11.1	4.6
Apr.	10.1	8.6	10.9	14.5	10.0	7.6	15.9	12.0	10.4
July	6.3	7.3	6.8	6.5	7.9	12.6	24.6	14.4	13.6
Oct.	2.0	4.3	9.3	16.1	11.4	16.4	20.4	7.8	12.3
Year	5.8	6.8	10.4	14.1	9.2	12.2	20.0	11.0	10.5
Wrocław (1951–60):									
Jan.	3.4	5.7	6.9	10.8	13.2	17.3	21.1	10.6	11.0
Apr.	9.6	8.9	10.0	9.4	9.1	8.5	19.3	15.1	10.1
July	5.9	5.7	4.4	8.1	8.5	10.4	30.9	15.3	10.8
Oct.	1.9	5.0	11.4	13.1	13.5	12.8	18.0	8.4	15.9
Year	5.1	6.8	10.5	11.1	10.6	10.9	21.9	11.6	11.6

TABLE V (*continued*)

	N	NE	E	SE	S	SW	W	NW	C
Kielce (1951–60):									
Jan.	1.1	4.7	6.9	8.6	18.0	20.8	20.2	6.9	12.8
Apr.	5.7	4.8	9.9	11.7	11.2	11.6	16.3	13.8	15.0
July	4.0	2.8	4.8	7.8	12.9	15.1	26.2	10.5	15.9
Oct.	1.8	4.1	7.1	11.4	15.4	19.1	18.1	4.0	19.0
Year	3.0	4.6	8.6	11.3	13.7	14.7	19.6	9.2	15.3
Zamość (1951–60):									
Jan.	6.0	5.2	11.6	9.1	6.3	24.5	20.7	11.8	5.8
Apr.	12.9	12.3	14.8	9.8	6.0	7.9	12.7	15.6	8.0
July	14.9	8.3	7.5	4.3	3.9	11.6	19.0	18.6	11.9
Oct.	5.8	7.7	14.3	8.1	6.7	18.0	17.2	12.7	9.5
Year	10.4	9.0	13.4	8.6	5.0	13.7	16.6	14.6	8.7
Prague (1946–53):									
Winter (D.–F.)	5.4	3.5	9.4	6.9	9.6	15.6	18.1	10.8	20.7
Summer (J.–A.)	8.0	4.2	6.6	3.5	6.0	11.2	17.0	16.0	27.5
Year	6.9	4.4	9.8	5.5	7.9	13.2	16.3	12.0	24.0
Cracow (1951–60):									
Jan.	4.0	10.6	13.0	2.0	1.1	10.0	31.1	9.0	19.2
Apr.	10.0	13.3	16.0	3.4	1.8	5.2	19.0	10.7	20.6
July	6.2	7.9	9.0	2.5	0.8	6.0	29.7	12.9	25.0
Oct.	3.0	9.8	17.0	2.5	0.9	5.6	23.4	6.9	30.9
Year	5.9	12.2	14.9	2.4	1.2	5.8	24.8	9.8	23.0
Mariánské Lázně (1946–54):									
Winter (D.–F.)	5.4	8.1	8.0	11.1	5.7	13.2	14.3	11.2	22.2
Summer (J.–A.)	7.3	6.0	3.8	5.0	3.9	10.3	12.9	15.4	35.4
Year	6.4	8.2	7.1	8.1	5.0	11.0	13.1	13.0	28.1
Ostrava (1946–54):									
Winter (D.–F.)	9.8	11.7	0.4	0.4	4.2	29.3	12.6	6.0	25.6
Summer (J.–A.)	10.9	6.8	2.3	1.1	3.5	15.9	8.7	8.0	42.8
Year	10.5	10.1	1.9	0.9	4.4	22.2	9.7	6.3	34.0
Kasprowy Wierch (1951–60):									
Jan.	16.7	10.9	2.2	7.1	21.9	13.6	12.2	11.1	4.3
Apr.	22.2	10.7	2.7	6.8	18.0	12.0	7.5	9.5	8.8
July	22.1	10.3	1.7	5.5	14.2	14.5	13.6	10.3	7.8
Oct.	11.4	8.6	2.9	9.1	25.4	19.4	9.7	6.5	7.0
Year	17.8	10.5	2.3	7.4	21.1	14.4	10.2	8.6	7.7
Brno (1946–54):									
Winter (D.–F.)	10.0	5.1	10.3	13.4	8.7	6.0	6.4	13.8	26.3
Summer (J.–A.)	17.5	6.8	7.0	5.8	8.6	7.3	6.6	16.6	23.8
Year	13.0	6.9	10.1	10.3	8.8	6.1	5.8	14.5	24.5

TABLE V (*continued*)

	N	NE	E	SE	S	SW	W	NW	C
Prešov (1946–53):									
Winter (D.–F.)	22.2	13.1	1.2	11.8	23.3	3.6	1.7	12.5	10.6
Summer (J.–A.)	25.3	13.3	2.3	6.8	13.9	4.8	1.8	22.4	9.4
Year	22.8	13.6	1.6	9.6	19.2	4.2	1.7	17.8	9.9
Bratislava (1946–53):									
Winter (D.–F.)	10.0	11.0	13.2	6.9	2.7	3.7	9.0	23.5	20.0
Summer (J.–A.)	13.4	7.8	4.8	3.7	5.9	4.3	10.4	27.3	22.4
Year	11.5	10.5	9.3	5.8	6.1	4.5	9.4	22.0	20.9
Eger (1901–50):									
Jan.	19.1	8.4	6.6	7.1	6.9	7.2	8.5	17.2	19.0
Apr.	14.4	10.3	7.8	8.2	8.0	9.6	10.8	17.6	13.3
July	13.3	7.6	9.0	6.4	4.6	8.7	14.1	24.1	12.2
Oct.	14.2	7.2	7.6	8.0	7.1	9.3	11.0	19.0	16.6
Year	14.2	9.0	8.0	7.9	6.8	8.6	10.8	19.3	15.4
Debrecen (1901–50):									
Jan.	9.6	19.4	11.3	9.4	15.2	17.4	5.1	2.0	10.2
Apr.	10.0	17.5	9.6	10.0	14.2	20.7	6.9	3.9	7.2
July	11.2	15.6	6.5	8.6	13.0	17.9	10.2	6.0	11.0
Oct.	8.7	15.4	9.0	10.5	15.5	16.2	6.6	3.4	14.7
Year	10.2	16.6	9.8	9.8	14.8	17.7	6.5	3.6	11.0
Budapest (1901–50):									
Jan.	13.0	10.7	5.4	7.8	6.9	5.1	7.8	24.0	19.3
Apr.	8.8	7.0	5.2	9.6	8.0	7.7	10.3	24.7	18.7
July	9.6	5.5	2.9	3.4	4.0	5.7	14.0	36.5	18.4
Oct.	8.2	6.8	5.6	9.1	6.1	6.3	9.2	20.9	27.8
Year	10.2	7.9	5.0	7.5	6.5	6.3	10.3	25.5	20.8
Szombathely (1901–50)									
Jan.	25.4	7.4	1.5	2.0	7.7	22.4	9.4	6.5	19.2
Apr.	30.5	7.9	2.3	2.7	8.3	18.2	5.3	7.7	17.1
July	30.0	9.8	2.5	1.6	5.5	14.0	6.7	9.7	20.8
Oct.	26.0	7.9	1.4	1.9	8.3	20.5	5.1	6.7	22.2
Year	27.4	8.9	2.0	2.1	7.7	18.6	6.5	7.2	19.6
Szeged (1901–50)									
Jan.	16.5	8.3	6.1	12.8	17.0	9.5	10.0	14.9	4.9
Apr.	15.4	7.8	6.0	11.0	18.6	10.4	11.7	15.6	3.5
July	17.0	7.5	4.9	5.8	11.1	8.1	15.3	24.1	6.2
Oct.	14.5	8.0	6.6	12.7	20.1	9.2	9.7	13.3	5.9
Year	15.5	8.0	6.1	11.3	16.6	9.6	11.4	16.2	5.3
Pécs (1901–15, 1926–50)									
Jan.	12.7	18.6	13.3	4.2	2.5	4.4	14.5	8.4	21.4
Apr.	15.2	18.8	9.3	4.7	3.6	6.5	15.2	9.3	17.4
July	21.0	10.8	5.6	3.7	4.1	6.7	15.0	13.4	19.7
Oct.	13.4	19.2	11.3	5.5	3.5	5.0	12.4	6.9	22.8
Year	15.2	17.5	9.7	4.9	3.4	5.4	14.5	9.2	20.2

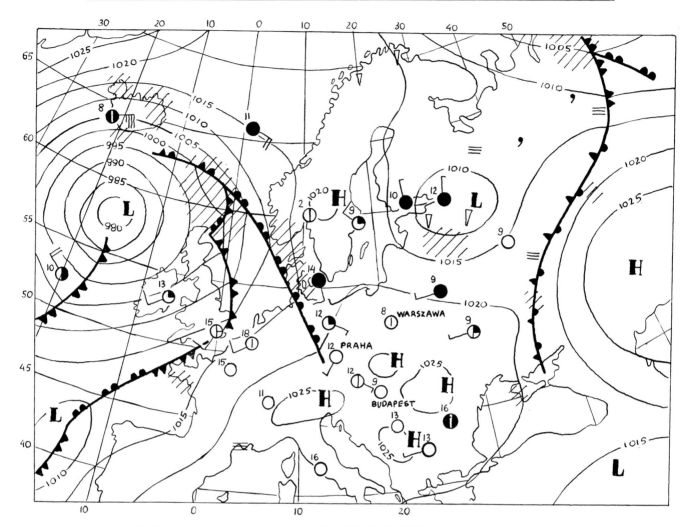

Fig.10. Synoptical situation, September 25, 1967, 00h00 GMT.

anticyclone develops encompassing the whole area from the Alps to the Romanian Carpathian Mountains. For this reason, the fall season in the mountainous regions of central Europe brings mostly continuous fine weather.

The high-pressure systems forming over eastern Europe at that time of the year and commonly affecting large parts of Poland, Czechoslovakia and Hungary, contribute to the formation of weather with small cloudiness and either warm (Fig.11) or cool with early frosts. Sometimes the first type changes into the second as a consequence of advancing cooling in the conditions of continuous anticyclonic weather in autumn. On October 4, 1966 (Fig.11) the air temperatures at midnight in Budapest, Prague and Warsaw were 14°, 15° and 14°C, respectively. The inflow of cold air from the east, producing early night frosts (in spite of high diurnal insolation) or facilitating their formation through radiation, is commonly associated with anticyclones whose centers are located farther northward.

Upper winds

The tropospheric circulation above the area under discussion is connected with the

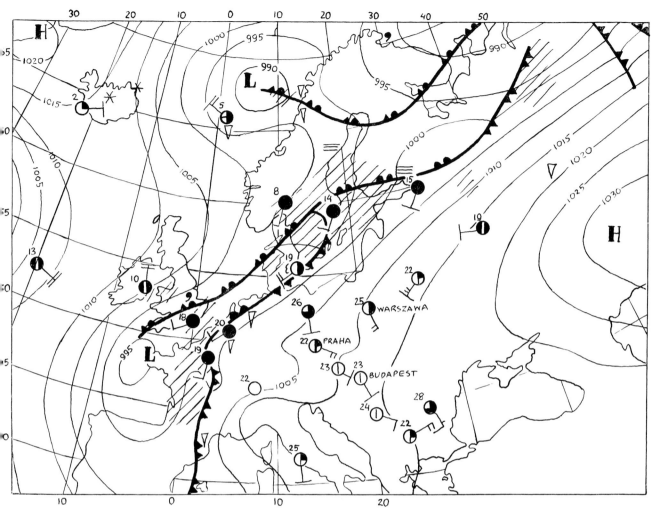

Fig.11. Synoptical situation, October 4, 1966, 12h00 GMT.

circulation appearing entirely over central Europe and adjacent territories and it is generally rather well known. In this paper attention is being paid only to some of its features and namely to strong-wind streams i.e., to long narrow passages of winds with velocities of 60 knots and more. Some of them have features typical of jet streams.

On the basis of an analysis of the working maps of baric 300 and 200 mbar topography, as well as of the distribution of velocities and other criteria, the occurrence frequency of such winds in the areas positioned between the parallels 48°–57°N and the meridians 10°–30°E for the period 1960–1967 has been defined (BUDZISZEWSKA, 1968). It has been found that the strong-wind streams in the upper troposphere are of western directions in 50%. In the summer season their frequency increases to 60%, whereas in February it decreases to its lowest frequency occurrence; the duration being 3–10 days on the average. Distinct periodicity is observed as regards the strong-wind streams from northern directions which blow most often in winter (in February they are more frequent than the western streams), in March and sometimes in October. Their average duration is 6–10 days in winter, whereas in transition periods of the year they may last 3–5 days. Streams from the southern directions, which are the most frequent in the transition period of the year, do not show distinct occurrence periodicity.

For the same area the occurrence of typical jet streams has been separately investigated —generally for the period of 1958–1971 and especially for the years of 1960–1969 (BUDZISZEWSKA, 1973). Their mean monthly frequency for winter and summer is given in Table VI and the number of cases and days, including persistence, is given in Table VII. The annual frequency course of jet streams is illustrated by Fig.12. Fig.13 illustrates several years' frequency variation of the jet stream appearance over the eastern part of central Europe.

When analyzing these tables and figures one can see that the most frequent jet streams are those blowing from the western quadrant, especially in autumn and winter.

Twice less frequent jet streams are those from the northern (the highest frequency also in autumn and winter) and the southern quadrants (the highest frequency in summer and autumn).

The least frequent jet streams are those from the eastern quadrant and from changeable directions.

Fig.12 illustrates clearly the short-term and Fig.13 the several-year periodicity of occurrence of those jet streams.

The minimum of jet-streams frequency which occurred in the years of 1967–1968 seems to be coincident with the 2–3 years' earlier period of minimum solar activity. Two per cent of all jet streams are shear-line cases.

TABLE VI

FREQUENCY OF JET STREAMS—MONTHLY MEANS ($\bar{x}$) AND STANDARD DEVIATIONS (σ)

(After BUDZISZEWSKA, 1973)

	Winter				Summer			
	S	N	W	Σ	S	N	W	Σ
$\bar{x}$	5	7	11	24	6	3	9	18
σ	3	3	5.5	5	4	2	5	6

Recording period: 1958–1971; height 300–200 mbar.

TABLE VII

NUMBER OF CASES AND NUMBER OF DAYS OF JET-STREAM OCCURRENCE IN RELATION TO DIRECTIONS AND PERSISTENCE

(After BUDZISZEWSKA, 1973)

	1–3 days		4–6 days		7 days		Total days
	cases	days	cases	days	cases	days	
W	215	396	76	362	54	520	1,278
E	11	22	2	9	—	—	31
N	135	223	33	160	15	125	508
S	140	235	34	167	11	109	511
Indef.	—	—	9	42	8	79	121
Total	—	876	—	740	—	833	2,449

Recording period: 1958–1971; height 300–200 mbar.

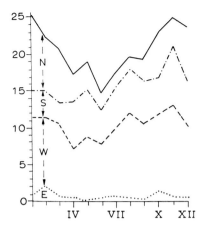

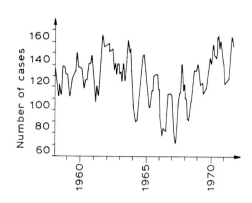

Fig.12. Mean frequency course of jet streams in relation to directions (after BUDZISZEWSKA, 1973).

Fig.13. Mean changes of 6-months totals of jet-stream occurrences (after BUDZISZEWSKA, 1973).

Usually dangerous phenomena are connected with them, for instance the flood caused by disastrous rainfalls on August 20–22, 1972 in the Silesian Beskid Mountains, especially in the spring area of the Vistula and Carpathian affluents of the Oder. During three days the precipitation at the meteorological station Lysa Hora (49°33′N 18°17′E, 1,327 m a.s.l.) was 415.5 mm.

The atmospheric situation of this period is shown by Fig.14. On the eastern side of the

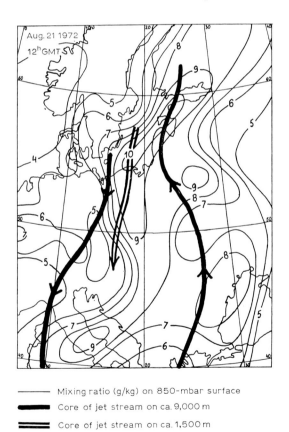

Mixing ratio (g/kg) on 850-mbar surface
Core of jet stream on ca. 9,000 m
Core of jet stream on ca. 1,500 m

Fig.14. Jet stream position and distribution of mixing ratio (g/kg)—850 mbar, 21.VIII.1972, 12h00 GMT (after BUDZISZEWSKA, 1973; and BUDZISZEWSKA et al., 1973).

shear-line system a very damp and warm air mass was flowing to the north. The velocity of flow in the core of the jet stream was reaching 50 knots (at 300 mbar surface). On the western side the air, also damp but much cooler, flew from the north. So the mixing ratio in the shear-line system in the area from the Gulf of Finland to Sardinia was above 9 g/kg. The western branch of the jet-stream core of the shear-line system directed to the south occurred much lower and caused very heavy rainfalls in the mountains where peaks reach 1,000–1,200 m above sea level: 3-days' sums of the precipitation in some stations of this region were above 200 and even 300 mm (after BUDZISZEWSKA et al., 1973).

The above mentioned data refer to the strong-wind streams of the upper troposphere. Somewhat different relations may appear in the lower parts of the troposphere and in connection with winds of smaller velocities. The records of the Observatory Lomnický Štit in the Tatra Mountains (Table VIII) may serve as a basis to formulate one's view on the winds of lower troposphere.

In the years 1960–1967 considerable variations in the frequency of the strong-wind streams in the troposphere were observed, reaching a maximum in 1961–1963 and a minimum in 1966–1967. In the period of maximum frequency the variation amplitudes were small whereas in the years 1964–1967, when a limitation of frequencies was generally observed, they were large. It would seem that since 1967 an increase of frequency of strong-wind streams in the troposphere has been observed again (BUDZISZEWSKA, 1968). It is worth adding that at the top of the Tatras jet streams are sporadically observed.

TABLE VIII

WIND DIRECTIONS FREQUENCY (%) AND MEAN WIND FORCE (Beaufort) IN THE TATRA MOUNTAINS (1941-1944)
(After ORLICZ, 1954)

		N	NE	E	SE	S	SW	W	NW	C
Kasprowy Wierch (1,988 m a.s.l.):										
Jan.	direction	18.7	9.5	2.7	7.3	22.8	9.4	12.1	12.4	5.1
	velocity	4.1	3.4	2.8	6.6	6.1	5.2	4.6	5.0	
Apr.	direction	22.2	14.3	2.2	6.4	17.4	14.7	11.4	9.5	1.9
	velocity	4.2	4 1	2.7	5.3	5.6	5.1	3.8	4.4	
July	direction	32.1	9.7	2.0	5.0	11.6	12.2	12.5	13.0	1.9
	velocity	2.7	2.5	3.2	4.7	4.7	4.1	3.4	2.7	
Oct.	direction	11.7	8.3	4.4	14.3	25.8	14.9	11.4	7.3	1.9
	velocity	3.0	2.6	3.2	4.6	5.3	4.5	3.4	3.1	
Year	direction	20.3	11.2	2.9	7.3	19.5	13.4	11.7	10.9	2.8
	velocity	3.3	2.9	2.9	4.9	5.6	4.9	3.9	3.7	
Lomnický Štit (2,635 m a.s.l.):										
Jan.	direction	17.3	4.7	2.6	3.6	5.1	15.1	15.7	31.3	4.6
	velocity	4.2	3.4	3.5	2.4	2.4	3.1	3.8	4.2	
Apr.	direction	15.1	3.6	2.5	1.8	4.0	10.1	21.9	33.5	7.5
	velocity	4.1	3.2	2.0	2.3	2.5	2.4	3.4	4.2	
July	direction	7.1	2.0	2.7	4.6	7.8	14.1	16.2	33.7	11.8
	velocity	3.2	2.5	2.1	2.2	2.2	2.3	3.4	3.5	
Oct.	direction	5.5	3.1	7.1	7.3	10.8	15.2	15.6	26.1	9.3
	velocity	3.0	2.1	2.2	2.2	2.6	2.7	3.2	3.2	
Year	direction	10.7	4.3	3.9	3.6	7.4	15.0	19.1	27.5	8.5
	velocity	3.4	2.9	2.6	2.4	2.5	2.7	3.4	3.8	

Surface winds

The frequency of different wind directions depends on the mean baric pressure field and on the relief of the ground. This can be observed not only in mountainous regions but even on plains intersected by large valleys or surrounded by mountains. Information on the anemometric conditions in the east part of central Europe is given in Table V. Generally it may be said that in Poland the directions west and southwest predominate in winter, while they are west and northwest in summer. In Czechoslovakia besides these same directions, also north winds and winds with a south and east component occur in some sites (primarily in Slovakia). In Hungary the north winds take first place, followed by those from south to northeast (in winter) and from southwest to northeast (in summer).

In mountainous regions, notably in Slovakia, many local winds act on and are conditioned by the temperature differences between day and night time on mountain tops, slopes and valleys. This phenomenon is reinforced by the fact that most mountain ranges run here approximately along the parallels and that their slopes, exposed primarily to the north and the south, have entirely different insolation. Also the aerodynamic conditions, affecting the regional and local wind directions and velocities, show very considerable differences. A good example is offered by the findings of the respective measurements performed at the observatories Kasprowy Wierch (1,988 m a.s.l.) and Lomnický Štit (2,635 m a.s.l.) (Table VIII).

Beside the frequent occurrence of calms, primarily in intermontane basins, another interesting and important fact is the occurrence of high-speed winds, most frequently in the cold half-year. The frequency of strong winds also differs at the afore-mentioned two observatories (Table IX). In principle, all winds from the south (and also some from southeast and southwest) with velocities in excess of 10 and 15 m/sec observed at Kasprowy Wierch are of föhn-type (local Polish name: "halny" wind = high-mountain meadows wind). This kind of wind is also noted on a number of Slovakian mountain ranges, in the Sudetes and even on the Matra Mountains and the Bakonya Forest in Hungary (ORLICZ, 1954; SZEPESI, 1963; OTRUBA, 1964, pp.130–136). Part of the airstreams of the same basic genesis is attributed to bora wind (at the Slovak side of the High Tatra Mountains and the southeast of the Bakonya Forest). This is explained by

TABLE IX

FREQUENCY OF WINDS OF DIFFERENT VELOCITY (1941–44)
(After ORLICZ, 1954)

	N	NE	E	SE	S	SW	W	NW
Kasprowy Wierch (1,988 m a.s.l.):								
⩾ 1 m/sec	20.9	11.5	3.0	7.5	20.0	13.8	12.1	11.2
⩾ 10 m/sec	8.7	3.5	1.1	11.1	37.0	19.8	10.5	8.3
> 15 m/sec	5.1	1.4	1.4	14.0	45.2	19.6	6.8	6.5
Lomnický Štit (2,635 m a.s.l.):								
⩾ 1 m/sec	11.7	4.7	4.3	3.9	8.1	16.4	20.9	30.0
⩾ 10 m/sec	12.9	3.5	0.4	0.5	2.2	7.5	23.9	49.1
> 15 m/sec	14.0	3.4	0.4	—	0.5	2.3	24.7	54.7

the fact that in advection of cold air from the north, the air flow over the mountains from the north is inhibited. Cold air can pass over an orographic obstacle only if it is accumulated in major masses. Such air undergoing föhn processes cannot attain such a high temperature as the air overlying precedingly the southern mountain slopes could, and therefore sweeps down as cold air (ORLICZ, 1954).

Beside the katabatic winds of föhn- or bora-type which cause sizeable damage in the mountains (forests), the storm winds which are observed on Lake Balaton, where they menace people practising aquatic sports, are also dangerous. During the May to September seasons of the period 1958–1963 at Siófok on Lake Balaton an average of 36 days with winds of speeds exceeding 15 m/sec was observed. In 30 % of the cases these winds rose abruptly, their speed increased within 10 minutes by 8 m/sec and reached 15 m/sec, in exceptional cases even 35 m/sec. Mainly they arrive from northwest or west, but the strongest winds came from west-southwest. The genesis of the Balaton winds is after GÖTZ and TÄNCZER (1966) related to: (*a*) the exchange of air masses and atmospheric fronts (56 %); (*b*) major changes in atmospheric pressure with great baric gradients (25 %); (*c*) great instability of the air with intense convection and a squall line (19 %). The afore-mentioned squall-line storms occur, for example, in the prefrontal area of cold fronts associated with secondary lows, which develop over the Po Valley (BODOLAI and BODOLAI, 1966).

Associated with a cold front and advection of continental air from the east (in the cold season) is the appearance of a harassing squall south or southeast wind in the south part of the Great Hungarian Plain. This wind is called the kossava wind (PÉCZELY, 1957). It blows when the continental mass, unable to conquer the Carpathian barrier, turns around it at the south, and streams to Serbia and Hungary through the Wallachian lowland.

The storms on the Baltic Sea may also be mentioned here. They occur mostly in the cold season of the year, particularly in fall and in mild winters. They are mostly associated with cyclones which come in from the west and often deepen over the Baltic. The storms related with these cyclones embrace the entire southern Baltic. Storms formed by advection of very cold air from the north seldom occur at an advanced stage of the winter when the southern waters of the Baltic are not covered by ice. When the differences between the air and sea surface temperatures are considerable and the air is therefore very unstable, very strong storms with north winds may form. Generally they do not embrace the west part of the southern Baltic (TARANOWSKA, 1966).

Sea breezes are sometimes noted on the Polish shore during the warm season (over 20 cases per season). One of the preconditions of their appearance is the presence of small gradients in the pressure field. The maximum landward extension of these sea breezes is about 70 km (MICHALCZEWSKI, 1965).

Spatial distribution and annual pattern of some meteorological elements

Air temperature

The spatial distribution of monthly mean temperatures in January and July is given in Fig.15 and 16. They show distinctly that, apart from the mountains, the lowest January

Fig.15. Mean temperatures (°C), January (after ANONYMOUS, 1958, 1960a, 1973a).

Fig.16. Mean temperatures (°C), July (after ANONYMOUS, 1958, 1960a, 1973a).

temperatures are observed in east Poland (Suwałki), the highest in southern Hungary and western Czechoslovakia (Prague), on the lower Odra and the Baltic Sea in Poland. The highest temperatures in July are found on the Hungarian Plain, the lowest in north Poland. The same conclusions can be derived from the climatic tables at the end of the chapter. The temperature differences between the mentioned regions amount to about 5°C in winter and in summer. Further characteristics of the thermal conditions are given in Tables X–XII. Compared with south Hungary (e.g. Szeged), the number of ice-days (maximum temperature below 0°C) is in north Poland (Chojnice, Suwałki) at least twice as high, whereas the number of hot days (maximum temperature above 30°C) is ten times lower.

The extreme temperatures recorded in a 30-year period were: in Poland −40.6°C (Żywiec, February 10, 1929) and +40.2°C (Prószków near Opole, July 29, 1921); in Czechoslovakia −42.2°C (Litvinovice, February 11, 1929) and +38.6°C (Bratislava, August 22, 1945); in Hungary −32.2°C (Kecskemét, February 11, 1929) and +39.8°C (Turkeve, July 1, 1950).

Cloudiness

Monthly mean cloudiness percentages are shown in the climatic tables, their annual spatial distribution in Fig.17, which figure, together with the map, explains the causes of the lower number of sunshine hours in Poland, compared with the number of sunshine hours in Hungary. Only in the southeast part of Poland (Zamość) the cloudiness ap-

TABLE X

MEAN NUMBER OF ICE-DAYS (max. temp. < 0°C; Hungary ≤ 0°C)

(After ANONYMOUS, 1958, 1960a, 1973a)

	Oct.	Nov.	Dec.	Jan.	Feb.	Mar.	Apr.	Year
Gdynia (a)[1]	—	0.7	5.3	10.5	8.9	3.7	0.1	29.2
Suwałki (a)	0.1	5.1	13.6	19.0	15.4	9.5	0.3	62.9
Chojnice (a)	0.1	2.7	10.2	15.3	11.8	5.0	0.1	45.2
Poznań (a)	0.0	1.7	8.6	13.5	10.6	3.3	—	37.4
Warsaw (a)	0.0	2.6	9.5	14.6	11.7	5.0	—	43.4
Wrocław (a)	—	0.9	6.5	11.6	8.7	2.1	0.0	29.8
Kielce (a)	0.1	2.8	11.4	16.4	12.3	4.9	—	47.7
Zamość (b)	0.1	2.8	7.9	13.8	12.9	8.2	0.1	45.8
Prague (c)	—	0.7	8.7	11.8	7.2	1.4	0.0	29.8
Cracow (a)	0.0	1.8	9.2	13.5	10.0	3.3	—	37.7
Mariánské Lázně (c)	0.1	2.3	12.6	13.7	8.3	2.8	0.1	39.9
Ostrava (c)	0.0	1.0	9.7	14.2	8.4	1.9	0.1	35.3
Brno (c)	—	0.7	9.3	14.3	6.8	1.2	—	32.3
České Budějovice (c)	0.0	1.0	10.0	12.9	7.2	1.5	0.1	32.7
Bratislava (c)	0.5	0.5	8.4	13.8	6.2	1.1	—	30.0
Eger (d)	—	1.7	8.4	12.7	6.6	1.0	—	30.4
Debrecen (d)	—	1.3	7.8	14.1	8.1	1.1	—	32.4
Budapest (d)	—	1.0	6.6	10.8	4.6	0.4	—	23.4
Szombathely (d)	0.0	1.4	7.1	11.9	6.2	0.5	—	27.3
Szeged (d)	0.1	1.3	5.5	10.4	5.0	0.4	—	22.7
Pécs (d)	—	0.9	6.0	10.5	5.4	0.4	—	23.2

[1](a) 1931–1960; (b) 1951–1960; (c) 1926–1950; (d) 1901–1950.

Fig.17. Mean annual cloudiness (%) (after ANONYMOUS, 1958, 1960a, 1973a).

TABLE XI

MEAN NUMBER OF FROST-DAYS (min. temp. $<0°C$; Hungary $\leqslant 0°C$)

(After ANONYMOUS, 1958, 1960a, 1973a)

	Sept.	Oct.	Nov.	Dec.	Jan.	Feb.	Mar.	Apr.	May	June	Year
Gdynia (a)[1]	—	0.9	6.8	16.0	22.4	19.9	19.4	5.5	0.5	—	91.3
Suwałki (a)	0.3	6.2	16.0	24.3	28.6	25.4	25.4	11.5	1.7	0.0	139.4
Chojnice (a)	—	3.9	13.5	22.2	27.0	23.3	23.1	10.3	1.9	—	125.2
Poznań (a)	0.1	3.6	12.3	21.4	25.3	22.1	19.9	6.9	1.0	—	112.4
Warsaw (a)	0.0	4.6	13.5	21.8	26.3	22.8	21.4	6.5	0.6	—	117.4
Wrocław (a)	0.4	4.5	10.8	21.1	25.0	21.8	18.5	6.8	1.2	—	110.1
Kielce (a)	0.1	5.9	14.6	23.9	27.5	24.5	22.6	9.2	1.6	0.0	129.9
Zamość (b)	0.5	6.9	13.9	21.7	26.9	23.8	24.3	10.1	0.5	—	128.6
Prague (c)	0.1	2.1	7.2	19.2	22.9	18.3	14.2	3.1	0.3	—	87.4
Cracow (a)	0.0	2.9	10.3	20.2	25.0	21.1	16.9	3.6	0.1	—	100.1
Mariánské Lázně (c)	0.9	5.3	16.2	24.8	26.2	23.7	21.4	9.7	2.0	—	130.2
Ostrava (c)	0.1	3.5	10.4	22.7	25.7	21.6	17.8	5.9	1.0	—	108.7
Brno (c)	0.1	3.2	9.6	21.5	26.8	21.8	17.0	4.6	0.6	—	105.2
České Budějovice (c)	0.7	5.0	11.8	22.9	24.9	21.5	18.3	6.5	1.9	0.1	113.6
Bratislava (c)	0.0	1.9	7.1	19.5	24.4	19.6	13.3	2.3	0.3	—	88.4
Eger (d)	0.1	2.9	12.5	21.6	26.0	22.3	14.1	3.8	0.4	—	104.6
Debrecen (d)	0.0	4.4	13.3	22.0	26.4	22.4	15.6	5.0	0.4	—	109.5
Budapest (d)	—	0.7	7.5	15.8	21.4	17.2	8.5	1.0	0.0	—	72.1
Szombathely (d)	—	2.3	12.2	22.0	26.0	22.5	14.8	3.8	0.2	—	103.7
Szeged (d)	0.1	0.9	8.8	18.2	23.5	20.0	11.5	1.6	0.0	—	84.7
Pécs (d)	—	0.9	8.3	17.4	22.9	18.3	9.6	1.4	0.1	—	78.5

[1] (a) 1931–1960; (b) 1951–1960; (c) 1926–1950; (d) 1901–50.

proaches that in Hungary. The differences in cloud cover between the lowland regions in the north and the south and the mountain areas separating them are also illustrated by the annual number of cloudy days (Fig.18). The number of clear days is, in some localities of Poland, three times lower than in Hungary where in some sites the annual mean exceeds 90 days.

Precipitation

The annual precipitation amounts on the lowland areas of Czechoslovakia, Poland and Hungary are relatively small, namely between 500 and a little above 600 mm (see the climatic tables). They only exceed 700 mm on the elevations of the northwestern Polish lake region, in the submountainous zones of the whole area, and in the mountains (Tatra 1,500 m). In some neighbouring lowlands, such as in the Tisza Basin in Hungary and the Labe Basin in Czechoslovakia and also in localities protected by relatively small elevations (e.g. on the lowlands of central Poland), the annual precipitation amounts are less than 500 mm (Fig.19). In analysing the isohyets pattern one is impressed by the surprisingly important role played by the exposure of elevations to the directions of advection of moist air masses of maritime origin. Thus, in spite of the relatively small altitude of the Pomorze lake district elevations in Poland (which is of the order 180–200, rarely 300 m a.s.l.), they receive precipitation amounts more than or equal to those received on the elevations and mountains located on the left bank of upper Vistula, although the altitudes there are much higher. The afore-mentioned lake-district elevations of north Poland form a shield sufficient to considerably reduce the precipitation in the environs of the lower Vistula Valley. The occurrence of larger precipitation amounts on one side of these elevations, and much smaller ones at their opposite side, may be due to föhn processes. Similar phenomena are observed in the Bakonya Forest and the Matra Mountains in Hungary (SZEPESI, 1963).

The increase in precipitation owing to orography is only partly due to hypsometric effects but is also a result of greater friction. This refers primarily to terrains of relatively small elevation but with strongly varied relief similar to those in many places of north Poland.

Major precipitation amounts observed south of the Carpathians occur, for example, in the case of cyclonic circulation of lows of Mediterranean origin. It contributes in the fall (October) to the formation of a secondary maximum of precipitation. The principal maximum on the Hungarian lowlands is shifted to May–June, while on the remaining area it is generally in July. The effect of the sea in north Poland can be felt almost uniquely in the larger amounts of precipitation in fall as compared to the spring values. Noteworthy is the fact that the smallest amounts of precipitation generally fall over areas with the highest summer temperatures and, as a consequence, strong evapotranspiration. Therefore these areas are very often exposed to drought.

TABLE XII

MEAN NUMBER OF HOT DAYS

(After ANONYMOUS, 1958, 1960a, 1973a)

	Maximum temperature $\geqslant 25°C$								Maximum temperature $\geqslant 30°C$							
	Mar.	Apr.	May	June	July	Aug.	Sept.	Oct.	Year	Apr.	May	June	July	Aug.	Sept.	Oct.	Year
Gdynia (a)[1]	—	0.1	1.0	2.2	3.6	3.8	1.0	—	11.7	—	0.1	0.2	0.5	0.4	0.1	—	3.1
Suwałki (a)	—	0.1	2.7	5.7	8.5	7.3	1.1	—	25.3	—	0.1	0.6	1.6	0.9	0.1	—	3.3
Chojnice (a)	—	0.1	2.8	6.6	9.3	7.0	1.8	—	27.0	—	0.3	0.7	1.2	0.8	0.0	—	3.0
Poznań (a)	—	0.2	3.8	8.6	12.5	10.7	3.5	0.0	39.3	—	0.3	1.3	2.8	2.1	0.2	—	6.7
Warsaw (a)	—	0.5	4.4	8.0	12.1	10.4	3.3	0.1	38.7	0.0	0.3	0.9	2.1	1.8	0.4	—	5.6
Wrocław (a)	—	0.7	4.6	8.8	13.8	11.7	4.6	0.2	44.2	—	0.3	1.4	2.8	2.5	0.7	—	7.7
Kielce (a)	—	0.5	4.2	7.3	12.0	10.5	3.4	0.1	37.9	0.0	0.3	1.0	2.7	2.0	0.3	—	6.3
Zamość (a)	—	0.4	2.4	10.3	15.1	11.7	4.0	0.1	44.0	—	0.4	0.8	3.3	1.7	0.6	—	6.9
Prague (b)	—	0.6	4.5	9.6	15.0	12.4	5.7	0.5	48.3	—	0.6	1.9	3.8	3.5	0.9	—	10.7
Cracow (a)	—	0.5	3.8	8.7	13.2	11.0	3.7	0.2	41.2	—	0.3	0.8	2.0	1.5	0.1	—	4.7
Mariánské Lázně (b)	—	0.1	2.6	6.2	9.3	7.8	3.8	0.1	29.9	—	0.2	0.6	1.5	1.5	0.2	—	4.0
Ostrava (b)	—	0.8	5.4	9.8	14.9	12.9	5.2	0.3	49.3	—	0.3	2.0	4.5	3.0	0.8	—	10.6
Brno (b)	—	0.7	4.8	10.7	17.0	15.2	6.1	0.2	57.4	—	0.1	1.8	4.2	3.2	0.9	—	10.2
České Budějovice (b)	—	0.4	3.9	8.8	13.3	12.1	5.7	0.6	44.8	—	0.3	1.5	3.1	2.6	0.9	0.0	8.4
Bratislava (b)	—	1.0	5.6	12.7	20.2	18.3	9.1	0.7	67.6	—	0.4	3.0	6.8	5.7	1.3	—	17.2
Eger (c)	0.0	1.3	8.5	14.9	22.7	20.2	9.3	0.9	77.9	0.1	1.0	3.1	7.5	6.6	1.3	0.0	19.6
Debrecen (e)	0.0	1.9	10.4	15.6	23.0	20.9	10.4	1.3	83.5	0.1	1.1	5.2	9.8	8.5	2.0	0.1	26.8
Budapest (c)	0.0	1.3	9.1	16.1	23.5	21.9	10.6	1.3	84.0	0.0	1.0	3.8	9.5	7.8	1.9	0.0	24.1
Szombathely (c)	—	0.7	5.3	12.2	18.8	16.2	6.0	0.3	59.5	—	0.2	1.4	4.8	3.5	0.6	—	10.5
Szeged (c)	—	1.7	11.7	18.6	25.0	23.1	11.8	1.6	93.6	0.0	1.3	5.4	11.3	9.1	3.5	0.0	30.7
Pécs (c)	—	1.5	8.0	14.6	22.2	20.1	10.6	1.5	78.5	—	0.7	3.2	8.5	8.0	2.2	—	22.5

[1] (a) 1931–1960; (b) 1926–1950; (c) 1901–1950.

Fig.18. Mean annual number of cloudy days, 80% of cloud cover (after ANONYMOUS, 1958, 1960a, 1973a).

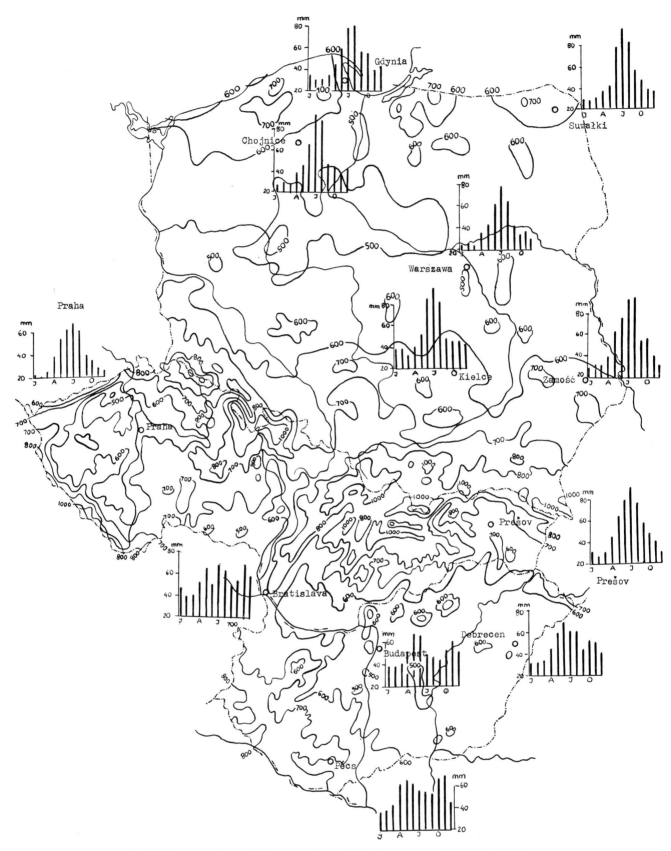

Fig.19. Mean annual precipitation sums in mm (after Anonymous, 1958, 1960a, 1973a).

The annual number of precipitation days north of the Carpathians is commonly over 150, reaching more than 170 days in some localities in the northern part of the country. South of the Carpathians, the corresponding figure (apart from highland regions and mountains) is as a rule not greater than 150 days (Prague) or even much less (Prešov, 126 days). North of the Carpathians the annual ratio of days with snowfall can exceed 30% of precipitation days, while south of these mountains it is 25–20% or less.

Air humidity and evaporation

More detailed data are missing regarding evaporation. In north Poland the amount of rainfall is 200 mm higher than the losses of water through evaporation whereas in central and south Poland the amount is only 100–180 mm. In Hungary this difference, which was observed at the stations situated on the lowlands, amounts to 80 mm and it sometimes falls to 0. In Hungary the evaporation generally surpasses the amount of rainfall in the summer season (see the climatic tables). (The annual amounts for the Polish stations are only approximate, being calculated from the equation of water balance, while for the Hungarian stations they reflect actual data as calculated by the Thornthwaite methods.)

Generally speaking, the volume of evaporation is related to air temperature and humidity. In Poland the relative humidity in the winter months amounts to nearly 90% and even higher (Suwałki) almost entirely over the country, the exception being in the south where it does not surpass 85%, as in Czechoslovakia. In Hungary it is ca 80% or less (e.g. in Debrecen). In the spring and summer months the percentages of humidity are as follows: in Poland ca 70% or less; in Czechoslovakia usually below 70%; and almost everywhere in Hungary, less than 50%. In Poland the humidity values are lowest in spring and they increase toward late summer and autumn, whereas in Hungary they are usually the lowest in mid- and late summer (see the climatic tables). This is due to the considerable loss of water reserves, exposed to evaporation, with the amount of precipitation being too low to meet the losses caused by evapotranspiration.

A considerable decrease of air humidity (and water contents in soils as well) in Poland and probably in other countries too, is connected with the intense water demand of plants from the very beginning of their vegetation. When precipitation amount is lower than normal (e.g. in May) agriculture may face detriment caused by drought.

The question of drought is, however, more complicated as it depends on the weather conditions which prevail in autumn (reserves of water in soils before it freezes), on the snow cover in winter and on the process of snow melting in spring.

Seasons

Winter

If we accept the start of winter to be on the day when the diurnal mean temperature falls below 0°C, and disregard mountains and highlands, then winter begins earliest in northeast Poland (November 20). Temperatures below the freezing point appear in eastern Czechoslovakia in the beginning of December. After December 15 they also occur in

Fig.20. Mean number of days with snow cover (after ANONYMOUS, 1958, 1960a, 1973a).

the west part of that country, but further south they occur not earlier than December 20. The maximum differences in the dates of winter onset are: between northeast and north-west Poland 20 days; between northeast Poland and southwest Slovakia over 30 days. In eastern Poland and Slovakia the number of days with snowfall in January represents over 80% of all precipitation days in that month. The warming which ends the winter (rise of diurnal mean temperature above 0°C) arrives from the south. It embraces the Hungarian plains and part of Slovakia, already between the 10th and 20th of February. The warming which approaches from the west takes place at the same time on the low-lands of the Labe Basin. The warming in Poland begins only in the last 8–9 days of February in the west part of the country. It finally reaches east Slovakia (in early March) and over the northeast areas of Poland (in the last 10 days of March). The difference between the extreme dates of the winter's end in Poland is more than 30 days, in Slovakia over 20 days.

The cold season of the year is characterized, among other factors, by the number of days with snow cover (see climatic tables and Fig.20). In the east, the cover is more stable and less disrupted by thawing. In the Tatra Mountains it persists over 250 days. Thawing which disrupts the snowcover occurs generally in connection with the activity of Atlantic–Baltic low-pressure systems (which encompass northwest Poland and north Czechoslovakia more often) or of Mediterranean lows which favour the Danubian regions. The lowest number of days with snowcover is noted in southern Hungary—almost three times lower than in northeast Poland (Pécs 26, Suwałki 99 days).

Summer

Taking as the characteristic feature the rising of daily mean temperature above 15°C, the summer begins earliest (prior to May 20) on the Hungarian plains, in southwest Slovakia and in the east of that country. On the first of June, mean summer temperatures are observed already on a large part of the lowlands of eastern Poland, also along the Odra River, on the sub-Carpathian depression, and in the Labe Basin in Czechoslovakia. A characteristic feature is that summer arrives in the east part of central Europe from three directions: the "Mediterranean summer" from the south, the "continental east-European" summer from the east and the "Atlantic west-European" summer from the west. Finally (after June 20) occurs the warming on the lowlands located on the northern peripheries of Poland.

Temperatures fall below 15°C earliest in the northeast part of the Polish lake districts and in western Czechoslovakia (September 1) and finally (after September 10) in western Poland, in Carpathian submontane depressions and also in east and south Slovakia.

The most characteristic feature of the warm season is the length of the vegetative period. It lasts in eastern Poland about 200 days or less, in south and west Poland 220 days or more, in the north of the country about 160 days. Duration of the growing season in Czechoslovakia is from 220 (České Budějowice) to approximately 240 days (Prague 239, Bratislava 242 days). In Hungary, the vegetative season ranges from about 240 days in the north to 250–260 days in the south (Szeged 259, Debrecen 242 days).

A typical feature of the warmer half of the year, notably in summer, are thunderstorms. Annual thunderstorm frequency in Poland is lowest in the north (15–20 days), reaches over 20 in some localities of central Poland and more than 30 days in the mountains

(Tatra). The analogous figures for Czechoslovakia are from 20 (Brno) to 27–30 (Bratislava, Prague). Corresponding data for Hungary are from 30 (Debrecen) to 50 days and over (Szeged, Pécs, Budapest). Maximum thunderstorm frequency in Poland (4–7 days) is observed in July, in Czechoslovakia (6–7 days) and in Hungary (12–15 days) generally in June. The number of thunderstorm days in the summer season is in Hungary two to three times greater than in Poland (after STOPA, 1962; ANONYMOUS, 1964, tables; GÖTZ and SZALAY, 1966, 1967).

Main climatic differences

If we take into consideration all the above mentioned climatic features, we arrive at a conclusion that the contrasts between lowland terrains are the greatest between north Poland and south Hungary on the one hand, and between the west and the east areas of Poland and of Czechoslovakia on the other. The climate of northwest Poland is the one which is most distinctly maritime in character. The influence of the Baltic Sea is evident in the lake districts zone. The annual amplitudes of air temperatures in Świno-ujście, Gdańsk–Gdynia and Suwałki reach to 19°, 20°, 23°C, respectively. A relatively distinct maritime climate is observed in Czechoslovakia in the northwest part of the country (temperature amplitude in Prague is 20°C).

A continental climate is most distinct in Poland in the southeast part of the country, where the annual amplitude of air temperatures reaches 23°C (Zamość), in east Czechoslovakia (Prešov 23°C) and in Hungary (annual air temperature amplitudes at Debrecen, Szeged and Pécs about 23° and 24°C, respectively). Those areas also differ in respect to their cloud and precipitation regimes. Types of circulation systems governing the weather are different too. The central lowland regions of Poland and of south Hungary belong to the driest and (in summer) warmest areas on both sides of the Carpathians; the Baltic regions of northwest Poland and south Hungary belong to the relatively most humid and warmest (in winter and fall). (Mountainous regions are not taken into account in this comparison.)

Oscillations and fluctuations of climate

One of the most difficult problems in climatology is to learn the regularities coherent to fluctuations, oscillations or changes of climate. The author is not going to deal with the changes in local climate and especially not in microclimate as they are the result of man's economic activities. Only some general remarks will be made in connection with the oscillations or fluctuations of macroclimate, i.e. of the climate of large regions.

Since the 19th century and up to the twenties of the present century a continuous diminishing of the annual range of air temperatures was observed in a fairly large part of Europe including its central part. As calculated per 100 years, it amounted to 5–10% of the average annual value. Very likely it excluded Hungary or at least its southern part. At the same time an increase in the annual amount of rainfall was observed, excluding, however, a part of the lowlands north of the Sudetes and north and east of the Carpathians. These phenomena were explained with the increasing frequency of advection of sea-air masses, occurring primarily in the winter season, as the diminution observed in the annual range of temperatures was mainly due to the increase of winter temperatures. Significant decrease in the amount of rainfall, as is observed beyond the Sudetes and Carpathians, suggested that the increased advection of air from over the Atlantic more often had the direction southwest than northwest. Both the changes in annual temperature amplitudes and in the amount of rainfall were not regular and they showed many oscillations, increases and decreases enduring sometimes some years and other times even several years (OKOŁOWICZ, 1948).

The recent studies of the changes of atmospheric circulation in the extratropical part of the Northern Hemisphere have proved that in the period of ca.1906–1915 to ca.1929–1938 continuous growth was observed in the frequency of zonal circulation whereas the meridional circulation showed a decrease. After 1929–1938 the frequency of zonal circulation again began to decrease and that of meridional began to increase until the years 1959–1968 (DZERDZEEVSKII, 1968).

The period of increase in the frequency of zonal circulation was conforming, to some extent, to the lowering of annual amplitudes of air temperatures.

Efforts have also been made to explain the diminution of the river outflow, observed in the north-European lowland, to be the result of the changes in atmospheric circulation and in the activeness of the sun. The results obtained suggest that the diminution in the river outflow is connected with the less frequent western advection, which is conditioned by the increased sun activeness. In the course of further analysis it is expected that considering the forecasts of average sun activeness (Wolf figures), in the period of 1965–1977 to 1978–1988 an increase in the river outflow will take place (Odra and Vistula rivers) due to the anticipated increase of the frequency of western circulation conditioned by the forecasted decrease of the sun activeness (STACHY, 1969).

The question of whether the above prognosis is true will be answered by the future or, maybe, by new research based on ample and more detailed observations to be made in the meantime.

References

ANONYMOUS, 1958. *Climatic Atlas of the Czechoslovak Republic*. Ústředni Správa Geodezie a Kartografie, Praha.
ANONYMOUS, 1960a. *Klima-Atlas von Ungarn*. Akademiai Kiadó, Budapest.
ANONYMOUS, 1960b. *The Climate of the Czechoslovak Socialist Republic.—Tables*. Hydromet. Ústav, Praha.
ANONYMOUS, 1964. *Atlas Mira. Fiziko-Geograficheskiĭ*. Akad. Nauk SSSR, Moskva.
ANONYMOUS. 1967. *Klima-Atlas von Ungarn, II. Tabellen*. Akademiai Kiadó, Budapest.
ANONYMOUS, 1973a. *Klimatyczny Atlas Polski*. Inst. Meteorologii i Gospodarki Wodnej, Warsawa.
ANONYMOUS, 1973b. *Narodowy Atlas Polski*. Polish Acad. of Sciences, Warsaw, in press.
BODOLAI, I. and BODOLAI, E., 1966. Instabilitätslinien und die Möglichkeiten deren Vorhersage im Raume des Balatonsees. Sturmwarnung am Balatonsee. *Veröff. Ung. Zentralanst. Meteorol. Budap.*, XXX: 128–139.
BUDZISZEWSKA, E., 1968. Jet streams over Poland, I. *Rev. Geophys. Warsaw*, XIII: 209–229.
BUDZISZEWSKA, E., 1973. Tropospherical jet streams over Poland, their duration and frequency in 1960–1969. In press.
BUDZISZEWSKA, E., HORAWSKA, M. and MOROZOWSKA, I., 1973. Meteorological conditions of flood rainfall occurrence in the southern part of Poland in August 1972. *Rep. Hydrol. Meteorol. Serv.*, IX (2–3): 51–60.

DZERDZEEVSKII, B. L., 1968. *Circulation Mechanisms in the Atmosphere of the Northern Hemisphere in the 20th Century.* Inst. of Geogr., Academy of Sciences of the USSR, Moscow.

FICKER, H., 1951. Die Zentralanstalt für Meteorologie und Geodinamik in Wien 1851–1951. *Denkschr. Österreich. Akad. Wiss.*, 109: 1–19.

FORGAČ, P., 1953. *Búrky na Slovensku.* Slovenska Akad. Vied., Bratislava, 86 pp.

GORCZYŃSKI, W., 1917. *Pression atmosphérique en Pologne et en Europe.* Cotty, Warsaw, 265 pp.

GÖTZ, G. and PAPAINÉ SZALAY, G., 1966. Thunderstorm activity during the summer half-year over Hungary. *Weather Budap.*, 70 (2): 106–116.

GÖTZ, G. and PAPAINÉ SZALAY, G., 1967. Thunderstorm activity in Hungary during the winter half-year. *Weather Budap.*, 71 (5): 302–309.

GÖTZ, G. and TÄNCZER, T., 1966. Statistische und synoptische Untersuchungen der Stürme am Balatonsee. Sturmwarnung am Balatonsee. *Veröff. Ungar. Zentralanst. Meteorol.*, XXX: 11–29.

GREGOR, A., 1966. Die Meteorologie in der ČSSR im historischen Überblick. *Meteorol.-Ergeb. Konf. Liblice Prag*, 1964, pp.13–22.

HRGIAN, A. H., 1959. *Ocherki Raswitija Meteorologii.* Gidrometeoisdat., Leningrad, pp.144–147.

KONČEK, M. and BRIEDOŇ, V., 1964. *Sneh a Snehová Pokrývka na Slovensku.* Slovenská Akad. Vied., Bratislava, 71 pp.

KONČEK, M., ORLICZ, M. and SLADKOVIČ, R., 1966. Uslovia pogody v Vysokich Tatrach v zavisimosti ot tipičnych sinoptičeskich položeniy. *Einfluss der Karpaten auf die Witterungserscheinungen. Konf. Karpatenmeteorol., 3, Beograd, 1965*, pp. 1–15.

KUCZMARSKA, L., and PASZYŃSKI, J., 1964. Distribution of global radiation in Poland. *Pol. Geogr. Rev.*, 36(4): 691–702.

LITYŃSKI, J., 1968. Liczbowa klasyfikacja typów cyrkulacji i typów pogody dla Polski. *Prace PIHM*, 97: 3–15.

MICHALCZEWSKI, J., 1965. Synoptic conditions of occurrence of sea breezes on the Polish coast of Baltic Sea. *Bull. Serv. Hydrol. Météorol., Warszawa,* I (XIII)–2: 29–40.

OKOŁOWICZ, W., 1948. About climatic changes. *Pol. Geogr. Rev.*, XXI: 205–224.

OKOŁOWICZ, W., 1962. Cloudiness in Poland. *Geogr. Stud. PAN, Warsaw*, 34: 1–107.

ORLICZ, M., 1954. Relations anémométriques sur les sommets des Tatra. *Bull. Serv. Hydrol. Météorol., Warszawa*, III (4): 38–59.

OTRUBA, J., 1964. *Veterné Pomery na Slovensku.* Slovenska Akad. Vied, Bratislava, 281 pp.

PASZYŃSKI, J., 1966. Materiały do bilansu cieplnego Polski—Atlas bilansu promieniowania w Polsce. *Dok. Geogr.*, 4: 8 pp. (+ tables and maps).

PÉCZELY, G., 1957. Grosswetterlagen in Ungarn. *Kleinere Veröff. Zentralanst. Meteorol. Budap.*, 30: 86 pp.

REIN, F., 1959. Weather typing with regard to dynamic climatology. *Stud. Geophys. Geod.*, 3: 177–194.

ROJECKI, A., 1966. The traditions of meteorology in Poland, from the 15th to 19th century. *Acta Geophys. Pol.*, XIV: 3–10.

STACHY, J., 1969. Wieloletnia prognoza odpływu rzek polskich. Sesja naukowa. . .19–20. VI.1969: 12–15 pp. State Hydrol.-Met. Inst., Warsaw.

STASZEWSKI, J. and UHORCZAK, F., 1966. *Geografia Fizyczna w Liczbach.* Państwowe Wydawnictwo Naukowe Warsawa, 518 pp.

STOPA, M., 1962. Thunderstorm in Poland. *Geogr. Stud. PAN, Warsaw*, 34: 109–185.

SZEPESI, D., 1963. Über die orographische niederschlagsbildende Wirkung der Gebirge des Karpatenbeckens. *Einfluss der Karpaten auf die Witterungserscheinungen. Konf. Karpatenmeteorol., 2., Budap.*, 1961, pp.37–43.

TARANOWSKA, S., 1966. *Stosunki Anemometryczne Południowego Bałtyku.* University of Wrocław, Wrocław (unpublished).

WARAKOMSKI, W., 1963. Cloudiness in Poland. *Rev. Geophys., Warsaw*, VIII (1–2): 21–35.

TABLE XIII

CLIMATIC TABLE FOR GDYNIA

Latitude 54°31′N, longitude 18°33′E, elevation 15 m

Month	Mean sta. press. (mbar)[1]	Temperature (°C)				Mean vapour press. (mbar)[4]	Relat. humid. (%)[4]	Precipitation (mm)	
		daily mean[2]	daily range[3]	extreme[3]				mean[2]	max. in 24 h
				max.	min.				
Jan.	1009.8	−1.3	5.1	10.8	−25.2	4.9	85	34	
Feb.	1013.3	−0.9	5.8	13.6	−28.8	5.1	84	30	
Mar.	1015.9	1.5	7.2	20.5	−12.9	5.8	81	30	
Apr.	1013.8	6.1	8.3	26.8	−4.3	7.4	76	34	
May	1015.0	11.0	8.8	29.9	−0.5	10.0	72	44	
June	1013.6	15.3	9.3	31.8	3.0	12.7	72	59	
July	1011.0	17.9	8.2	36.0	8.2	15.3	73	77	
Aug.	1010.8	17.6	8.3	32.5	8.2	14.8	75	79	
Sept.	1013.2	14.0	8.3	30.7	1.7	12.4	77	55	
Oct.	1014.6	9.1	7.8	22.4	−6.5	9.3	81	54	
Nov.	1014.8	4.3	4.8	14.7	−8.2	6.9	85	39	
Dec.	1011.0	1.1	2.5	12.2	−14.8	5.6	87	42	
Annual	1013.1	7.9	7.0	36.0	−28.8	9.2	79	576	

Month	Mean evapor. (mm)	Number of days with				Mean cloudiness (%)[7]	Mean sunshine (h)[1]	Wind[3]	
		precip. ⩾1 mm[2]	thunderstorm[5]	fog	snow cover[6]			preval. direct.	mean speed (m/sec)
Jan.		15.4	0.1		16.6	81	38	S	4.2
Feb.		13.6			15.5	76	70	S	3.8
Mar.		12.7	0.1		5.7	66	134	S	4.0
Apr.		12.5	1.4		1.3	65	163	N	3.9
May		10.5	2.8			61	224	N	3.6
June		11.4	3.8			60	259	N	3.0
July		12.7	4.5			63	236	N	3.2
Aug.		13.1	4.0			59	225	W	3.0
Sept.		13.1	1.8			59	174	W	3.3
Oct.		14.0	0.1			66	105	S	3.4
Nov.		12.3			1.4	81	45	S	3.4
Dec.		15.2			5.9	81	32	S	3.9
Annual	±350	156.5	18.4		46.4	68	1,705	S	3.6

[1] 1951–1960; [2] 1931–1960; [3] 1951–1960, from Gdańsk; [4] 1931–1960, from Gdańsk; [5] 1946–1955, from Gdańsk; [6] 1950/51–1959/60; [7] 1950–1959.

TABLE XIV

CLIMATIC TABLE FOR SUWAŁKI

Latitude 54°06′N, longitude 22°75′E, elevation 170 m

Month	Mean sta. press. (mbar)[1]	Temperature (°C)				Mean vapour press. (mbar)	Relat. humid. (%)[2]	Precipitation (mm)	
		daily mean[2]	daily range[1]	extreme[1]				mean[2]	max. in 24 h
				max.	min.				
Jan.	991.5	−5.6	5.2	7.0	−32.0		89	29	
Feb.	992.9	−4.8	6.4	6.5	−32.0		86	28	
Mar.	996.6	−1.3	8.0	16.2	−21.9		80	30	
Apr.	994.7	5.7	9.3	25.4	−8.3		75	37	
May	994.9	12.2	11.1	30.8	−4.2		67	42	
June	994.8	15.8	11.1	31.2	1.4		69	77	
July	992.5	17.7	10.4	35.3	6.2		74	93	
Aug.	993.1	16.8	10.3	32.9	3.9		76	81	
Sept.	994.8	12.3	9.4	29.8	−0.5		80	56	
Oct.	996.1	6.5	7.4	22.1	−14.2		86	47	
Nov.	997.0	1.5	4.5	14.5	−20.7		90	38	
Dec.	992.5	−2.4	4.2	10.3	−22.9		91	36	
Annual	994.3	6.2	8.1	35.3	−32.0		80	594	

Month	Mean evapor. (mm)	Number of days with				Mean cloudiness (%)[5]	Mean sunshine (h)[1]	Wind[1]	
		precip. ⩾1 mm[2]	thunderstorm[3]	fog	snow cover[4]			preval. direct.	mean speed (m/sec)
Jan.		16.9			26.7	82	28	SW	4.9
Feb.		14.4			24.4	80	55	SW	4.3
Mar.		12.9	0.1		20.0	63	147	SE	4.3
Apr.		12.5	1.6		3.4	66	162	SE	4.1
May		11.3	3.2			62	227	NW	4.0
June		12.6	4.7			61	254	NW	3.5
July		15.6	5.0			64	234	W	3.5
Aug.		13.5	3.1			62	217	W	3.4
Sept.		12.3	1.4			64	158	W	3.5
Oct.		13.3			0.4	72	86	SW	4.0
Nov.		14.0			6.8	87	31	SE	4.5
Dec.		16.4			17.0	86	22	SW	4.6
Annual	±350	165.9	19.1		98.7	71	1,621	SW	4.1

[1] 1951–1960; [2] 1931–1960; [3] 1946–1955; [4] 1950/51–1959/60; [5] 1950–1959.

TABLE XV

CLIMATIC TABLE FOR CHOJNICE

Latitude 53°42'N, longitude 17°33'E, elevation 172 m

Month	Mean sta. press. (mbar)	Temperature (°C)				Mean vapour press. (mbar)	Relat. humid. (%)	Precipitation (mm)	
		daily mean*1	daily range*2	extreme*2				mean*1	max. in 24 h
				max.	min.				
Jan.		−3.4	4.8	9.4	−26.7			27	
Feb.		−2.7	6.0	13.0	−29.8			29	
Mar.		0.4	7.4	18.2	−15.2			29	
Apr.		6.2	9.5	26.2	−6.0			39	
May		11.9	11.1	30.5	−3.5			46	
June		15.5	11.1	32.6	0.4			65	
July		17.2	10.0	35.5	4.9			89	
Aug.		16.6	9.8	30.7	5.0			85	
Sept.		12.7	9.3	29.3	0.0			46	
Oct.		7.4	7.1	21.7	−8.3			42	
Nov.		2.5	4.5	14.8	−10.1			35	
Dec.		−0.8	4.1	11.2	−15.2			40	
Annual		7.0	8.0	35.5	−29.8			572	

Month	Mean evapor. (mm)	Number of days with				Mean cloud-iness (%)*5	Mean sun-shine (h)*2	Wind*2	
		precip. ⩾ 1 mm*1	thunder-storm*3	fog	snow cover*4			preval. direct.	mean speed (m/sec)
Jan.		14.4			22.9	81	29	W	4.2
Feb.		14.0			21.6	76	58	W	3.5
Mar.		11.1	0.2		7.6	63	136	E	3.5
Apr.		11.8	1.0		0.8	67	150	W	3.1
May		11.5	3.4			62	220	NW	3.2
June		11.8	4.1			60	230	NW	2.8
July		15.1	4.9			65	207	W	2.7
Aug.		13.8	4.1			62	198	W	2.5
Sept.		11.9	1.1			59	169	W	3.0
Oct.		13.2			0.1	69	96	SW	3.0
Nov.		13.5			1.8	85	31	SW	3.1
Dec.		15.0	0.1		12.8	83	21	SW	3.8
Annual	± 350	157.1	18.9		67.6	69	1,545	W	3.2

*1 1931–1960; *2 1951–1960; *3 1946–1955; *4 1950/51–1959/60; *5 1950–1959.

TABLE XVI

CLIMATIC TABLE FOR POZNAŃ

Latitude 52°25'N, longitude 16°50'E, elevation 92 m

Month	Mean sta. press. (mbar)*1	Temperature (°C)				Mean vapour press. (mbar)*2	Relat. humid. (%)*2	Precipitation (mm)	
		daily mean*2	daily range*1	extreme*1				mean*2	max. in 24 h
				max.	min.				
Jan.	1002.8	−2.2	5.4	11.2	−22.7	5.1	87	30	
Feb.	1003.2	−1.6	6.8	13.7	−28.0	5.3	85	29	
Mar.	1006.2	2.3	8.1	18.8	−14.8	6.2	81	28	
Apr.	1005.0	8.0	10.4	28.5	−6.2	7.8	73	38	
May	1005.7	13.6	11.5	31.8	−3.0	10.6	68	59	
June	1004.9	17.0	11.6	34.4	0.6	12.6	67	60	
July	1003.4	18.8	10.7	38.2	3.8	14.3	69	73	
Aug.	1003.5	17.9	11.0	34.2	5.2	13.9	73	64	
Sept.	1005.9	14.0	10.6	32.4	−1.6	11.8	77	41	
Oct.	1006.5	8.5	9.0	24.8	−6.0	9.3	83	40	
Nov.	1006.7	3.7	5.2	17.0	−10.1	7.2	88	34	
Dec.	1002.9	0.2	4.8	12.5	−14.7	5.7	89	33	
Annual	1004.7	8.3	8.7	38.2	−28.0	9.2	78	528	

Month	Mean evapor. (mm)	Number of days with				Mean cloud-iness (%)*5	Mean sun-shine (h)	Wind*2	
		precip. ⩾ 1 mm*2	thunder-storm*3	fog	snow cover*4			preval. direct.	mean speed (m/sec)
Jan.		16.4	0.1		16.9	75		SW	4.6
Feb.		14.6			13.8	72		W	4.2
Mar.		12.3	0.4		4.3	60		W	4.5
Apr.		13.1	1.1		0.2	64		W	4.0
May		12.1	3.6			61		W	4.0
June		12.4	4.3			62		W	3.6
July		14.9	4.8			65		W	3.8
Aug.		13.2	4.2			59		W	3.4
Sept.		11.6	1.3			57		W	3.5
Oct.		14.1	0.2			61		SW	3.5
Nov.		14.6			1.1	79		SW	3.7
Dec.		15.9			6.8	78		SW	4.2
Annual	±400	165.2	20.0		43.1	66		W	3.9

*1 1951–1960; *2 1931–1960; *3 1946–1955; *4 1950/51–1959/60; *5 1950–1959.

TABLE XVII

CLIMATIC TABLE FOR WARSAW (Okęcie)

Latitude 52°09′N, longitude 20°59′E, elevation 107 m

Month	Mean sta. press. (mbar)[1]	Temperature (°C)				Mean vapour press. (mbar)	Relat. humid. (%)[1]	Precipitation (mm)	
		daily mean[2]	daily range[1]	extreme[1] max.	extreme[1] min.			mean[2]	max. in 24 h[1]
Jan.	1001.3	−3.5	5.0	10.7	−27.1		86	23	8.4
Feb.	1002.1	−2.5	6.6	12.0	−26.1		85	26	13.0
Mar.	1004.6	1.4	7.7	18.3	−19.0		77	24	11.9
Apr.	1002.8	8.0	9.8	27.0	−6.9		73	36	24.6
May	1002.9	14.0	10.5	30.8	−3.0		68	44	22.4
June	1002.9	17.5	10.9	32.1	2.3		69	62	49.1
July	1000.9	19.2	10.5	35.1	5.2		74	79	39.6
Aug.	1001.6	18.2	10.8	35.1	5.4		74	65	27.4
Sept.	1003.9	13.9	10.3	31.4	−0.3		77	41	22.2
Oct.	1005.1	8.1	9.0	25.2	−8.0		82	35	50.3
Nov.	1005.4	3.0	5.2	16.7	−10.0		86	37	28.6
Dec.	1001.6	−0.6	4.4	11.9	−18.9		88	30	17.4
Annual	1002.9	8.1	8.5	35.1	−27.1		78	502	50.3

Month	Mean evapor. (mm)	Number of days with				Mean cloud- iness (%)[5]	Mean sun- shine (h)[1]	Wind	
		precip. ≥ 1 mm[2]	thunder- storm[3]	fog	snow cover[4]			preval. direct.[1]	mean speed (m/sec)[1]
Jan.		14.3		5.0	20.9	78	21	W	5.2
Feb.		14.1	0.1	5.9	18.6	73	51	W	4.7
Mar.		9.4	0.1	2.7	10.3	63	131	E	4.9
Apr.		11.4	1.7	2.3	0.9	62	138	W	4.2
May		11.0	4.3	0.8		61	192	W	4.0
June		11.6	5.2	1.5		58	198	W	3.4
July		14.1	5.5	0.9		61	203	W	3.3
Aug.		12.1	4.9	1.3		56	208	W	3.1
Sept.		12.4	1.5	2.5		57	171	W	3.5
Oct.		10.4	0.2	5.5		62	106	W	3.6
Nov.		13.0		7.6	2.1	81	36	SE	4.3
Dec.		14.8		7.1	9.5	80	10	W	4.7
Annual	± 400	148.6	23.5	43.1	62.3	64	1,465	W	4.1

[1] 1951–1960; [2] 1931–1960; [3] 1946–1955; [4] 1950/51–1959/60; [5] 1950–1959.

TABLE XVIII

CLIMATIC TABLE FOR WROCŁAW

Latitude 51°08′N, longitude 16°59′E, elevation 119 m

Month	Mean sta. press. (mbar)[1]	Temperature (°C)				Mean vapour press. (mbar)[2]	Relat. humid. (%)[2]	Precipitation (mm)	
		daily mean[2]	daily range[1]	extreme[1] max.	extreme[1] min.			mean[2]	max. in 24 h
Jan.	1000.3	−2.0	6.6	13.0	−23.3	4.9	83	31	
Feb.	1000.8	−1.1	8.1	16.0	−32.0	5.0	80	30	
Mar.	1002.5	2.8	9.2	22.1	−19.0	6.0	75	31	
Apr.	1001.7	8.3	10.8	27.6	−6.8	7.4	69	39	
May	1002.1	13.6	11.5	30.0	−2.4	10.4	66	60	
June	1001.8	17.0	11.8	32.8	1.4	12.2	66	72	
July	1000.3	18.8	11.0	36.6	4.3	14.1	67	81	
Aug.	1000.8	17.9	11.8	35.6	4.6	13.4	69	73	
Sept.	1002.9	14.1	11.3	31.6	−1.8	11.6	73	48	
Oct.	1003.6	8.7	10.5	25.7	−6.2	9.0	78	42	
Nov.	1003.4	4.1	6.9	18.4	−10.3	6.8	83	34	
Dec.	1000.2	0.3	6.3	15.4	−17.6	5.4	84	33	
Annual	1001.7	8.5	9.7	36.6	−32.0	8.9	74	574	

Month	Mean evapor. (mm)	Number of days with				Mean cloud- iness (%)[5]	Mean sun- shine (h)[1]	Wind[1]	
		precip. ≥ 1 mm[2]	thunder- storm[3]	fog	snow cover[4]			preval. direct.	mean speed (m/sec)
Jan.		15.8	0.1		15.2	73	48	W	4.0
Feb.		13.5			15.1	70	67	W	3.6
Mar.		14.0	0.3		6.2	63	124	W	3.7
Apr.		13.5	1.6		0.5	65	146	NW	3.2
May		14.1	4.9			61	200	NW	3.1
June		13.4	5.3			62	204	NW	2.9
July		14.7	5.3			59	202	NW	3.1
Aug.		13.2	4.7			61	199	W	2.7
Sept.		12.1	1.0			58	156	NW	3.0
Oct.		13.0	0.1		0.3	59	112	SE	2.8
Nov.		14.3			1.1	76	48	SE	3.3
Dec.		16.5			8.1	75	38	SE	3.6
Annual	± 460	168.1	23.3		46.5	65	1,644	W	3.2

[1] 1951–1960; [2] 1931–1960; [3] 1946–1955; [4] 1950/51–1959/60; [5] 1950–1959.

TABLE XIX

CLIMATIC TABLE FOR KIELCE

Latitude 50°51'N, longitude 20°37'E, elevation 268 m

Month	Mean sta. press. (mbar)	Temperature (°C)				Mean vapour press. (mbar)	Relat. humid. (%)	Precipitation (mm)	
		daily mean[*1]	daily range[*4]	extreme[*4]				mean[*1]	max. in 24 h
				max.	min.				
Jan.		−4.1	6.2	8.6	−28.4			38	
Feb.		−3.0	7.8	11.3	−31.0			39	
Mar.		1.0	9.5	20.2	−19.9			32	
Apr.		7.1	11.0	26.8	−9.4			40	
May		13.1	11.6	33.4	−3.6			52	
June		16.5	11.5	31.9	−0.2			84	
July		18.2	11.4	35.0	4.8			95	
Aug.		17.3	12.2	35.2	3.7			81	
Sept.		13.3	11.1	31.4	−1.5			49	
Oct.		7.7	9.8	25.2	−6.7			45	
Nov.		2.7	5.8	16.6	−13.1			46	
Dec.		−1.1	5.2	11.6	−15.4			43	
Annual		7.4	9.5	35.2	−31.0			644	

Month	Mean evapor. (mm)	Number of days with				Mean cloud-iness (%)[*3]	Mean sun-shine (h)	Wind[*4]	
		precip. ⩾1 mm[*1]	thunder-storm[*2]	fog	snow cover			preval. direct.	mean speed (m/sec)
Jan.		16.2				79		E	3.2
Feb.		15.2				72		W	3.1
Mar.		13.1	0.1			66		E	3.2
Apr.		12.6	1.1			63		E	2.7
May		11.4	2.6			64		E	2.9
June		13.3	4.1			61		W	2.6
July		14.1	5.5			59		W	2.5
Aug.		13.0	3.0			56		W	2.5
Sept.		11.1	1.2			58		W	2.6
Oct.		12.0	0.3			63		W	2.6
Nov.		14.2				78		E	2.8
Dec.		16.6				81		E	3.0
Annual	±460	162.8	17.9			66		W	2.8

[*1] 1931–1960; [*2] 1946–1955; [*3] 1950–1959; [*4] 1951–1960.

TABLE XX

CLIMATIC TABLE FOR ZAMOŚĆ

Latitude 50°44'N, longitude 23°15'E, elevation 219 m

Month	Mean sta. press. (mbar)[*1]	Temperature (°C)				Mean vapour press (mbar)	Relat. humid. (%)	Precipitation (mm)	
		daily mean[*2]	daily range[*1]	extreme[*1]				mean[*2]	max. in 24 h
				max.	min.				
Jan.	988.6	−4.4	5.9	11.6	−27.2			29	
Feb.	988.6	−3.3	6.8	14.3	−27.6			32	
Mar.	991.1	0.7	7.0	19.0	−21.8			31	
Apr.	989.6	7.6	10.3	26.7	−5.6			39	
May	989.5	13.5	11.3	31.6	−1.4			63	
June	989.5	16.9	11.8	32.2	0.3			76	
July	988.3	18.6	11.5	35.7	5.1			93	
Aug.	988.9	17.6	11.7	35.7	5.0			94	
Sept.	991.3	13.3	11.1	31.2	−1.1			55	
Oct.	992.6	7.6	8.5	25.4	−6.4			58	
Nov.	992.6	2.6	5.5	17.5	−13.9			42	
Dec.	989.3	−1.3	5.0	14.1	−16.2			32	
Annual	990.7	7.4	9.1	35.7	−27.6			644	

Month	Mean evapor. (mm)	Number of days with				Mean cloud-iness (%)[*3]	Mean sun-shine (h)[*1]	Wind[*1]	
		precip. ⩾1 mm	thunder-storm[*3]	fog	snow cover[*4]			preval. direct.	mean speed (m/sec)
Jan.					24.7	77	42	SW	3.9
Feb.					20.3	71	60	W	3.4
Mar.			0.1		15.7	61	128	E	3.4
Apr.			1.6		1.9	57	157	NW	2.8
May			3.9			55	201	NW	2.7
June			4.5			50	232	NW	2.4
July			5.1			49	237	W	2.3
Aug.			4.5			46	213	W	2.3
Sept.			1.1			53	168	W	2.7
Oct.					0.3	56	120	SW	2.8
Nov.					3.1	77	37	SE	3.1
Dec.			0.1		13.3	76	36	SW	3.6
Annual			20.9		79.3	61	1,631	W	3.0

[*1] 1951–1960; [*2] 1931–1960; [*3] 1946–1955; [*4] 1950/51–1959/60; [*5] 1950–1959.

TABLE XXI

CLIMATIC TABLE FOR CRACOW

Latitude 50°05′N, longitude 20°01′E, elevation 213 m

Month	Mean sta. press. (mbar)[*1]	Temperature (°C)					Mean vapour press. (mbar)	Relat. humid. (%)	Precipitation (mm)	
		daily mean[*2]	daily range[*1]	extreme[*1]					mean[*2]	max. in 24 h
				max.	min.					
Jan.	989.1	−2.9	5.3	11.3	−22.8				34	
Feb.	989.3	−1.4	7.0	15.8	−26.6				34	
Mar.	990.8	2.6	8.5	21.2	−18.1				35	
Apr.	989.8	8.6	10.1	27.1	−7.1				42	
May	990.0	14.1	10.4	32.6	−2.2				57	
June	990.0	17.5	10.5	32.5	3.0				86	
July	988.8	19.3	11.2	35.2	6.2				95	
Aug.	989.4	18.4	11.1	35.7	4.2				83	
Sept.	991.8	14.4	10.4	30.7	−1.5				56	
Oct.	992.7	8.8	9.6	26.4	−5.0				46	
Nov.	992.5	3.8	6.2	20.9	−13.0				42	
Dec.	989.3	−0.2	5.5	16.6	−17.1				34	
Annual	990.3	8.6	8.8	35.7	−26.6				645	

Month	Mean evapor. (mm)	Number of days with				Mean cloud-iness (%)[*5]	Mean sun-shine (h)[*1]	Wind[*1]	
		precip. ⩾ 1 mm[*2]	thunder-storm[*3]	fog	snow cover[*4]			preval. direct.	mean speed (m/sec)
Jan.		16.5			21.5	74	43	W	3.2
Feb.		14.8	0.1		18.1	72	55	W	3.2
Mar.		13.4	0.2		11.0	65	113	W	3.5
Apr.		13.9	1.0		0.6	61	147	W	2.7
May		13.3	3.1			61	189	W	2.8
June		15.3	5.1			59	199	W	2.4
July		14.9	5.2			57	220	W	2.4
Aug.		13.8	4.5			53	210	W	2.2
Sept.		11.9	1.8			56	151	W	2.5
Oct.		13.3	0.2		0.2	61	111	W	2.3
Nov.		16.0			1.8	76	48	W	2.7
Dec.		15.2			8.7	77	36	W	3.0
Annual	± 500	172.3	21.2		61.9	64	1,522	W	2.7

[*1] 1951–1960; [*2] 1931–1960; [*3] 1946–1955; [*4] 1950/51–1959/60; [*5] 1950–1959.

TABLE XXII

CLIMATIC TABLE FOR PRAGUE

Latitude 50°05′N, longitude 14°25′E, elevation 197 m

Month	Mean sta. press. (mbar)[*1]	Temperature (°C)					Mean vapour press. (mbar)[*4]	Relat. humid. (%)[*1]	Precipitation (mm)	
		daily mean[*2]	daily range[*3]	extreme[*5]					mean[*1]	max. in 24 h[*5]
				max.	min.					
Jan.	1018.2	−2.6	4.7	13.3	−21.2	4.6	86	23	16
Feb.	1018.2	−1.6	5.6	13.2	−27.1	4.7	83	24	17
Mar.	1018.4	2.7	7.7	19.8	−21.2	5.6	77	23	25
Apr.	1017.1	7.8	8.8	27.6	−5.5	7.0	70	32	40
May	1016.4	12.9	9.7	32.4	−1.2	9.1	69	61	56
June	1016.7	16.2	10.0	37.2	4.0	11.7	70	67	61
July	1015.9	17.9	9.8	35.5	8.5	13.4	72	82	87
Aug.	1015.9	17.4	9.3	35.0	6.9	13.1	71	66	58
Sept.	1018.4	13.9	8.8	32.1	0.9	11.3	75	36	46
Oct.	1019.1	8.2	6.8	27.0	−5.4	8.7	81	42	29
Nov.	1018.2	3.1	4.6	19.5	−7.2	6.8	87	26	25
Dec.	1018.4	−0.8	4.0	13.6	−20.4	4.8	89	26	44
Annual	1017.6	7.9	7.5	37.2	−27.1	8.4	78	508	87

Month	Mean evapor. (mm)	Number of days with				Mean cloud-iness (%)[*3]	Mean sun-shine (h)[*4]	Wind	
		precip. ⩾ 1 mm[*5]	thunder-storm[*6]	fog[*6]	snow cover[*7]			preval. direct.[*8]	mean speed (m/sec)
Jan.		13.1		6.4	10.9	75	55	W	
Feb.		12.0	0.1	5.8	9.4	70	86	W	
Mar.		11.9	0.3	6.3	4.0	60	153		
Apr.		13.0	1.8	1.1	0.3	58	189		
May		12.7	6.3	0.7		57	242		
June		12.7	6.3	0.2		56	264	W	
July		13.4	6.6	0.2		56	265	W	
Aug.		13.1	4.8	1.2		51	245	W	
Sept.		10.4	2.3	3.4		52	191		
Oct.		12.9		7.6	0.1	65	117		
Nov.		11.7		6.5	0.9	78	53		
Dec.		13.4	0.1	9.4	7.1	79	42	W	
Annual		150.3	28.6	48.8	32.7	63	1,902	W	

[*1] Praha Ruzyně, 1938–1960; [*2] Praha Ruzyně, 1931–1960; [*3] Praha Klementinum, 1926–1950; [*4] Praha Karlov, 1926–1950; [*5] Praha Klementinum, 1901–1950; [*6] Praha Karlov, 1946–1955; [*7] Praha Klementinum, 1920/21–1949/50; [*8] Praha Karlov, 1946–1953.

TABLE XXIII

CLIMATIC TABLE FOR MARIÁNSKÉ LÁZNĚ

Latitude 49°58'N, longitude 12°42'E, elevation 581 m

Month	Mean sta. press. (mbar)	Temperature (°C) daily mean*1	daily range	extreme max.	extreme min.	Mean vapour press. (mbar)*1	Relat. humid. (%)*1	Precipitation (mm) mean*2	max. in 24 h*3
Jan.		-3.1	5.7			4.4	88	53	35
Feb.		-2.1	7.2			4.7	84	46	35
Mar.		1.6	8.9			5.6	80	44	31
Apr.		5.9	10.0			7.0	75	54	35
May		11.4	10.8			10.0	72	63	73
June		14.3	11.1			11.7	72	73	57
July		16.0	11.4			13.3	74	82	64
Aug.		15.1	11.1			13.0	76	78	64
Sept.		11.8	10.6			11.2	79	54	36
Oct.		6.6	7.9			8.3	84	51	38
Nov.		1.4	5.1			6.5	90	51	36
Dec.		-2.0	5.0			4.8	90	53	30
Annual		6.4	8.7			8.3	80	702	73

Month	Mean evapor. (mm)	Number of days with precip. ≥ 1 mm*2	thunderstorm*4	fog*4	snow cover*5	Mean cloudiness (%)*1	Mean sunshine (h)	Wind preval. direct.*6	mean speed (m/sec)
Jan.		15.7		4.6	21.8	78		W	
Feb.		13.8		3.8	18.0	70		W	
Mar.		13.0	0.2	2.2	10.0	59			
Apr.		13.7	1.1	1.3	1.8	62			
May		13.2	4.6	0.8		59			
June		13.6	5.5	1.3		58		NW	
July		14.1	6.3	2.0		57		NW	
Aug.		13.5	4.7	3.0		55		NW	
Sept.		11.8	1.7	4.6		54			
Oct.		12.3	0.1	6.0	0.7	69			
Nov.		13.5		7.5	4.1	82			
Dec.		15.8		8.2	17.5	81		W	
Annual		164.0	24.2	45.3	73.9	65		W	

*1 1926–1950; *2 1901–1950; *3 1901–1942, 1948–1950; *4 1946–1955; *5 1920/21–1949/50; *6 1946–1954.

TABLE XXIV

CLIMATIC TABLE FOR OSTRAVA

Latitude 49°51'N, longitude 18°18'E, elevation 212 m

Month	Mean s.l. press. (mbar)*1	Temperature (°C) daily mean*2	daily range*3	extreme*3 max.	extreme*3 min.	Mean vapour press. (mbar)*3	Relat. humid. (%)*2	Precipitation (mm) mean*2	max. in 24 h*4
Jan.	1017.6	-2.9	6.3	14.0	-27.5	4.3	85	30	30
Feb.	1016.6	-1.4	6.9	15.2	-32.5	4.7	80	29	21
Mar.	1017.5	2.9	9.2	22.2	-25.0	5.7	77	35	39
Apr.	1015.9	8.0	11.2	28.6	-15.5	7.7	71	39	41
May	1015.5	13.1	12.2	31.2	-1.8	10.4	72	79	68
June	1015.2	16.4	12.4	34.6	0.1	12.9	72	91	68
July	1014.7	18.2	12.5	35.0	5.1	14.7	74	99	86
Aug.	1014.8	17.4	12.2	35.9	3.5	14.4	78	92	68
Sept.	1017.5	13.9	11.5	33.2	-1.1	12.1	79	59	59
Oct.	1019.4	8.2	9.0	28.2	-8.0	9.0	82	51	47
Nov.	1017.5	3.8	6.2	20.5	-9.5	6.8	85	39	36
Dec.	1016.4	0.0	5.1	15.2	-27.9	5.1	89	32	23
Annual	1016.5	8.1	9.5	35.9	-32.5	9.0	79	675	86

Month	Mean evapor. (mm)	Number of days with precip. ≥ 1 mm*4	thunderstorm	fog	snow cover*5	Mean cloudiness (%)*3	Mean sunshine (h)	Wind preval. direct.*6	mean speed (m/sec)
Jan.		17.5			18.6	76		SW	
Feb.		14.0			15.4	75		SW	
Mar.		14.0			7.1	65			
Apr.		14.6			0.8	63			
May		13.7				59			
June		15.1				58		SW	
July		14.8				57		SW	
Aug.		16.0				56			
Sept.		12.5				54			
Oct.		14.8			0.3	65			
Nov.		14.1			2.6	76			
Dec.		15.9			12.3	79		SW	
Annual		177.0			57.1	65		SW	

*1 1946–1960; *2 1931–1960; *3 1926–1950; *4 1901–1950; *5 1920/21–1949/50; *6 1946–1954.

TABLE XXV

CLIMATIC TABLE FOR BRNO TUŘANY

Latitude 49°12′N, longitude 16°34′E, elevation 223 m

Month	Mean sta. press. (mbar)[1]	Temperature (°C)					Mean vapour press. (mbar)[3]	Relat. humid. (%)[2]	Precipitation (mm)	
		daily mean[2]	daily range[3]	extreme[3]					mean[2]	max. in 24 h[4]
				max.	min.					
Jan.	1017.9	−2.7	6.0	14.4	−21.6	4.4	86	26	26
Feb.	1016.6	−1.0	7.1	14.5	−30.4	4.8	83	24	20
Mar.	1017.4	3.7	9.3	21.8	−20.2	6.0	77	21	32
Apr.	1015.6	8.7	10.8	27.0	−7.9	7.5	70	33	29
May	1015.1	14.2	11.5	31.6	−2.6	10.7	69	55	58
June	1015.1	17.8	11.7	35.4	0.8	13.0	70	81	95
July	1014.6	19.3	12.0	36.1	5.7	14.8	72	73	65
Aug.	1014.8	18.5	12.0	35.4	5.1	14.4	71	67	90
Sept.	1017.4	14.8	11.8	32.0	−2.7	12.0	75	37	49
Oct.	1019.2	9.1	9.2	27.6	−7.2	9.0	81	41	41
Nov.	1017.8	4.0	5.5	19.8	−9.1	7.0	87	39	34
Dec.	1017.0	−0.2	4.9	14.0	−22.7	5.2	89	30	38
Annual	1016.5	8.8	9.3	36.1	−30.4	9.1	78	527	95

Month	Mean evapor. (mm)	Number of days with				Mean cloud-iness (%)[3]	Mean sun-shine (h)[3]	Wind	
		precip. ⩾ 1 mm[4]	thunder-storm[5]	fog[5]	snow-cover[6]			preval. direct.[7]	mean speed (m/sec)
Jan.		12.2	0.1	2.4	18.2	72	49	NW	
Feb.		10.9		2.0	13.1	66	77	NW	
Mar.		10.2		1.7	3.9	54	140		
Apr.		11.7	1.2	0.5	0.3	54	175		
May		12.6	3.8	0.1		51	233		
June		12.7	5.2	0.3		50	252	N	
July		12.8	5.9	0.2		49	264	N	
Aug.		12.0	3.1	0.1		47	237	N	
Sept.		10.0	1.1	1.6		45	183		
Oct.		12.2	0.1	4.4	0.1	58	114		
Nov.		13.0		5.5	0.9	77	46		
Dec.		14.5		5.2	10.0	76	36	NW	
Annual		144.8	20.5	24.0	46.5	58	1,806	NW	

[1] 1946–1960; [2] 1931–1960; [3] 1926–1950; [4] Brno Pisárky, 1901–1950; [5] 1946–1955; [6] Brno Pisárky, 1920/21–1949/50; [7] 1946–1954.

TABLE XXVI

CLIMATIC TABLE FOR PREŠOV

Latitude 49°00′N, longitude 21°15′E, elevation 270 m

Month	Mean sta. press. (mbar)	Temperature (°C)					Mean vapour press. (mbar)[2]	Relat. humid. (%)[2]	Precipitation (mm)	
		daily mean[1]	daily range	extreme					mean[1]	max. in 24 h[3]
				max.	min.					
Jan.		−3.9				3.6	84	31	23
Feb.		−2.1				3.9	81	27	20
Mar.		3.2				5.3	73	31	25
Apr.		8.8				7.5	67	46	57
May		14.3				10.5	66	65	44
June		17.2				12.9	67	80	48
July		19.1				14.6	68	91	58
Aug.		18.2				14.4	70	77	65
Sept.		14.3				11.8	72	59	68
Oct.		8.7				8.7	78	49	68
Nov.		3.1				6.6	82	42	35
Dec.		−1.2				4.4	84	33	18
Annual		8.3				8.7	74	631	68

Month	Mean evapor. (mm)	Number of days with				Mean cloud-iness (%)[2]	Mean sun-shine (h)[2]	Wind	
		precip. ⩾ 1 mm[1]	thunder-storm	fog[4]	snow cover[5]			preval. direct.[6]	mean speed (m/sec)[7]
Jan.		9.9		13.1	22.8	73	61	N	3.8
Feb.		9.1		10.1	20.3	70	79	N	4.1
Mar.		9.0		5.9	6.8	60	158	N	4.2
Apr.		10.7		1.8	0.3	60	190	N	3.9
May		11.2		0.5		56	247	N	3.5
June		12.6		0.3		56	256	N	3.2
July		12.8		0.7		52	274	N	3.0
Aug.		11.8		0.4		51	247	N	2.9
Sept.		8.9		2.1		50	203	S	2.9
Oct.		9.7		5.5	0.1	61	136	S	3.2
Nov.		10.2		11.4	2.4	77	58	S	3.6
Dec.		10.6		16.4	13.0	78	47	N	3.6
Annual		126.5		68.2	65.7	62	1,956	N	3.5

[1] 1901–1950; [2] 1926–1950; [3] 1901–1918, 1921–1950; [4] 1946–1955; [5] 1921/22–1950/51; [6] 1946–1953; [7] 1934–1944, IV.1945–III.1954.

TABLE XXVII

CLIMATIC TABLE FOR ČESKÉ BUDĚJOVICE

Latitude 48°59′N, longitude 14°28′E, elevation 383 m

Month	Mean sta. press. (mbar)	Temperature (°C)				Mean vapour press. (mbar)*2	Relat. humid. (%)*2	Precipitation (mm)*1	
		daily mean*1	daily range*2	extreme*2				mean	max. in 24 h
				max.	min.				
Jan.		−2.1	6.8	15.1	−31.3	4.3	83	25	23
Feb.		−1.1	8.2	18.2	−39.7	4.7	80	28	33
Mar.		3.1	9.7	21.0	−27.3	5.7	76	29	30
Apr.		7.5	11.0	27.6	−18.2	8.1	73	46	34
May		12.8	12.2	31.2	−4.8	10.7	73	67	37
June		15.8	12.5	36.8	−0.2	13.0	73	85	67
July		17.4	12.5	36.7	3.0	14.4	74	102	57
Aug.		16.6	12.6	36.5	2.9	13.9	74	73	128
Sept.		13.0	12.4	33.5	−3.4	11.8	76	54	36
Oct.		7.8	9.1	30.1	−12.2	8.7	80	46	43
Nov.		2.9	6.1	21.6	−13.2	6.6	84	33	39
Dec.		−0.7	5.4	13.5	−29.9	4.8	85	32	30
Annual		7.8	9.9	36.8	−39.7	8.8	78	620	128

Month	Mean evapor. (mm)	Number of days with				Mean cloud-iness (%)*2	Mean sun-shine (h)*2
		precip. ≥ 1 mm*1	thunder-storm*3	fog*3	snow cover		
Jan.		10.3		2.9		74	46
Feb.		9.4		3.5		67	82
Mar.		9.6		1.8		61	136
Apr.		11.2	1.5	1.0		63	164
May		12.3	4.2	2.8		61	207
June		12.9	4.4	3.2		57	226
July		13.3	5.6	1.6		56	238
Aug.		12.8	4.2	3.4		52	219
Sept.		10.3	1.0	5.7		52	174
Oct.		10.1		7.2		63	108
Nov.		9.9		5.2		76	55
Dec.		10.6		6.0		77	36
Annual		132.7	20.9	44.3		63	1,691

*1 1901–1950; *2 1926–1950; *3 1946–1955.

TABLE XXVIII

CLIMATIC TABLE FOR BRATISLAVA VAJNOVY

Latitude 48°12′N, longitude 17°12′E, elevation 133 m

Month	Mean sta. press. (mbar)	Temperature (°C)				Mean vapour press. (mbar)*2	Relat. humid. (%)*2	Precipitation (mm)	
		daily mean*1	daily range*2	extreme*2				mean*2	max. in 24 h*3
				max.	min				
Jan.		−1.6	5.5	15.0	−26.3	4.3	84	46	38
Feb.		0.1	6.7	16.6	−31.8	4.7	80	39	31
Mar.		4.9	8.8	22.0	−17.8	6.1	73	40	40
Apr.		9.8	10.6	27.6	−8.0	8.3	69	53	40
May		15.0	11.3	32.0	−1.4	11.4	68	65	64
June		18.1	11.6	36.2	2.0	13.5	67	51	65
July		20.1	12.0	38.6	6.5	15.5	66	70	62
Aug.		19.2	11.8	37.7	5.6	15.1	68	64	44
Sept.		15.3	11.6	34.6	−1.2	12.4	70	50	64
Oct.		9.9	9.1	28.2	−7.5	9.1	77	54	51
Nov.		4.4	5.6	20.6	−10.3	7.2	84	69	61
Dec.		0.6	4.8	14.4	−22.8	5.1	85	56	45
Annual		9.6	9.1	38.6	−31.8	9.4	74	657	65

Month	Mean evapor. (mm)	Number of days with				Mean cloud-iness (%)*2	Mean sun-shine (h)*2	Wind	
		precip. ≥ 1 mm*1	thunder-storm*4	fog*5	snow cover*6			preval. direct.*7	mean speed (m/sec)*8
Jan.		12.7		6.2	14.3	73	62	NW	3.4
Feb.		11.2		2.2	10.6	67	93	NW	4.2
Mar.		10.9	0.1	2.4	2.7	58	157	NW	4.2
Apr.		12.1	2.6	0.4	0.3	58	214	NW	4.0
May		11.7	6.3			56	271	NW	3.2
June		10.9	6.4			54	286	NW	3.3
July		11.9	7.1			49	307	NW	3.3
Aug.		10.4	2.9			46	284	NW	2.8
Sept.		9.9	1.2	0.8		46	226	NW	2.7
Oct.		12.4	0.1	3.1	0.3	59	147	NW	2.8
Nov.		13.4		6.8	0.9	77	71	NW	3.4
Dec.		15.4	0.1	8.8	7.4	77	50	NW	3.3
Annual		142.9	26.8	30.7	36.4	60	2,168	NW	3.4

*1 1901–1950; *2 1926–1950; *3 Bratislava, výskumne ústavy, 1901–1918, 1921–1950; *4 1946–1955; *5 Bratislava, výskumne ústavy, 1946–1955; *6 1921/22–1950/51; *7 1946–1953; *8 1934–1951.

TABLE XXIX

CLIMATIC TABLE FOR EGER

Latitude 47°53′N, longitude 20°23′E, elevation 173 m

Month	Mean sta. press. (mbar)[1]	Temperature (°C)[1]				Mean vapour press. (mbar)[1]	Relat. humid. (%)[1]	Precipitation (mm)[1]	
		daily mean	daily range	extreme				mean	max. in 24 h
				max.	min.				
Jan.	998.5	−2.3	6.6	13.1	−27.0	4.3	80	27	24
Feb.	996.5	−0.1	8.2	17.4	−28.0	4.8	78	30	47
Mar.	994.7	4.7	10.3	24.2	−15.6	6.2	74	33	26
Apr.	992.6	10.2	11.6	32.0	−6.2	8.5	68	45	80
May	993.9	16.1	11.9	35.0	−2.5	12.1	70	65	51
June	993.9	18.8	12.5	38.1	1.0	14.4	68	73	72
July	993.7	20.8	13.4	39.3	4.0	15.9	65	59	56
Aug.	994.5	20.0	13.2	38.0	4.0	15.1	67	57	76
Sept.	997.0	15.9	12.7	34.2	−2.0	12.7	73	48	53
Oct.	997.1	10.1	11.1	30.0	−12.6	9.5	80	52	43
Nov.	997.0	4.1	7.2	20.0	−16.5	7.2	84	53	39
Dec.	997.1	0.1	6.1	13.7	−28.0	5.5	85	40	42
Annual	995.5	9.9	10.4	39.3	−28.0	9.8	75	582	80

Month	Mean evapor. (mm)[1]	Number of days with				Mean cloudiness (%)[1]	Mean sunshine (h)[1]	Wind	
		precip. ⩾ 1 mm[1]	thunderstorm[2]	fog[3]	snow cover			preval. direct.[1]	mean speed (m/sec)
Jan.		6.2		6.0		67	73	N	
Feb.		6.0	1	4.2		63	89	NW	
Mar.	18	6.3	1	1.8		56	141	NW	
Apr.	50	,7.6	4	1.1		57	183	NW	
May	96	8.9	10	0.3		52	268	NW	
June	109	9.2	15	0.3		53	267	NW	
July	108	7.6	13	0.4		47	303	NW	
Aug.	87	6.8	8	0.1		45	288	NW	
Sept.	61	6.0	3	0.7		45	208	NW	
Oct.	40	7.6	1	2.1		54	143	NW	
Nov.	12	8.5		6.4		67	72	NW	
Dec.	1	8.5	1	8.6		73	50	NW	
Annual	582	89.2	60	32.8		57	2,085	NW	

[1] 1901–1950; [2] 1956–1965; [3] 1940–1954.

TABLE XXX

CLIMATIC TABLE FOR DEBRECEN

Latitude 47°33′N, longitude 21°37′E, elevation 123 m

Month	Mean sta. press. (mbar)[1]	Temperature (°C)				Mean vapour press. (mbar)[1]	Relat. humid. (%)[2]	Precipitation (mm)	
		daily mean[2]	daily range[1]	extreme[1]				mean[2]	max. in 24 h[3]
				max.	min.				
Jan.	1004.3	−2.7	6.0	13.8	−30.2	4.4	85	35	23
Feb.	1002.2	−0.6	7.0	17.9	−26.0	4.8	82	36	29
Mar.	1000.5	4.5	10.2	25.8	−17.8	6.2	74	30	23
Apr.	998.2	11.0	12.3	33.6	−7.1	8.3	67	36	36
May	999.3	16.5	14.0	32.7	−3.0	11.8	67	61	65
June	999.1	19.8	13.4	37.0	−0.4	11.1	69	80	45
July	998.7	21.8	13.7	38.5	5.2	15.7	67	59	62
Aug.	999.5	20.8	13.6	39.0	2.7	15.9	70	64	57
Sept.	1002.5	16.4	12.9	36.0	−2.9	12.5	73	41	34
Oct.	1002.7	10.2	10.8	29.5	−14.9	9.5	78	49	49
Nov.	1002.7	4.9	7.3	21.4	−19.0	7.1	85	53	31
Dec.	1002.9	0.5	5.3	16.0	−28.0	5.5	87	40	30
Annual	1001.0	10.3	10.6	39.0	−30.2	9.6	75	584	65

Month	Mean evapor. (mm)[1]	Number of days with				Mean cloudiness (%)[1]	Mean sunshine (h)[1]	Wind	
		precip. ⩾ 1 mm[1]	thunderstorm[4]	fog[5]	snow cover[6]			preval. direct.[1]	mean speed (m/sec)[7]
Jan.		7.9		5.3	15.9	69	61	SW	3.3
Feb.		6.8		5.0	8.8	65	82	SW	3.2
Mar.	18	6.9	1	1.8	2.7	57	145	SW	3.5
Apr.	51	7.9	2	0.5		57	191	SW	3.5
May	93	9.2	6	0.3		52	258	NE	3.2
June	110	9.9	8	0.2		54	271	SW	2.8
July	110	8.1	6	0.2		50	297	SW	2.7
Aug.	90	7.1	5	0.4		44	268	SW	2.5
Sept.	60	6.7	2	0.7		47	201	NE	2.5
Oct.	40	7.9	1	2.0	0.1	55	144	SW	2.6
Nov.	12	9.1	1	5.3	0.3	70	67	SW	2.5
Dec.	1	8.6		6.9	7.9	78	45	NE	3.6
Annual	585	96.1	28	28.6	35.6	58	2,030	SW	3.0

[1] 1901–1950; [2] 1931–1960; [3] 1901–1944, 1947–1950; [4] 1956–1965; [5] 1940–1954; [6] 1929/30–1943/44; [7] 1958–1962.

TABLE XXXI

CLIMATIC TABLE FOR BUDAPEST (Met. Intézet)

Latitude 47°31'N, longitude 19°02'E, elevation 120 m

Month	Mean sta. press. (mbar)[1]	Temperature (°C)				Mean vapour press. (mbar)[1]	Relat. humid. (%)[2]	Precipitation (mm)	
		daily mean[2]	daily range[1]	extreme[1]				mean[2]	max. in 24 h[1]
				max.	min.				
Jan.	1004.1	−1.1	4.9	15.1	−21.7	4.7	81	42	34
Feb.	1002.1	1.0	6.5	18.0	−23.4	4.9	76	44	40
Mar.	1000.2	5.8	8.8	25.4	−13.6	6.2	67	39	37
Apr.	997.9	11.8	10.1	30.2	−4.2	8.2	60	45	44
May	999.0	16.8	11.3	32.4	0.0	11.7	62	72	94
June	999.3	20.2	11.7	39.5	3.0	13.9	62	76	55
July	999.0	22.2	11.8	38.4	8.9	15.1	60	54	64
Aug.	999.8	21.4	12.5	39.0	7.0	14.6	62	51	62
Sept.	1002.3	17.4	11.3	35.2	1.2	12.5	65	34	62
Oct.	1002.5	11.3	9.0	30.8	−9.5	9.8	74	56	46
Nov.	1002.3	5.8	5.5	22.6	−11.9	7.3	81	69	40
Dec.	1002.6	1.5	4.2	15.7	−19.1	5.7	83	48	37
Annual	1000.9	11.2	9.0	39.5	−23.4	9.5	69	630	94

Month	Mean evapor. (mm)[1]	Number of days with				Mean cloud-iness (%)[1]	Mean sun-shine (h)[1]	Wind	
		precip. ⩾1 mm[1]	thunder-storm[3]	fog[4]	snow cover[5]			preval. direct.[1]	mean speed (m/sec)[6]
Jan.		7.6		10.0	17.1	70	59	NW	2.1
Feb.	2	6.8	1	7.5	10.5	65	83	NW	2.5
Mar.	22	7.3	1	3.4	2.9	59	136	NW	2.7
Apr.	52	7.4·	4	0.6		58	186	NW	2.6
May	97	8.5	11	0.2		54	252	NW	2.5
June	114	8.0	14	0.1		52	269	NW	2.5
July	107	6.5	12	0.0		46	297	NW	2.5
Aug.	89	6.3	7	0.1		43	270	NW	2.5
Sept.	59	6.2	2	0.6		46	195	NW	2.0
Oct.	43	7.5	1	2.5		57	134	NW	1.8
Nov.	14	8.8	1	7.6	0.5	71	67	NW	1.7
Dec.	2	9.1	1	10.8	9.7	77	40	NW	2.2
Annual	601	90.0	54	43.4	40.7	58	1,988	NW	2.3

[1] 1901–1950; [2] 1931–1960; [3] 1956–1965; [4] 1940–1954; [5] 1929/30–1943/44; [6] 1958–1962.

TABLE XXXII

CLIMATIC TABLE FOR SZOMBATHELY (Gazd isk.)

Latitude 47°15'N, longitude 16°36'E, elevation 218 m

Month	Mean sta. press. (mbar)[1]	Temperature (°C)[1]				Mean vapour press. (mbar)[1]	Relat. humid. (%)[1]	Precipitation (mm)[1]	
		daily mean	daily range	extreme				mean	max in 24 h
				max.	min.				
Jan.	993.9	−1.6	6.2	15.2	−24.0	4.7	83	34	27
Feb.	991.8	−0.1	7.6	19.2	−29.3	5.1	80	36	29
Mar.	990.1	4.7	9.5	22.4	−17.8	6.6	76	38	31
Apr.	988.2	9.5	11.2	29.6	−5.6	8.2	71	53	33
May	989.5	14.5	11.7	32.2	−3.5	11.8	72	68	50
June	990.3	17.6	12.0	36.4	2.6	14.2	71	80	69
July	990.2	19.8	12.3	38.3	6.8	15.9	72	89	59
Aug.	990.9	19.1	11.8	36.2	5.5	15.2	73	79	74
Sept.	992.9	15.3	11.1	33.0	−1.6	12.6	77	69	62
Oct.	992.3	9.7	9.3	27.9	−11.7	9.6	81	56	50
Nov.	991.9	4.1	6.3	23.5	−14.6	6.8	84	54	47
Dec.	991.8	0.3	5.3	18.0	−19.2	5.6	86	44	29
Annual	991.1	9.4	9.5	38.3	−29.3	9.7	77	700	74

Month	Mean evapor. (mm)[1]	Number of days with				Mean clous-iness (%)[1]	Mean sun-shine (h)[1]	Wind	
		precip. ⩾1 mm[1]	thunder-storm[2]	fog[3]	snow cover[4]			preval. direct.[1]	mean speed (m/sec)[5]
Jan.		5.8		4.6	12.5	74	63	N	3.6
Feb.		5.5	1	2.9	7.1	68	83	N	3.6
Mar.	19	6.1	1	1.7	3.1	65	129	N	4.1
Apr.	49	8.5	3	0.6	0.1	65	162	N	4.2
May	90	9.3	6	0.7		61	218	N	3.3
June	110	9.3	10	0.4		61	224	N	3.2
July	121	10.1	11	0.2		57	243	N	2.7
Aug.	102	8.9	8	0.5		53	239	N	2.7
Sept.	73	7.3	2	1.7		55	165	N	2.6
Oct.	41	7.9	1	4.0	0.4	65	113	N	2.8
Nov.	13	6.9	1	6.8	0.5	76	59	N, S	3.3
Dec.	1	6.8		7.2	10.3	79	47	N	3.7
Annual	619	92.4	47	31.3	34.0	65	1,745	N	3.3

[1] 1901–1950; [2] 1956–1965; [3] 1940–1954; [4] 1929/30–1943/44; [5] 1958–1962.

TABLE XXXIII

CLIMATIC TABLE FOR SZEGED

Latitude 46°15′N, longitude 20°09′E, elevation 79 m

Month	Mean sta. press. (mbar)[1]	Temperature (°C)				Mean vapour press. (mbar)[1]	Relat. humid. (%)[2]	Precipitation (mm)	
		daily mean[2]	daily range[1]	extreme[1]				mean[2]	max. in 24 h[3]
				max.	min.				
Jan.	1008.3	−1.4	6.1	15.4	−29.1	4.9	87	34	25
Feb.	1006.2	0.7	7.1	18.3	−27.2	5.2	83	38	62
Mar.	1004.1	5.8	9.3	24.5	−16.8	6.5	74	35	40
Apr.	1001.8	11.9	10.5	31.5	−5.0	8.5	66	41	51
May	1002.6	17.3	11.0	34.0	−4.5	11.7	65	63	43
June	1002.9	20.8	10.6	38.8	5.7	14.2	64	63	45
July	1002.6	23.0	11.5	38.7	7.2	14.8	61	51	51
Aug.	1003.4	22.2	11.4	39.0	6.4	14.3	63	47	68
Sept.	1005.9	18.2	10.7	38.2	0.1	12.4	66	42	45
Oct.	1006.2	12.0	9.7	31.2	−10.0	9.9	75	46	38
Nov.	1006.3	6.1	6.6	24.9	−11.9	7.5	85	59	33
Dec.	1006.6	1.5	4.9	21.1	−21.7	5.8	88	39	31
Annual	1004.7	11.5	9.3	39.0	−29.1	9.6	73	558	68

Month	Mean evapor. (mm)[1]	Number of days with				Mean cloud- iness (%)[1]	Mean sun- shine (h)[1]	Wind	
		precip. ≥ 1 mm[1]	thunder- storm[4]	fog[5]	snow cover[6]			preval. direct.[1]	mean speed (m/sec)[7]
Jan.		6.8		7.6	15.1	71	64	S	3.3
Feb.	1	6.6	1	6.4	9.6	65	90	S	3.4
Mar.	22	6.9	1	2.1	5.2	59	143	S	4.0
Apr.	52	8.0	3	0.8		59	187	S	3.7
May	90	8.5	8	0.2		53	258	S	3.2
June	106	8.2	12	0.4		51	271	N, W	2.9
July	100	6.4	9	0.0		42	309	NW	2.9
Aug.	79	˙6.4	6	0.3		39	286	NW	2.7
Sept.	59	5.9	2	1.2		42	211	NW	2.6
Oct.	46	7.1	1	1.7		54	152	S	3.0
Nov.	16	8.1	1	6.6		69	79	S	3.0
Dec.	2	7.7	1	9.8	5.1	75	52	S	3.7
Annual	573	86.6	44	37.1	33.0	57	2,102	S	3.2

[1] 1901–1950; [2] 1931–1960; [3] 1901–1944, 1946–1950; [4] 1956–1965; [5] 1940–1954; [6] 1929/30–1943/44; [7] 1958–1962.

TABLE XXXIV

CLIMATIC TABLE FOR PÉCS

Latitude 46°05′N, longitude 18°15′E, elevation 141 m

Month	Mean sta. press. (mbar)[1]	Temperature (°C)				Mean vapour press. (mbar)[1]	Relat. humid. (%)[2]	Precipitation (mm)	
		daily mean[2]	daily range[1]	extreme				mean[2]	max. in 24 h[3]
				max.	min.				
Jan.	1001.4	−0.7	6.0			4.7	83	41	40
Feb.	999.4	1.3	7.6			5.1	78	46	38
Mar.	997.5	6.1	9.4			6.5	70	41	54
Apr.	995.3	11.9	10.6			8.7	64	58	39
May	996.3	16.9	10.6			11.8	66	66	45
June	996.9	20.4	10.9			14.4	64	69	72
July	996.9	22.6	11.4			15.2	62	64	41
Aug.	997.5	21.9	11.9			14.7	63	55	58
Sept.	999.8	17.9	11.3			12.9	67	47	59
Oct.	999.7	11.8	9.4			10.1	75	64	69
Nov.	999.5	6.2	6.5			7.5	81	71	59
Dec.	999.8	1.8	5.5			5.7	83	45	37
Annual	998.3	11.5	9.3			9.8	71	667	72

Month	Mean evapor. (mm)[1]	Number of days with				Mean cloud- iness (%)[1]	Mean sun- shine (h)[1]	Wind	
		precip. ≥ 1 mm[1]	thunder storm[4]	fog[5]	snow cover[6]			preval. direct.[7]	mean speed (m/sec)[8]
Jan.		7.5		5.7	8.6	70	67	NE	3.2
Feb.	2	7.2	1	4.2	8.5	64	93	NE	3.5
Mar.	23	7.5	1	1.5	2.5	60	136	NE	4.0
Apr.	53	8.9	4	0.6		60	179	NE	4.1
May	94	9.6	10	1.1		53	249	W	3.2
June	112	8.3	12	0.7		51	266	N	3.0
July	114	7.8	10	0.3		44	299	N	2.9
Aug.	94	7.0	6	0.8		40	278	N	2.8
Sept.	66	7.1	2	1.3		46	193	NE	2.7
Oct.	44	8.2	1	1.7		56	141	NE	3.0
Nov.	16	8.5	1	5.3	0.1	70	73	NE	3.1
Dec.	3	8.7		7.9	6.7	74	51	NE	3.4
Annual	621	96.3	48	31.1	26.4	57	2,025	NE	3.3

[1] 1901–1950; [2] 1931–1960; [3] 1901–1918, 1922–1950; [4] 1956–1965; [5] 1940–1954; [6] 1929/39–1943/44; [7] 1901–1915, 1926–1950; [8] 1958–1962.

Chapter 4

The Climate of Italy

V. CANTÙ

Meteorological observations in Italy

The fervid interest in experimental science shown by Florence in the 17th century led to an early use of fundamental meteorological instruments and to the establishment of a network of stations taking systematical observations similar to the basic observations of a modern synoptic network.

In 1639 Benedetto Castelli, a disciple of Galileo, invented a pluviometer. In 1641, after a few decades of experience with thermoscopes, the Accademia del Cimento established the first liquid thermometer on the initiative of Grand Duke Ferdinand II of Tuscany; the mercury barometer followed after five years of experiments with water.

The observations of the "Tuscan network" began in 1654 when, on the initiative of Ferdinand II, thermometers were distributed; barometers, anemoscopes and hygroscopes were added in 1657. Synoptic observations of pressure, temperature, humidity, wind direction and atmosphere were carried out. This service lasted for 13 years; unfortunately, almost all these observations have been lost. "As so many things in Italy brightly begun and afterwards abandoned, so ended also this first meteorological service of the world" (R. BILANCINI, 1950). For nearly two centuries no synoptic observations were made in Italy.

However, the beginnings of a modern synoptic network started in the Pontifical States on 20th June 1855. The observations were carried out in Bologna, Ferrara, Urbino and Ancona at noon and were sent to Rome via telegraph. Not until 1866 was there a real meteorological service. On 10th April of that year the collection of data from 20 Italian stations, all of which belonged to the international network then existing, began. A daily weather report, sometimes with a forecast, was issued by telegraph.

In 1879 the number of stations had already increased to 35. In the same year the Ufficio Centrale di Meteorologia in Rome was established. This Office still exists under the denomination of Ufficio Centrale di Meteorologia ed Ecologia Agraria and is still located in its original place: the Collegio Romano, which since 1572 was the seat of the astronomical and meteorological activities of the Jesuits; weather observations have been carried out there since 1782.

The successive decades saw a growth of meteorological services, with various re-organizations and partial unifications, as well as a remarkable increase in the number of synoptic stations. There were 55 synoptic stations in 1925, 89 in 1932 and well over 210 in 1939. At present their number is still about 160. The reduction has been made since new methods of analysis of synoptic charts have been introduced which do not require so many surface observations.

The climatological stations, too, have always been numerous, as a consequence of some very advanced initiatives. Around 1908 the number of such stations already amounted to 888 (ROSTER, 1909). Around 1925 Italy took a pride in having the important station system of the "Osservatori del Monte Rosa", which included an observatory at Gressoney (1,385 m), a thermo-pluviometric station at Lake Gabiet (2,340 m), an observatory at the Col d'Olen (2,901 m) and an observatory at the "Capanna Regina Margherita" (4,560 m). The observations had begun in 1893 but were discontinued around 1950.

Since World War II around 5,000 climatological stations have existed in Italy and their number has changed little in the latest decades.

Unfortunately, however, the coordination of all stations under one agency and the systematical and uniform publication of the results of the observations have not been realized so far.

The Ministry for Public Works publishes the *Annali Idrologici* which includes the data of its more than 3,000 dependable stations most of which are pluviometric only.

Since 1959 the Istituto Centrale di Statistica has published data in its *Annuario di Statistiche Meteorologiche* on a great number of elements from approximately 1,000 stations, belonging to different public services.

There are more long series of meteorological observations in Italy than in most countries. In compiling their answer to a WMO investigation on this matter, CANTÙ and NARDUCCI (1967) found 44 series more than 80 years long, of which 9 began before 1800 and 5 before 1750. With certain series it has been possible to obtain only the indication that they exceed 80 (> 80) years, or 100 (> 100) years.

Although a remarkable increase in the number of stations is not to be expected in the near future, it is to be hoped that the IFA (Istituto Fisica dell' Atmosfera, formerly CENFAM: Centro Nazionale per la Fisica dell' Atmosfera e la Meteorologia) reaches agreement with the above-mentioned agencies to recuperate and evaluate the very numerous data which they collect and which are spread over the country.

Geographical factors influencing the climate of the Apennine Peninsula

The two major geographical factors influencing the Italian climate are: the barrier effect of the Alpine–Apennine Range and the unstabilizing effect of the Mediterranean Sea which tends to facilitate cyclogenesis in the vicinity of the peninsula.

The Alps protect the Italian Peninsula from cold air flows and represent an especially important obstacle for the northwestern quadrant. However, if these currents are strong enough, they move out into the Mediterranean, at first through the Rhone Valley then through the Carnic Gate north of Trieste, and finally, when they gain a considerable thickness, they overflow into the southern parts of the country through the various gates of the Alps (Fig.1). From the Rhone Valley the cold air travels to Sicily often giving rise to the violent northern air stream, known as *mistral* in southern France and *maestrale* in Italy. Such northern air currents cause the severe storms in the Golfe du Lion, legendary among Italian sailors under the name of "lionate". Under these conditions depressions frequently develop in the Gulf of Genoa; it is the well-known lee-cyclogenesis phenomenon recognized by meteorologists all over the world as the *Genoa cyclone*.

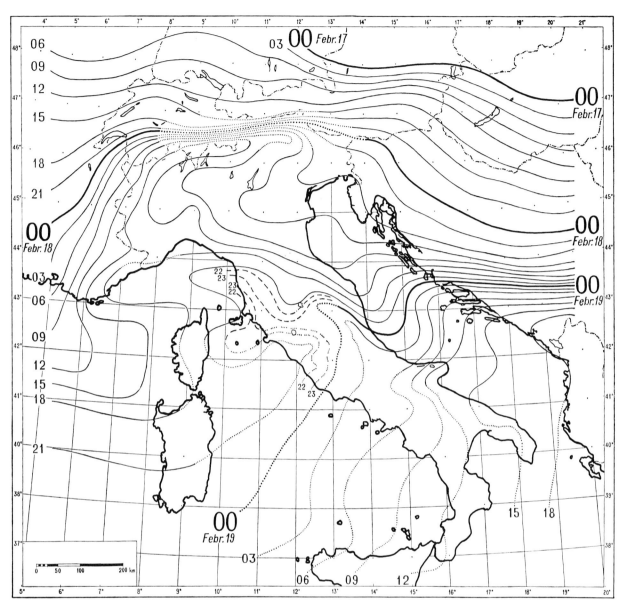

Fig. 1. Isochrones of a cold front that crossed Italy during the period February 17–19, 1958 (GMT time).

The cold air bursting through the Carsic Gate causes a very violent wind to occur on the northern Adriatic, which is locally called *bora*. This wind is particularly dangerous because of its sudden appearance. The air bursting over the Alpine Chain only rarely causes important foehn phenomena noticeable at the earth's surface (in fact no special name for such a phenomenon exists in the Italian language). Often cold air at upper levels overflows warm air at the lower level causing sudden and imposing instability manifestations in the Po Valley.

The Apennines too are a barrier with some importance especially in connection with the cold air from the northeast. In fact, the winter on the Tyrrhenian coast is warmer than the winter on the Adriatic coast at the same latitude. The mean annual isotherm of 15°C is found entirely along the Tyrrhenian coast but appears on the Adriatic coast only south of Pescara.

At the same latitude the Mediterranean Sea is permanently warmer than the Atlantic Ocean and has a considerable unstabilizing effect on all arriving air masses, with the exception of those carried by the sirocco from Libya. Maritime air-masses are frequent over all Italy and are quite common even in the Po Valley, due to the vicinity of the Adriatic Sea.

Meteorological phenomena over Italy

The meteorological phenomena which are most decisive for Italian weather and which have to be considered by the weather forecaster in a typical city of central Italy such as Rome are: (*1*) the upper troughs; (*2*) the cold cut-off lows; (*3*) the afternoon convective activity; and (*4*) the lee depressions. Over the regions to the north and south of central Italy the depressions are of greater importance and have greater impact upon the weather so they should be considered in the second or third place of importance rather than the fourth.

The fronts, as interpreted according to the classical Norwegian schools of thought, have little practical importance over Italy since they are generally not well defined and even if so, they bear little significance upon the "rain/good weather" dilemma; in particular the fronts are hard to recognize in Genoa-cyclones.

In general the surface depressions over Italy, as well as over the whole Mediterranean Sea, are less deep, and move less regularly than those over the Atlantic Ocean. It is often difficult to find their centre and to track their motions on the ordinary synoptic chart. For instance, in the Gulf of Genoa, it is easy to confuse a main moving low-pressure system with a derived low-pressure system that remains stationary in the same area. Over the central Mediterranean and the Ionian Sea the cyclones sometimes move very slowly, causing very heavy rainfall on the eastern coasts of Calabria and Sicily. 1,500 mm of rain in three days has been observed in the Calabria Mountains.

The forecasting of the motion of the system on the basis of the isallobaric centres, is almost impossible; but trajectories can sometimes be useful. For example, the lee depressions forming in the Gulf of Genoa, which constitute the majority of the depressions affecting Italy, tend to move southeastwards along the Tyrrhenian during the winter time while in summer they tend to move across the Po Valley and then southwards into the Adriatic Sea or directly towards the east.

The synoptic climatology of Italy

Climatology of depressions

URBANI (1955, 1956a,b, 1957, 1961) has extensively studied the weather types and the depressions which affect Italy. The weather types have been divided into seven groups, and each type is represented by the 700-mbar mean pressure pattern, both for the warm and the cold season. Such patterns, however, imply different developments in the two seasons, and this requires a separate classification of sub-types. URBANI (1961) arrived at 22 such sub-types[1] for each one of the seasons. Urbani has also studied the winter and summer depressions of the decade 1946–1955 with the purpose of establishing the relation of cyclogenesis in the Mediterranean with the weather types. During winter time the principal cyclogenetic regions are the Gulf of Genoa, with 17.9% of the Mediterranean cyclogenesis; the western Mediterranean with 14.4%; the Golfe du Lion with 13.9%; the central and southern Tyrrhenian with 11.7%. The other areas in Fig.2A cause less than 10% of all cyclogenesis. In summer the principal cyclogenetic regions are northern Italy and the Gulf of Genoa, with 24.3% and 22.1%, respectively (Fig.2B). During winter-time the principal cyclogenetic weather types are those of family VII (Fig.3A) with 15% of the cyclogenesis and a frequency of 8%. In summer, family II (Fig.3C) is the most typical with 34% of the cyclogenesis and 19% frequency.

URBANI (1956b) also studied the winter cyclones coming from the Atlantic Ocean and made in this connection a special classification of weather types. The results that have most general interest appear to be those regarding trajectories and durations, as follows:

(*1*) Depressions which form in the northern Atlantic Ocean and cross the British Islands, normally reach as far as Turkey; if passing on a southern track they cross northern Africa entirely, and if passing south of Gibraltar they fill up rapidly.

(*2*) Depressions forming over the British Islands cross the Tyrrhenian and fade over the Balkans, the Black Sea or the eastern Mediterranean.

(*3*) Depressions forming in the Bay of Biscay have a mean development cycle of 3–4 days, those which form west of Portugal, of 1–2 days.

(*4*) Depressions which form near the Azores have a long development cycle and they often reach as far as Russia.

This information agrees substantially with, and is completed by the information published by ZENONE (1959) and by the METEOROLOGICAL OFFICE (1962–1964).

Synoptic climatology of the upper air

The analysis of quarterly mean contour lines of the 500-mbar isobaric surface enables one to represent the dynamic characteristics of the Italian upper air climate quite effectively. If from one year to the other the differences between corresponding quarters are plotted, various types of isallohypse patterns are obtained which make it possible to classify the behaviour of seasonal weather. Such patterns are described in the research work of LECCE and DEL TRONO (1957, 1958, 1960).

The main anomaly types resulting from an analysis of such isallohypse patterns are the following (DEL TRONO, 1964, 1965).

(*1*) An anticyclonic type characterized by positive pressure anomalies over Italy, from one year to the next, gives precipitation about 10% below normal and temperatures about 1°C above normal.

(*2*) A type with a general airflow from west, which is characterized by negative pressure anomalies over the Mediterranean and Italy, from one year to the next, gives precipitation above normal and temperatures slightly below normal.

---

[1] This study is an up-dating of his earlier work of 1955, in which he then called "families" those which are now called "types", and "types" those that in the later study are called "sub-types". The order of the "families" and "types" is the same so that family I corresponds to type A, family II to type B and so on.

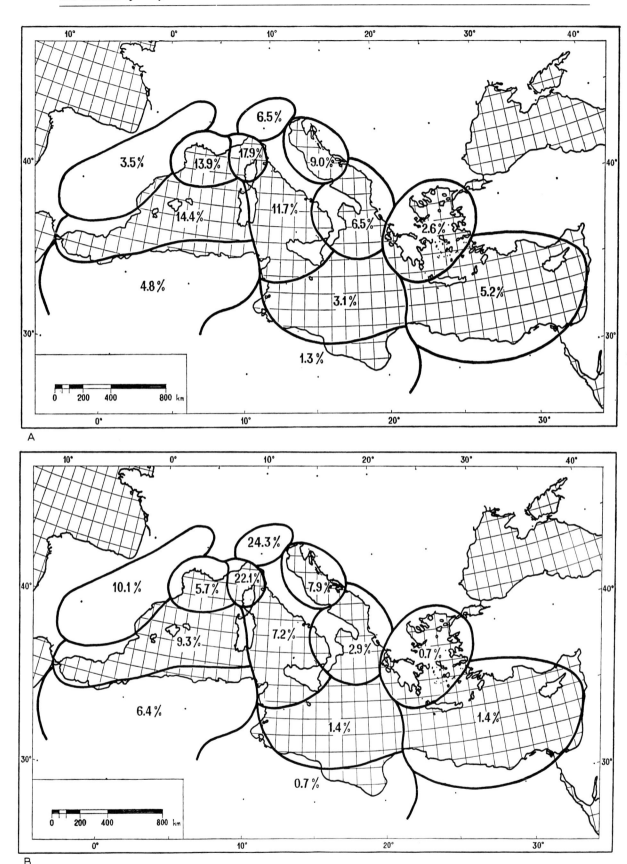

Fig.2. Percentages of cyclogenesis in the different regions of the Mediterranean: A. in winter; B. in summer (ZENONE, 1959). Conic projection; the scale is precise at 30° and 60°N.

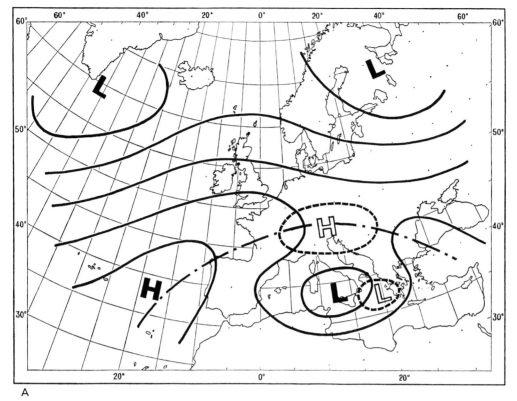

A

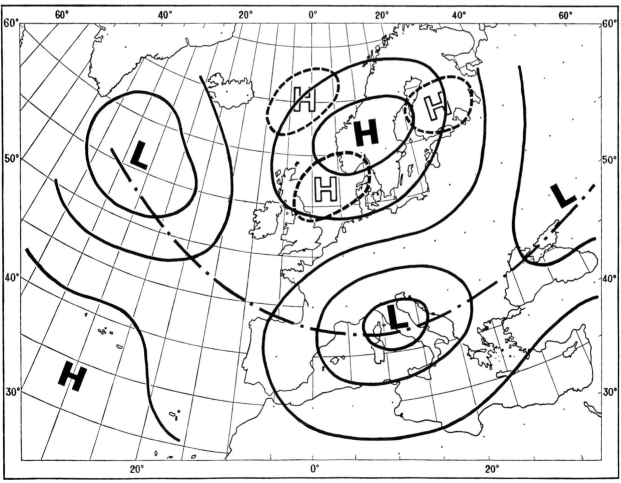

B

Fig.3. The principal cyclogenetic weather types. In winter: A. family VII; and B. family VI. In summer: C. family II; and D. family I. (From URBANI, 1955). The broken lines represent other frequently occurring situations of the same type.

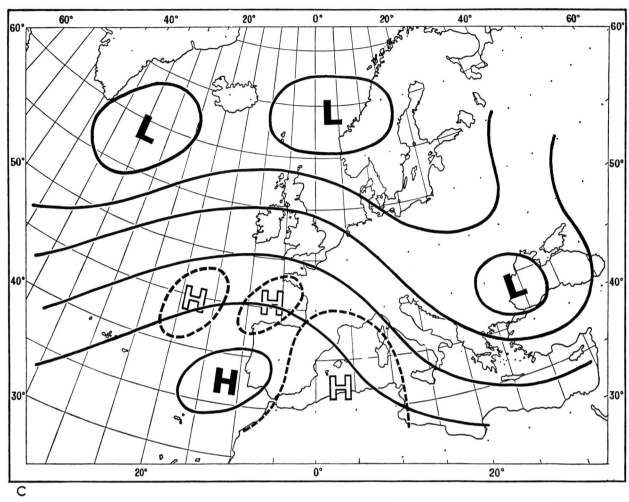

C

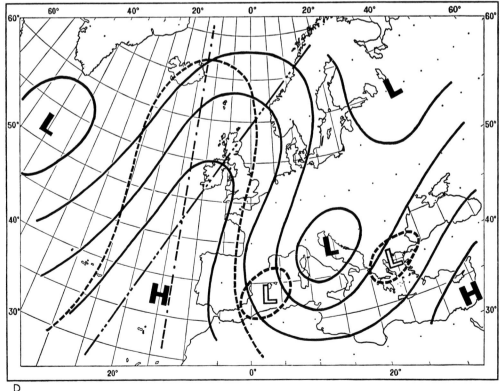

D

Fig.3C,D. (legend see p. 133).

(*3*) A type with a general airflow from northeast characterized by a belt of positive pressure anomalies spreading from the central Atlantic to northern Europe, and negative variations over the central Mediterranean. Precipitation with this type is remarkably below normal over central-northern Italy and slightly above normal over southern Italy; temperature is below normal, particularly over southern Italy.

The application of the above types of surface-pressure patterns has proved to be of great usefulness in long-range forecasting of weather over Italy.

Dynamic climatology of the cyclogenesis

Research in the climatology of cyclogenesis has been carried out by Lecce (see Lecce, 1964; Lecce in CENFAM, 1964, 1965) within the framework of investigations on circulation in the lee of the Alps during the decade 1952–1961, organized by the Centro Nazionale per la Fisica dell' Atmosfera e la Meteorologia (CENFAM now IFA).

For each surface-pressure pattern has been computed the number of days with Genoa cyclones, and with large-scale weather situations[1]; then the frequency of days with a Genoa cyclone has been put in relation to the persistence of the situation itself. The results of more general interest appear to be the following.

(*1*) For each large-scale weather situation, the frequency of Genoa cyclones rapidly decreases after the first two days.

(*2*) Maritime northern weather types present the maximum frequency of days with Genoa cyclones (37.5% compared with 12% of the western types).

(*3*) Ground situations characterized by ill-defined centres of action, of relatively small area (at a continental scale) and rapidly changing, present the maximum frequency of Genoa cyclones and also the maximum variability of large-scale situations.

The relationship between the behaviour of cyclogenesis and the behaviour of temperatures and precipitation has also been studied, and it appeared that to a greater frequency of the Genoa cyclones correspond below-mean temperatures and a precipitation which deviates little from the mean.

Weather in climatic regions

In the following the territories, which belong to each climatic region, are indicated (Fig.4, *1–8*) and the meteorological phenomena which characterize the various seasons are described.

The Alps (Fig.4,*1*)

In the Alpine region (> 1,000 m above m.s.l.), autumn, winter and spring are characterized by a succession of Atlantic, Genoa or Mediterranean depressions; Genoa

---

[1] "A day with a Genoa cyclone" means here a day in which either cyclogenesis in the Gulf of Genoa occurs or a cyclone which originated in the Gulf of Genoa is found in the central-northern Tyrrhenian. "A large-scale weather situation" has here the same meaning as Baur (1951) gave it: the mean topography of a period during which the positions of the major centres of action and the direction of the air streams over a given region remain substantially unchanged. In Italy, the term "synoptic weather type" is preferred instead.

The climate of Italy

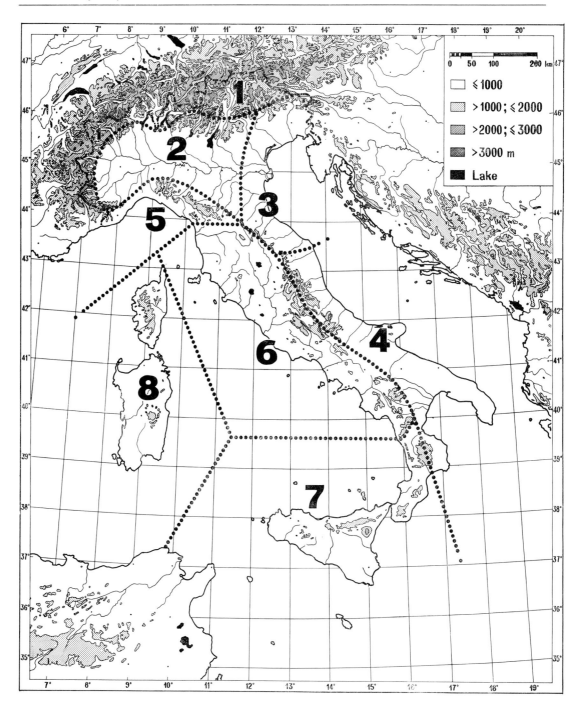

Fig.4. Italian climatic regions. The numbers are explained in the text.

depressions are typical for the region in the autumn and in the spring, whilst during winter-time they move more southwards; the Mediterranean depressions characterize the region in the spring.

During the mid-winter, the central-European anticyclone gives periods of stable weather, however, with inversion clouds around 1,000 m above m.s.l. In summer, mainly anti-cyclonic weather prevails due to the Azores high; frequent instability phenomena with showers and thunderstorms occur however.

Po Valley (Fig.4,*2*)

This is a wide basin with a flat bottom, almost symmetrical, surrounded by reliefs in the north, west and south but open to the sea in the east. As a climatic region it may be regarded as limited by the 1000-m isohypse on the slope of the Alps side; it extends as far as the Apennine watershed (north of the 44° parallel) and loses its characteristics at a distance of around 50 km from the sea.

In autumn, winter and spring lee and Mediterranean depressions occur, alternating with periods of no perturbations due to the influence of the central-European anticyclone; in winter-time the whole valley is occupied by a cold air layer near the ground, which has a thickness of about 1,000 m and causes heavy and very frequent fogs; perturbations often surmount this cold air cushion.

In summer, fair weather prevails due to the Azores high, however, with very frequent instability manifestations, favoured by the weak thermal depression that forms over the area during daytime.

Northern Adriatic (Fig.4,*3*)

This region includes the Adriatic coast subject to the bora. The region is constituted by the maritime belt of the Po Valley from the 43° parallel to the Alps, by the calcareous lowlands of the Carso, and by the Istria flats. In autumn and winter, lee depressions and consequently the bora dominate. In spring, lee depressions and with less frequency Mediterranean depressions are succeeding each other.

In summer, this region has fair anticyclonic weather with frequent instability phenomena.

Central and southern Adriatic (Fig.4,*4*)

This region is defined as the belt enclosed between the Apennine watershed and the Adriatic Sea from the 43° to the 39° parallel. The Apennines protect it from the western flows, hence the easterly winds are of a special importance for this region.

In autumn and winter, lee depressions from the Gulf of Genoa cross the Po Valley and reach the area, while in spring weather is usually imperturbated due to the influence of the Balkan anticyclone; the northern part of the region is sometimes influenced marginally by lee depressions. In summer there is usually fair anticyclonic weather with some instability phenomena.

Liguria and northern Tuscany (Fig.4,*5*)

The belt is enclosed by the Alps, the Ligurian Apennines and the sea, north of the 44° parallel. The watershed is found at an altitude of approximately 1,000 m and up to 50 km off the coast. Between Savona and Genoa it lowers to 500 m or even less and approaches the coast (minimum distance 8 km). The 500-m isohypse is generally found at a distance not longer than 10 km from the shore.

In autumn, lee and Mediterranean depressions prevail becoming more frequent as the season proceeds. Winter is characterized by easterly winds bringing clear sky, due to the influence of the central-European high pressure; some slow-moving lee depressions may occur. During the winter, the very privileged position of the Ligurian Riviera becomes apparent as mean temperatures between $+8°$ and $+9°C$ are commonly observed (see p.142). In spring, the lee depressions become less frequent and anticyclonic weather gradually prevails transforming from the winter type (European high pressure) into the summer type (Azores high). In summer, this area has fair anticyclonic weather with some instability phenomena.

Tyrrhenian versants (Fig.4,*6*)

These territories are situated between the 39° and the 44° parallel, comprised between the Apennine watershed and the Tyrrhenian Sea. In autumn, lee and Mediterranean depressions are becoming more frequent as the season proceeds. In winter easterly winds giving rise to cold weather and clear sky are prevailing. Hence levelled pressures with clear sky at higher levels and fogs in the valleys dominate but lee depressions may occur causing perturbated weather for several consecutive days. In spring, the lee depressions become less frequent and fair anticyclonic weather becomes dominant; it turns the winter type (European high pressure) into the summer type (Azores high). In summer a flat pressure field takes over with generally clear sky.

Calabria and Sicily (Fig.4,*7*)

Calabria is the southernmost part of the Italian Peninsula ("the toe of the boot"), south of the 39° parallel. Sicily prolongs it towards the southwest. In autumn, winter and spring there is often perturbated weather caused by depressions marginally affecting the area. Instability phenomena with numerous thunderstorms chiefly occur during the autumn season. In summer, a flat pressure field with clear sky and extraordinarily stable weather dominates.

Sardinia (Fig.4,*8*)

In autumn, depressions become more frequent when the season proceeds and fair anticyclonic weather of the summer type (Azores high) transforms into the winter type (European high). Perturbated weather is normal in winter due to passages of depressions or a persistent upper trough over the Tyrrhenian Sea with a surface low over the Gulf of Genoa. In spring depressions become less frequent and fair anticyclonic weather from the winter type gradually turns into the summer type. In summer, the weather is normally not perturbated.

Climate and life in Italy

How does "the man in the street" experience the climate of an Italian city? What are the psychological effects? Let us take Rome as an example as this is no doubt the most well-known Italian city. Towards the end of October the weather deteriorates. A lee depression arrives which brings heavy, long-lasting rains, followed by other rains which fall in an almost uninterrupted succession until Christmas. The impression is of a "rainy season", very damp and not cold, psychologically depressive. This is the most intense "city-life" period of the year. Any activity in the open air is a priori disregarded (the Italians are not used, for example, to going on excursions in rain), and people enjoy the sense of protection offered by the city and the home intimacy.

Around the beginning of the year a period of easterly streams commences and the sky is very clear and the air transparent. The cooling by nocturnal radiation is considerably important but during the day time the temperature is quite high. In the morning it is common to see ice on the ground but by midday overcoats are unnecessary. This is the most "tonic" period of the year and there is very much open air activity, especially since winter sports have become more popular.

During March and April gloomy and rainy days alternate with sunny days, though the latter are frequently interrupted by showers. This is the time when the Italian weather is most similar to that of Atlantic Europe. The Italian, who is used to a marked and long-lasting persistency of the weather, feels uncomfortable with such a regime which he calls "tempo matto" (crazy weather). In Italy one does not have the sensation of a slow, but marked and continuous fading of the winter which can be noticed in trans-Alpine Europe, where the atmosphere and landscape change colour day-by-day. After an atavic experience, there is rather a period of expectancy for the "outburst" of summer, that occurs in May. At a certain time the Azores high protrudes into the Mediterranean and the summer regime begins to dominate rather abruptly: nice, stable weather, high temperatures, and very regular breezes.

This regime persists until the end of September, with a few interruptions caused by cold-air outbreaks near the beginning and at the end. The summer weather is not so heavy as one might imagine; in the early morning the weather is in fact rather cool. Around midday there are one or two very hot hours, then the well-known Roman "ponentino" (little westerly) rises, a rather cool and fresh sea breeze (about 10 knots). Until quite recently human activity in these summer days was very limited; a short morning's activity was followed by a prolonged idleness in the afternoon and then by some work in the listless evenings after the air cools: "dopo l'infrescata" according to the old popular idiom. However, the necessities of today's life, new habits, modern attire and means of transportation are changing this picture.

The only real period of transition is the month of October. A progressive decrease in the temperature is perceived, cold air outbreaks follow one after another, each becoming more active; at this time there is a gradual passage into winter.

It is a very beautiful period, the most romantic month of the year in Rome. The transparent air, the mild temperature, and appealing lights are very inviting to go out and merge into nature. In fact, during the past centuries too, October was the month for going to the country for a day to celebrate the Roman "ottobrate".

The comprehensive impression is that the Roman climate is very pleasant, not very stimulating, but only seldom debilitating or too fatiguing to bear.

The climatic elements

Temperature distribution

As may be expected, the mean temperature in winter in Italy (Fig.5A) generally decreases

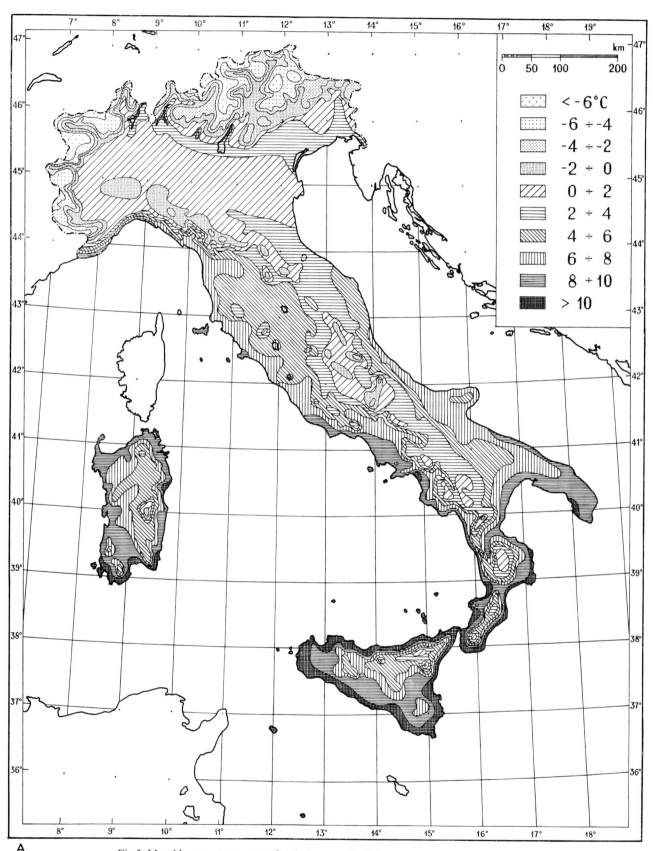

A

Fig.5. Monthly mean temperature in: A. January; B. July.

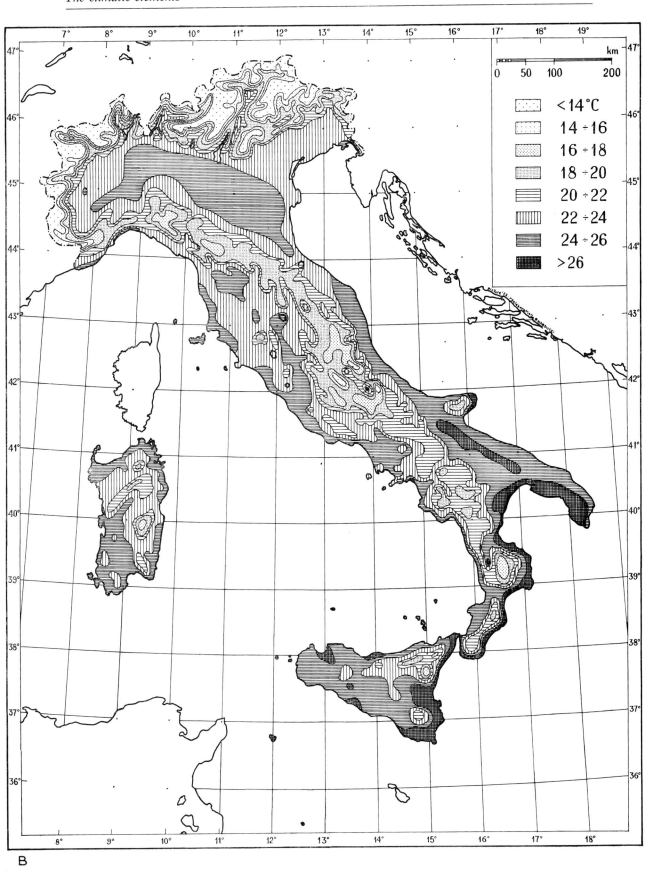

B

when latitude, altitude and the distance from the sea increase. One important exception from this rule is found in the earlier mentioned differences in winter temperatures between the Adriatic and the Tyrrhenian coasts the former being more exposed to northerly winds and the latter more to southerly winds. The Adriatic Sea has a mean temperature of 5°C on the coast and 1°C in the eastern slopes of the Apennines, while the corresponding values on the Tyrrhenian side are 8° and 1°C, respectively. This difference disappears south of Pescara, which represents the southern limit of the region hit by the bora.

Another difference in winter temperatures exists between the Po Valley which is occupied by a cold-air cushion and the Ligurian Riviera which is well protected from cold-air outbreaks, especially in the western parts. The Ligurian Riviera shows mean temperatures in January as high as +8° to +10°C whilst the Po Valley at approximately the same latitude and altitude experiences temperatures around 0° to +2°C. The station at San Remo on the Riviera (43°49′N, 113 m) has mean maximum and minimum temperatures for January over a decade of +11.8° and +6.0°C, respectively. The absolute maximum and minimum temperatures are +19.2° and −1.9°C, respectively. For Novi Ligure (44°47′N, 118 m) in the Po Valley the mean maximum and minimum values are +4.1° and −2.9°C, respectively, while the absolute maximum and minimum values are +17.2° and −18°C for January.

The 0°C isotherm for January runs along the southern slopes of the Italian Alps from southwest to east-northeast. In the area north of this isotherm temperature decreases rapidly with altitude and in January mean temperature may be −5°C or even lower. The highest temperatures in January are found in Sicily, Calabria and southern Sardinia where they are above +10°C indicating that these areas are slightly subtropical in character.

In July (Fig.5B) again the lowest temperatures are found in the high Alps where they are below 20°C. The highest, above +25°C, are experienced in Sicily and the southernmost part of the country in general. The distribution of mean temperatures in January and July is given in Fig.5. The pattern of the temperature extremes has been studied by Coccia (1961) for Milan, Rome and Palermo, on the basis of six years of observations. The results of this comparison are given in Table I.

In this table $\overline{T_{03}} - \overline{T_n}$ is the mean difference between 03h00 GMT temperature and minimum temperature; f_{3n} is the frequency of differences between 03h00 GMT temperature and minimum temperatures higher than 3°C; d_n is the mean difference between sunrise hour and minimum temperature hour, $\overline{T_x} - \overline{T_{12}}$ is the difference between maximum temperature and 12h00 temperature; f_{3x} is the frequency of differences between maximum temperature and 12h00 temperature higher than 3°C; d_x is the mean difference between maximum temperature hour and local midday.

The main conclusion from this comparison is that Milan represents a much more continental type of climate than both Rome and Palermo.

Precipitation and aridity

Fig.6 presents the yearly mean precipitation for the 30-year period 1921–1950, plotted by using the data of 2,372 stations.

From Fig.6 it is obvious that rainfall increases both with increasing latitude and altitude.

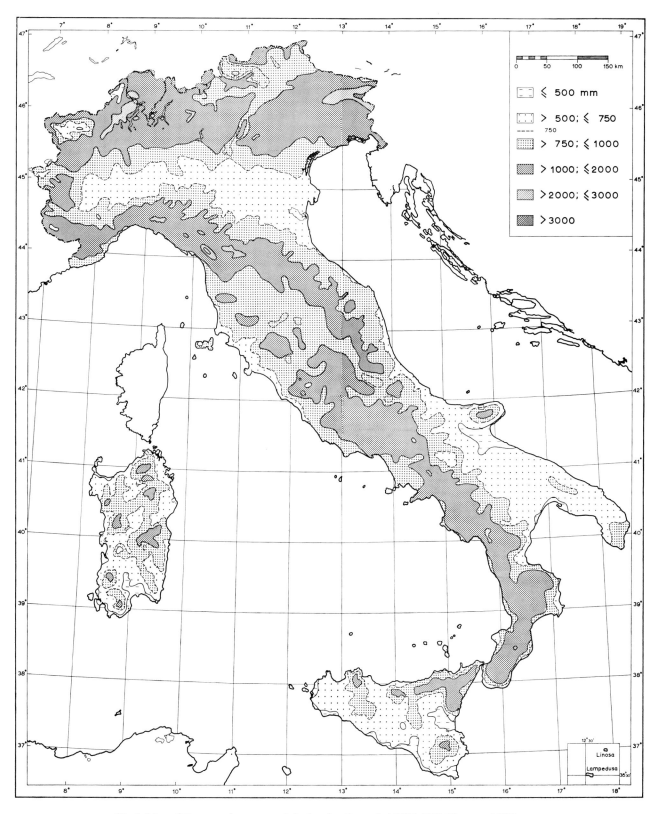

Fig.6. Map of the annual mean precipitation for the period 1921–1950 (FROSINI, 1961).

TABLE I

PATTERN OF TEMPERATURE EXTREMES FOR MILAN, ROME AND PALERMO DURING A PERIOD OF 6 YEARS[1]
(After Coccia, 1961)

	January			April		
	Milan	Rome	Pal.	Milan	Rome	Pal.
$\overline{T_{03}} - \overline{T_n}$	1.8°	1.6°	1.8°	1.6°	1.4°	1.5°
f_{3n}	12%	14%	13%	12%	6%	11%
d_n	−1 h	−30 min	−75 min	+30 min	−30 min	+30 min
$\overline{T_x} - \overline{T_{12}}$	2.1°	1.5°	1.1°	2.1°	1.5°	1.3°
f_{3x}	22%	8%	4%	17%	6%	8%
d_x	+3 h	+150 min	+2 h	+3 h	+90 min	+90 min

	July			October		
	Milan	Rome	Pal.	Milan	Rome	Pal.
$\overline{T_{03}} - \overline{T_n}$	1.5°	1.1°	0.9°	1.6°	1.5°	1.3°
f_{3n}	10%	1%	3%	12%	7%	10%
d_n	+1 h	0	0	+20 min	0	−15 min
$\overline{T_x} - \overline{T_{12}}$	2.3°	1.2°	1.2°	2.1°	1.5°	1.0°
f_{3x}	19%	3%	5%	19%	17%	4%
d_x	+4 h	+1 h	+2 h	+3 h	+2 h	+1 h

[1] For explanation of symbols see text.

The most conspicuous exceptions from this rule are the differences found between regions more or less exposed to the high frequency of southerly to southwesterly winds as well as the small "islands" of large amounts of rainfall which occur due to windward effects in various regions of the Alps and the Apennines.

The highest amounts of precipitation (between 2,000 — 3,000 mm a year) are found, first of all, on the southern slopes of the northernmost Apennines which are exposed to the southwesterly winds from the Gulf of Genoa. Other areas with a similar amount are found in the southern Apennines inland of the Gulf of Salerno and in the highest areas of Sicily. The lowest amounts of annual precipitation are experienced in the lee-ward areas of the Apennines particularly north of Bari where the amount is only between 300–500 mm. The leeward areas otherwise receive between 500 and 750 mm. The eastern Po Valley which also is a region of fairly low precipitation, receives 750–1,000 mm annually.

The annual distribution of rainfall is in general typical of the Mediterranean regime with a maximum in autumn and a minimum in summer. Only the Tyrrhenian coast shows a secondary maximum in spring too. The Po Valley and the Alpine area show a rather continental regime of distribution with a maximum in summer and a minimum in winter.

Fig.7 presents a chart of maximum intensities of precipitation (in 1 h and in $^1/_4$ h) observed during the 30-year period 1930–1959 at 21 stations of the southernmost regions of the peninsula, known as being subject to very intense precipitations. Fig.8 shows the per cent ratio between the maximum precipitations in 24 h of November 1959 and those of the 30-year period 1931–1960. This ratio represents an eloquent index of how variable the maximum intensity of precipitation is in Basilicata and Calabria, and gives

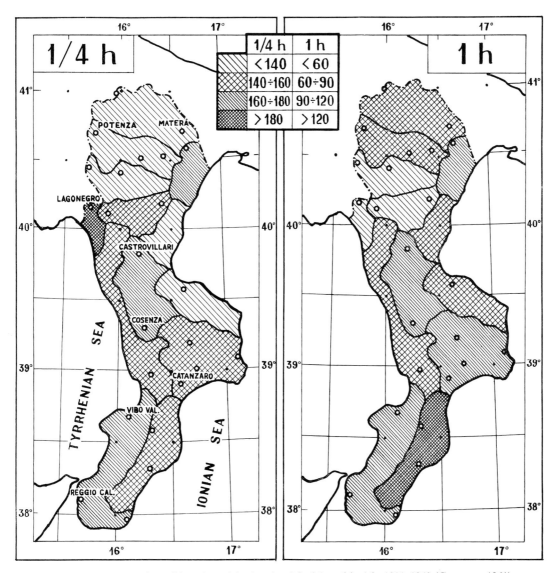

Fig.7. Maximum intensities of precipitation (mm) in ¼ h and in 1 h, 1931–1960 (CAVAZZA, 1961).

an indication of the difficulties to be met in hydrological studies and in the planning of works of civil engineering.

PINNA (1957) has calculated the aridity index of De Martonne for 473 Italian stations for which precipitation data were available relating to the decade 1921–1930 and temperature data relating to the period 1926–1935. The fact that practically no observations exist on evaporation has excluded the possibility of taking this into account. It has therefore been necessary to consider an index based on temperature and precipitation to give an indication of aridity. Of these indexes Pinna chose the second index of De Martonne, because it defines rather well the regions where irrigation is necessary for agriculture. The index is obtained through the formula:

$$\frac{1}{2}\left(\frac{P}{T+10}+\frac{12p}{t+10}\right)$$

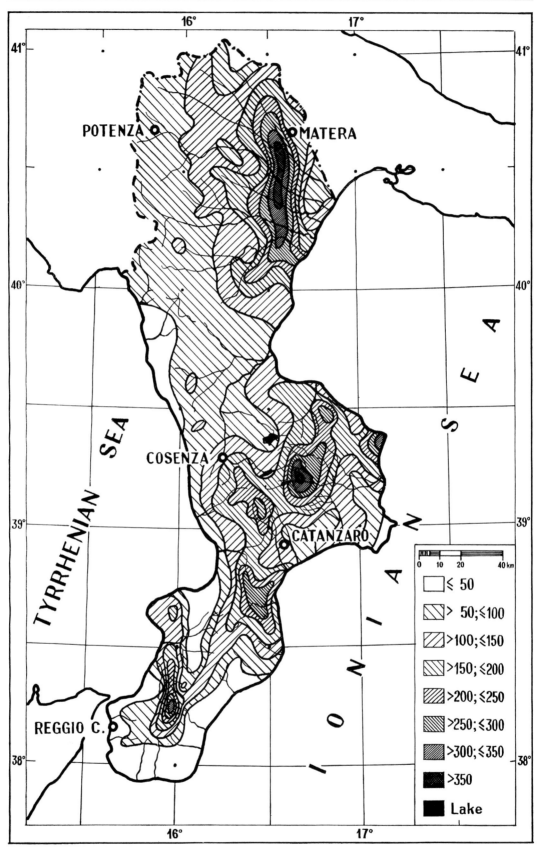

Fig.8. Percent ratios between maximum precipitations in 24 h of November 1959 and those of the period 1931–1960 (CAVAZZA, 1961).

where T and P are the annual temperature and precipitation, and t and p are the temperature and precipitation of the most arid month, which is singled out by the second term of the above-indicated formula. Index values smaller than 10 are considered to correspond to an arid climate. Such climates are found in Italy only in the coastal areas of Apulia, Basilicata, Sicily and Sardinia (Fig.9). Values between 10 and 20 correspond to a sub-arid climate, which is the climate of the typical Mediterranean vegetation. The Po Valley, too, includes sub-arid areas, but not of the Mediterranean vegetation type: a belt around the Po from Mantua to the delta; the Langhe and small portions, that cannot be represented on the chart, of the Aosta and Susa valleys; the Valtellina and the central valley of the Adige. Values between 20 and 30 represent a sub-humid climate

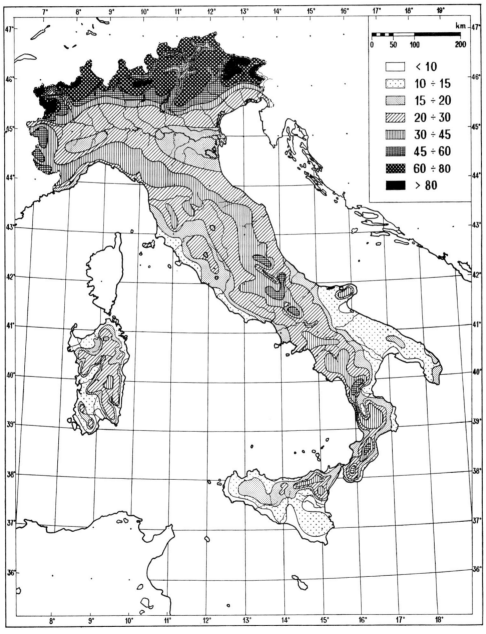

Fig.9. Chart of the aridity indexes of De Martonne.

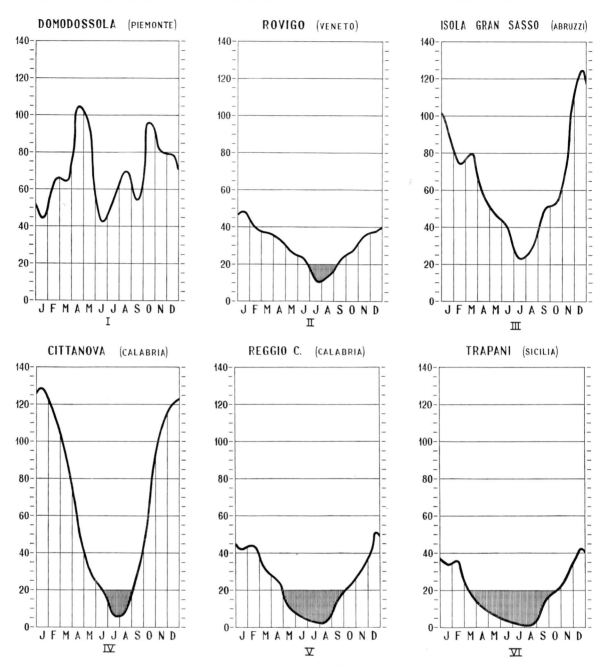

Fig.10. Annual variations of the aridity index at six different places.

which, however, does not exclude the necessity for irrigation for some types of cultivation. Humid climates correspond to a value > 30 where arboreal vegetation is abundant. Fig.10 presents the annual variation of the aridity index relating to a few Italian localities.

Snow

Human activities in Italy are more greatly affected by the imposing accumulations of fresh snow than by persistence of the mantle of snow. The greatest thicknesses of snow

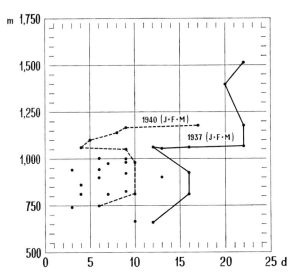

Fig.11. Variation with elevation of number of days with snow in the first quarter of the years 1937 and 1940.

occur in relatively restricted areas with a low level of population in the Alpine chain as well as in relatively more populous regions of the central-southern Apennine massifs. These areas are subjected to very intense snowfall, carried in by moisture-rich maritime air, overlying a layer of cold air, which usually originates in the Balkans or Russia.

We have preferred to reproduce here the map of average annual snow precipitation (Fig.12) which was drawn up by GAZZOLO and PINNA (1973) based on data from 1,507 stations (1,379 of which with more than 20 years of observations and 859 with more than 30 years) and which represents fundamental progress in our knowledge of snow in Italy.

The explanatory monograph contains (even though in a somewhat fragmentary way) numerous other data which permit of establishing synthetic comparisons in Tables IIA and IIB.

TABLE IIA

COMPARISON OF SNOWFALLS

	h(m)	No	$SDm(y)$	$SDn(y)$	$SDx(y)$	Mean snowfall (cm): Dec.	Jan.	Feb.	year	$SNx(y)$	1928–29: Jan.	Feb.	H.Y.	1950–51: Dec. Feb.	H.Y.	1956: Feb.
Cortina d'Ampezzo	1,275	1	124	–	–	40	46	41	203	–	–	–	–	443	542	–
Madonna di Campiglio	1,553	3	157	104	209	–	–	–	327	600	–	–	–	330	640	–
Lago Goillet	2,526	4	218	157	272	150	103	90	689	1,106	–	–	–	–	1,204	–
Valsavaranche	1,545	5	130	59	170	–	–	–	207	435	–	–	–	–	–	–
Bardonecchia	1,275	6	96	43	154	–	–	–	197	426	–	–	–	–	–	–
Crissolo	1,410	7	113	–	–	60	54	74	292	–	–	–	–	–	–	–
Pescocostanzo	1,395	10	82	–	–	60	65	61	260	–	–	–	–	–	–	–
Capracotta	1,400	11	79	15	128	70	75	72	309	705	150	135	–	–	–	365
Fonni	992	13	–	–	–	–	–	–	–	–	230	153	–	–	–	131
Floresta	1,250	14	52	–	–	–	–	–	–	–	144	349	–	–	–	142

Explanation of symbols: h = height of the station; No = number of the station in Fig.12; $SDm(y)$ = mean yearly duration of snow cover (total number of days with snow lying); $SDn(y)$ = minimum yearly duration of snow cover; $SDx(y)$ = maximum yearly duration of snow cover; $SNx(y)$ = maximum yearly snowfall; H.Y. = hydrological year (Sep.–Aug.).

149

TABLE IIB

SNOW FREQUENCIES FOR MILAN (1763–1955) AND ROME (1901–1960)

	Days with snow								Years with snow (%)							
	Oct.	Nov.	Dec.	Jan.	Feb.	Mar.	Apr.	Year	Oct.	Nov.	Dec.	Jan.	Feb.	Mar.	Apr.	Year
Milan	0.02	0.5	2.1	3.3	2.1	1.0	0.1	9.1	2	30	70	86	65	46	9	99
Rome	0	0.03	0.3	0.7	0.5	0.3	0.02	1.8	0	3	17	35	27	17	2	65

The incoherences between data in the columns $SNx(y)$ and H.Y. are probably due to the fact that the first refer to the civil year and the second to the hydrologic year.

The annual maximum totals have been measured for the Alps at the Col du Tonale (2; 1,850 m), with 1,693 cm during the hydrologic year 1950–51 and for the Apennines at Roccacaramanico (9; 1,050 m) with 1,049 cm in 1928–29. The greatest daily snowfalls were at Bagnara (8; 620 cm) in the central Apennines (123 cm), at Montevergine (12: (12; 1,270 m) in the southern Apennines (138 cm) and at Floresta in Sicily (14; 1,250 m) where 130 cm was recorded on 22 February 1929.

In any event, snowfall is only really significant in the mountainous regions. It is not significant in the case of the principal cities and not even in the plain of the Po, as is shown in Table IIB.

Table IIB gives some information on the number of days with snow and the frequency (%) of years with snow in Milan and in Rome.

Moreover, it is interesting to remark, as BOSSOLASCO (1948) does, that around the Po Valley a snow frequency minimum is verified at about the 1,000-m level (Fig.11). At this altitude the peplopause (so called by SCHNEIDER-CARIUS, 1947, p.82; 1948, p.6) is found, which is the transition layer between the cushion of cold air remaining at low level and the air masses conveyed by synoptic changes. When it snows an inversion at the peplopause height is verified, and it seems logical to propose this inversion as an explanation for the above-mentioned frequency minimum, in the sense that precipitation is sometimes constituted of rain above and snow below this layer.

Thunderstorms

Thunderstorms occur in northern Italy almost exclusively in summertime, whereas in central Italy they occur throughout the year. In Sicily they prevail in winter. This is an aspect of the thunderstorm distribution in Italy that mostly concerns the man in the street. To someone who lives in the northern areas, a thunderstorm is a summer characteristic, and the first time he moves southwards and observes a wintry thunderstorm, he no doubt will be surprised. Fig.13 and 14 represent the distribution of the number of days with thunderstorm activity annually and in the single seasons.

Fog

In Italy fog is traditionally considered as a phenomenon exclusive of the Po Valley because it is only there that fog covers very extensive areas and persists all day or throughout series of several consecutive days. Elsewhere, in fact, night and morning fogs have such a frequency that they often hinder the complex and delicate system

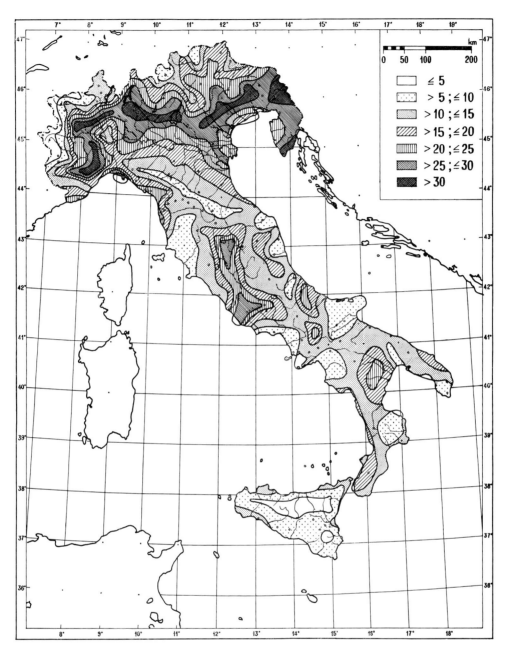

Fig.13. Annual mean number of days with thunderstorm for the period 1880–1948 (Bossolasco, 1949).

of modern transport. However, the fog phenomenon takes conspicuous aspects only in the Po Valley, and there it has been studied more thoroughly than in other areas. The general features of the fog distribution in Italy are shown in Fig.15. They refer to the months of January, April and October for the 06h00 GMT observation[1]. The map of April shows fog principally in mountain regions, while in October and in January fog appears also in the lower region, especially in the northern parts of Italy. In the Po Valley fog exhibits a distribution much more uniform than in the rest of Italy,

---

[1] Corresponding to 07h00 local time.

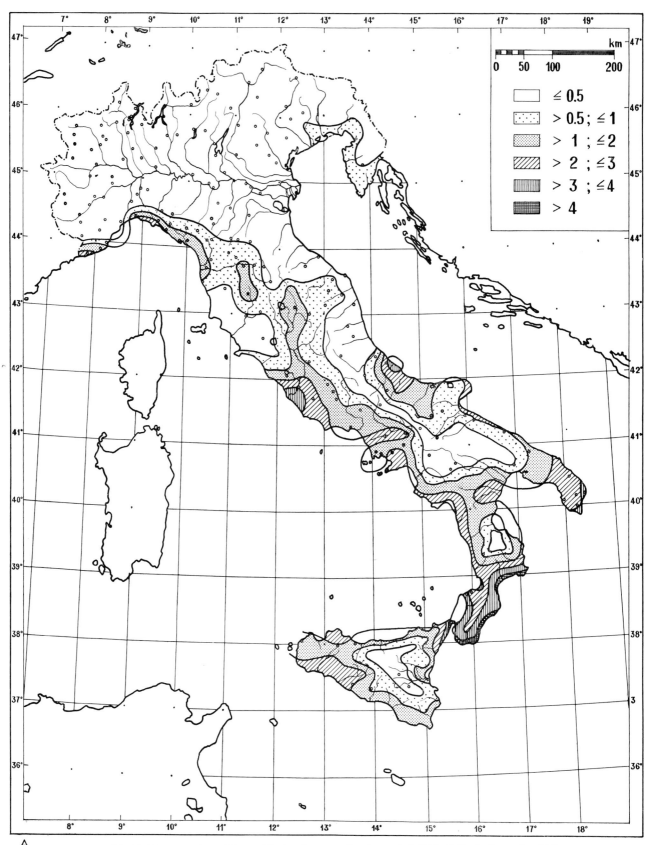

A

Fig.14. Mean number of days with thunderstorm for the period 1880–1948: A. December–February; B. March–May; C. June–August; D. September–November. From Bossolasco, 1949.

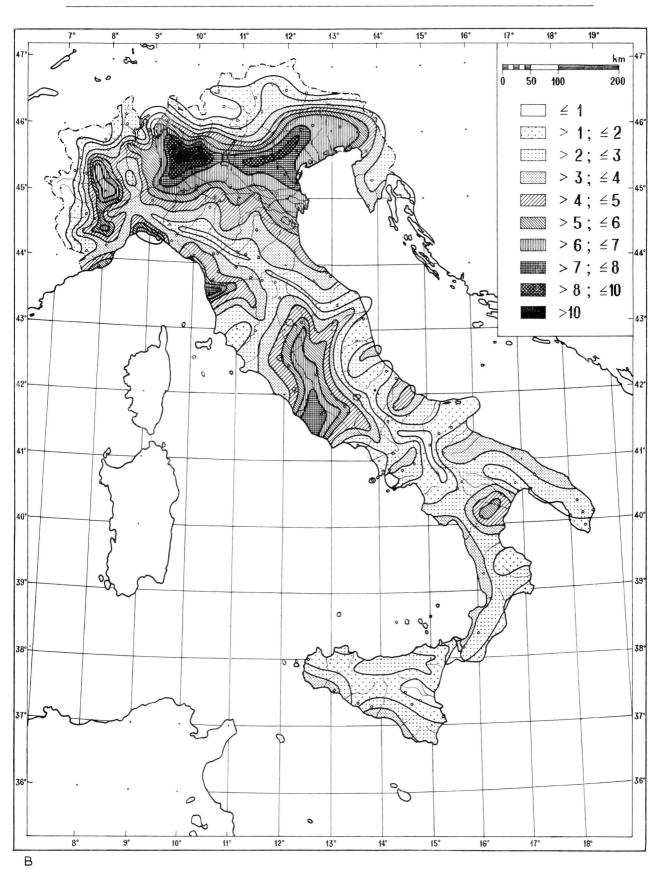

B

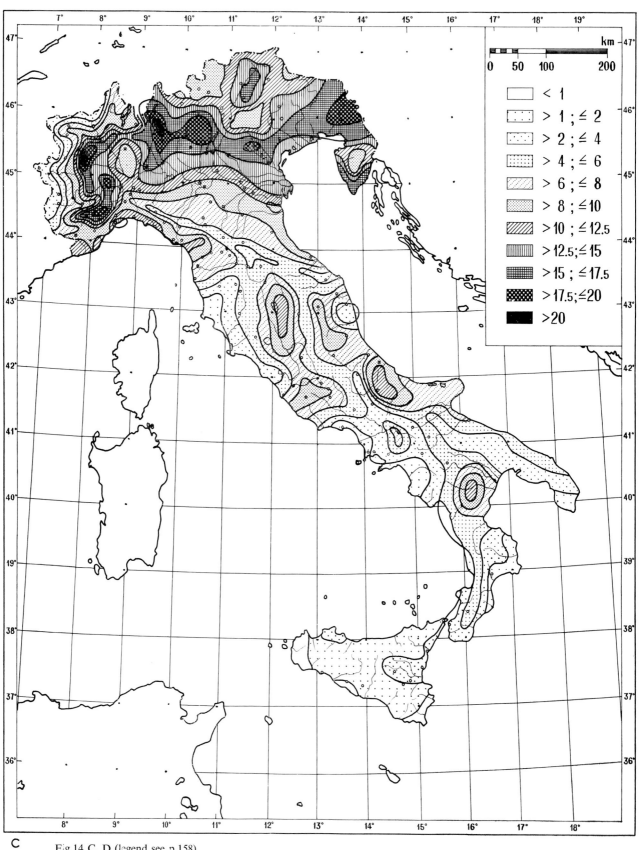

	km
	0 50 100 200

	< 1
	> 1 ; ≦ 2
	> 2 ; ≦ 4
	> 4 ; ≦ 6
	> 6 ; ≦ 8
	> 8 ; ≦ 10
	> 10 ; ≦ 12.5
	> 12.5 ; ≦ 15
	> 15 ; ≦ 17.5
	> 17.5 ; ≦ 20
	> 20

C   Fig.14 C, D (legend see p.158)

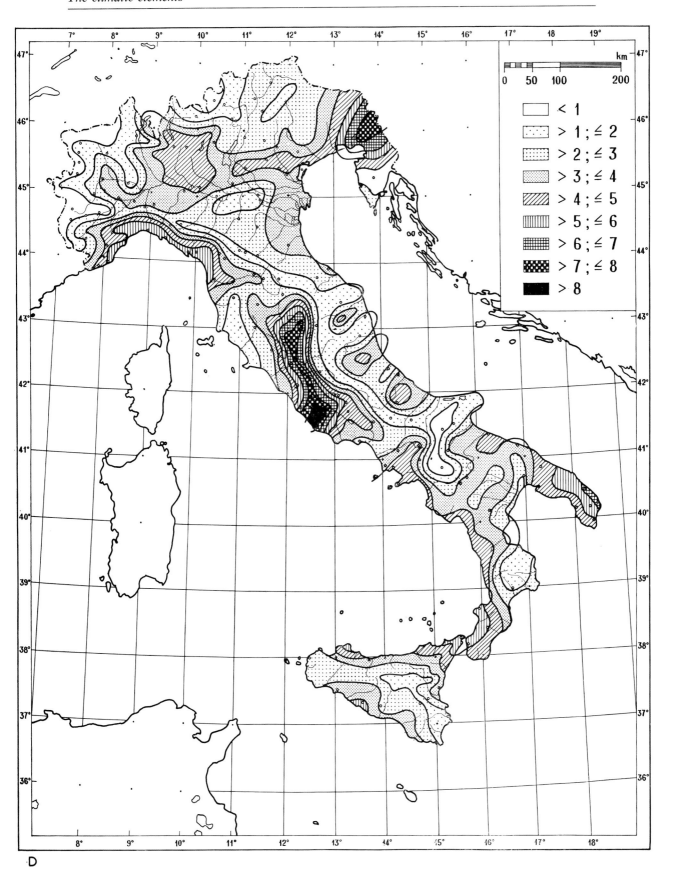

Fig.15. Fog frequency (days/month) based on observations from 100 Italian stations in the period 1951–1960 at 06h00 GMT. A. January; B. April; C. October.

Fig.15 C (legend see p.162)

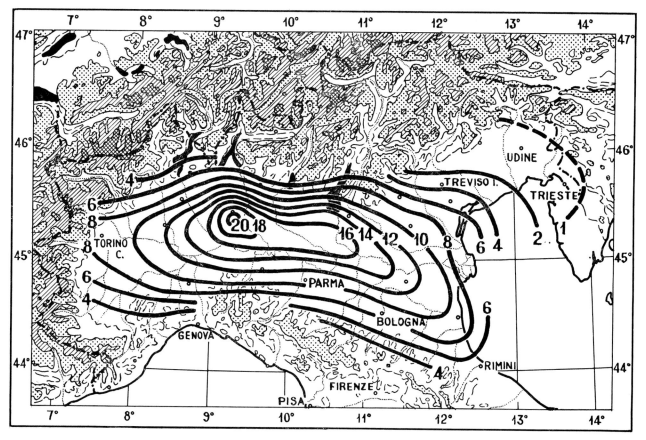

Fig.16. Mean number of days with visibility below 1 km in the Po Valley, in December at 06h00 GMT (PALMIERI and TODARO, 1958).

and for this reason it has been possible to draw the isolines of frequencies as number of days per month (EREDIA, 1916; PALMIERI and TODARO, 1958). From Palmieri and Todaro a map of the number of days in December with visibility less than 1,000 m is reproduced (Fig.16). The diurnal and annual variations of fog frequency at the Milan airport are given in Fig.17.

Diurnal variations of pressure

These variations have been studied for 20 Italian stations by ZANOLI (1957) on the basis of the synoptic observations in the months of January, April, July and October. The main results are the following:

(*1*) The more continental stations, which are also the northernmost ones, present one maximum only (in the summer, around 03h00 GMT; in the winter around 06h00).

(*2*) The more maritime stations, which are also the southernmost ones, present two maxima (around 09h00 and 21h00 GMT, and two minima (in the summer, around 03h00 and 18h00 GMT; in the winter, around 06h00 and 15h00 GMT).

(*3*) The intermediate stations gradually pass from the first to the second type because of the appearance, and the accentuation, of the morning minimum. The oscillation

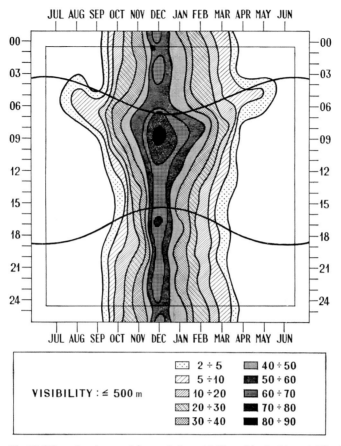

JUL AUG SEP OCT NOV DEC JAN FEB MAR APR MAY JUN

VISIBILITY : ≤ 500 m

☐ 2 ÷ 5	▨ 40 ÷ 50
▨ 5 ÷ 10	▧ 50 ÷ 60
▨ 10 ÷ 20	▤ 60 ÷ 70
▨ 20 ÷ 30	■ 70 ÷ 80
▨ 30 ÷ 40	■ 80 ÷ 90

Fig.17. Diurnal and annual fog variation at Milan, Linate Airport. The two lines in horizontal direction indicate time of sunrise and sunset.

amplitude decreases with continentality (Bolzano 1.12 mbar; Milan 0.48 mbar; Naples 0.26 mbar).

Pressure, temperature, humidity and wind at upper levels

The network of radiosonde observations has reached only recently enough uniformity, density and reliability to allow for climatological conclusions. The first detailed studies have been carried out (ROSINI and RISPOLI, 1962) on the basis of data of the five-year period 1956–1960, and some results are presented here in maps of mean contour lines and isotherms of the same standard surfaces (see Fig.18).

Climatic variations

The energy devoted in Italy to the study of the long series of meteorological data—longer and more numerous series than in most other countries—largely was absorbed by *period analysis*. The development of the period analysis is due to the geophysicist Francesco Vercelli. Both the choice of the period and the judgement on the reality of the results are, to a great extent, subjective in this method (POLLI, 1954). The technique was developed during World War I and, as happens frequently, at first seemed successful. A school of researchers on periodicities was established at the Osservatorio Talassografico of Trieste, of which Vercelli was for a long time the director. This school survived after its founder's death and its studies are still followed by many geophysicists of the Italian universities. The works indicated in the bibliography (VERCELLI, 1915, 1916, 1930, 1940; POLLI, 1942, 1943, 1945, 1954, 1955; DEPIETRI, 1955) constitute only a part of the many publications on this subject. According to the above-mentioned authors, the most important periodicities found are those of 11, 16, 22 and 35 years.

The period analysis was not entirely popular, mainly because it did not render a useful technique for the practical meteorologist. However, it is not a completely negligible scientific effort since it was one of the early examples of an attempt to abandon the traditional, rigid, causalistic conceptions. It implied the application of a calculation technique built on a mechanical analyser of oscillating curves. A first purposeful study on climatic variations was undertaken by the Italian geographers who, since 1934, have been publishing a series called *Richerche sulle Variazioni Storiche del Clima Italiano* (Researches on historical variations of the Italian climate). Later, studies have been carried out using more refined and more up-to-date methods, by the meteorologists of the Italian Air Force Meteorological Service (ROSINI, 1955, 1963; TODARO, 1964).

The remarks made by periodicity researchers and the conclusions of geographers and meteorologists appear to agree with a statement of Giacobbe in 1961: that there are in Italy indications of a tendency for an increase of temperature and temperature ranges at the same time as for a decrease of precipitation; generally speaking this means that there has been a trend towards a continentalization of the climate. Beginning from ecological considerations relating to the *habitat* of some arboreal species, Giacobbe made a comparison between temperature means of the period 1892–1906 and those of the period 1936–55, for 91 Italian stations. The maximum temperature increased between these periods in 73 stations by an average of 1.4°C, and decreased at 16 stations by an average of 0.3°C, whilst it remained unchanged at 3 stations. The minimum temperature decreased at 68 stations by an absolute average value of 1.0°C; it increased at 22 stations by an absolute mean value of 0.7°C, and it remained unchanged at 2 stations. The annual range increased at 76 stations by a mean increment of 2.0°C, decreased at 15 stations by an absolute average value of 0.6°C and remained unchanged at 2 stations.

In Milan an increase in the mean summer temperature of 1° C/century during the period 1838–1963 (TODARO, 1964) has been found, and during the period 1850–1950 (POLLI, 1955) an increase of the mean annual temperature of 2°C/century occurred, as well as a decrease in precipitation of 200 mm/century. Todaro, after using modern processes of sequential analysis, indicates that 1866 and 1915 have been years with a climatic turn. Likewise, precipitation seems to have decreased by 140 mm/century at Modena during the period 1830–1954 (DEPIETRI, 1955); by about 50 mm/century at Padua during the period 1727–1940, and at Trieste during the period 1841–1940 (POLLI, 1943); by a few dozen mm at Foggia during the period 1881–1950 (BIANCHINI, 1954); by about 100 mm at Genoa during the period 1833–1947 (BORETTI and DE AMICIS, 1948), and probably the same amount at Torino during the period 1806–1951 (MENNELLA, 1956).

On the contrary, Mantua during the period 1840–1940 (POLLI, 1943); Florence during

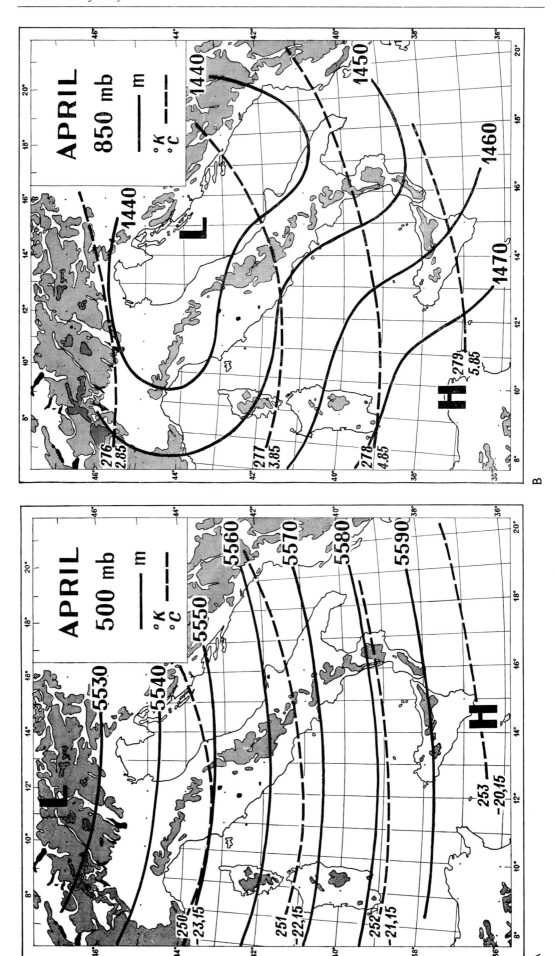

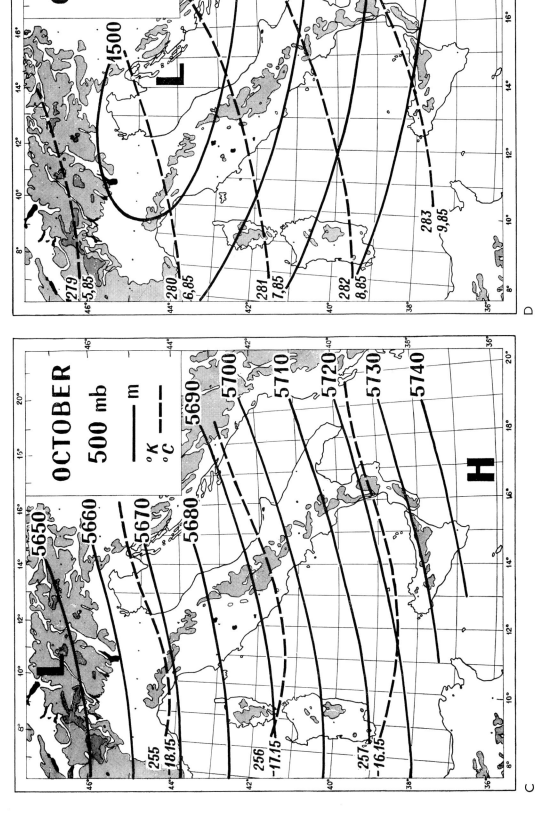

Fig.18. Charts of mean contour lines and mean isotherms for April and October.

the period 1820-1900 (BRAMANTI, 1946); Bologna (CAPRA, 1952) Perugia and Rome during the period 1831-1930 (MELICCHIA, 1942) do not show a clear tendency to variation in precipitation, while Citania seems to show an increase.

MELICCHIA (1939, 1942) studied the regime behaviour of the rivers and precipitation in northern and central Italy until 1930. The mean hydrometric heights of the Po result appear to have decreased, those of the Tiber appear to have markedly decreased in the upper course, to be almost unchanged in the medium, and to have slightly increased in the lower part. This corresponds, more or less, to the secular variation of precipitation. From all the above-mentioned studies it is obvious that there exist important modifications of the atmospheric circulation. However, it shows even more clearly that so far this problem has not been studied adequately.

Indirect, but in our opinion substantially reliable, deductions on the climate change were derived by MONTERIN (1937) and BULI (1949). The former examined information about the transit through the Alpine passes, the existence of vegetal residues at altitudes at which growth is now impossible and the old, very dense network of irrigation canals in the Aosta Valley. He concluded that a relatively warm and dry period endured until the last decades of the 16th century, followed by a cold and rainy period that lasted until the middle of the 19th century. As a consequence the maximum expansion of glaciers in historical ages occurred at the time. Since then there was a return to a warmer and drier climate until about 1940, when the climate changed again towards a colder and rainier type. BULLI (1949) examined the annual meristematic increase (i.e. the rings) of old trees in the Ravenna pine forests and found that the minimum meristematic increment corresponds with an important minimum of the mean temperature at Bologna (CAPRA, 1939, p.20) and Milan (TODARO, 1964b, p.29); with a precipitation maximum at Bologna (CAPRA, 1952, table III) and Mantua (POLLI, 1945, p.18); and with an absolute maximum of the hydrometric height of the Po at Pontelagoscuro (MELICCHIA, 1939, Fig.5). On the contrary, Florence (BRAMANTI, 1946), Foggia (BIANCHINI, 1954) and the Tiber (MELICCHIA, 1942) record behaviour almost opposite, even if less marked.

Final considerations

The most probable explanation for the modest interest which the Italians show in climatology and meteorology is that the climate in Italy is less severe than in northern countries of Europe and in the United States. The Italian climate is perhaps the least systematically investigated European climate. However, as far as the density of the station network and the number of the long series of observations concerns, Italy is probably above the European standard. Groups of researchers, trained to cooperate and organized to utilize the data available at local archives, therefore may attain quite interesting results. Along these lines, and with this purpose, the IFA (Istituto per la Fisica dell' Atmosfera, formerly CENFAM) operates. It was established at the Consiglio Nazionale delle Ricerche after an agreement with various interested agencies, of which the Air Force Meteorological Service is the most important. The Istituto Centrale di Statistica (ISTAT) is making a vigorous effort to collect and publish data of meteorological observations from very different sources, in its *Annuari di Statistiche Meteorologiche*, published since 1959. Finally the Servizio Idrografico del Ministero del Lavori

Publici and the Ufficio Centrale di Meteorologia Ecologia Agraria should be mentioned as maintaining important archives and as publishers of observation data.

References

ALIVERTI, G. and FEA, G., 1968. Lineamenti meteorologici e climatologici delle regioni italiane colpite dalle alluvioni del novembre 1966. *Acc. Naz. dei Lincei, Atti del Convegno sul tema: Le Scienze della Natura di Fronte agli eventi idrologici, Roma, 1967.* Quaderno 112, pp. 13–19 + table XV.

BIANCHINI, M., 1954. Il clima di Foggia. *CNR, Ricerche sulle Variazioni Storiche del Clima Italiano*, 12.

BIEL, E. R., 1944. *Climatology of the Mediterranean Area.* University of Chicago Press, Chicago, Ill. (Institute of Meteorology of the University of Chicago, Misc. Rep. 13.)

BILANCINI, R., 1956–57. Tavole climatologiche di alcune localitá italiane. *Riv. Meteorol. Aeronaut.*, XVI(4) 31–37; XVII(1): 43–49; XVII(2): 34–39; XVII(3): 22–27; XVII(4): 30–37.

BORETTI, L. and DE AMICIS, V., 1948. Analisi periodale della serie pluviometrica di Genova. *Geofis. Pura Appl.*, XII(5–6): 261–285.

BOSSOLASCO, M., 1948. Le precipitazioni nevose nell' Italia settentrionale. *Geofis. Pura Appl.*, XIII (5–6) 213–233. (See also XII(5–6): 287–292.)

BOSSOLASCO, M., 1949. La distribuzione dei temporali in Italia. *Energ. Elettr.*, XXVI(11–12): 3–15.

BRAMANTI, L., 1946. Analisi di una serie pluviometrica (Firenze). *Geofis. Pura Appl.*, VIII: 167–179.

BULI, U., 1949. Ricerche climatiche sulle pinete di Ravenna. *CNR, Ricerche sulle Variazioni Storiche del Clima Italiano*, 10: 1–77.

CANTÙ, V., 1971. Sulla distribuzione della nebbia in Italia. *Riv. Meteorol. Aeronaut*, XXIX(4): 14–22.

CANTÙ, V., 1971. Periodicità e variazioni climatiche: il punto di vista di un meteorologo. *Riv. Meteorol. Aernonaut.*, XXXI(3): 289–296.

CANTÙ, V. and NARDUCCI, P., 1967. Lunghe serie di osservazioni meteorologiche. *Riv. Meteorol. Aeronaut.*, XXVII(2): 71–79.

CANTÙ, V. and NARDUCCI, P., 1973. *Bibliographia Climatologica Italiana.* CNR-IFA, Roma, Str. 22, 24 pp.

CAPRA, A., 1939. Variazioni periodiche della temperatura media a Bologna dal 1814 al 1933. *CNR, Ricerche sulle Variazioni Storiche del Clima Italiano*, 6: 1–31.

CAPRA, A., A., 1952. Andamento delle precipitazioni a Bologna dal 1813 al 1942. *CNR, Ricerche sulle Variazioni Storiche del Clima Italiano*, 11: 1–24.

CAVAZZA, S., 1961. Precipitazioni brevi ed intense in Lucania e Calabria. *Energ. Elettr.*, 1961(8): 746–748.

CENFAM, 1964–1966. *Cyclone Development in the Lee of the Alps.* Res. Reports, Nos. 4 and 5, of the Centro Nazionale per la Fisica dell' Atmosfera e la Meteorologia, Roma.

COCCIA, L., 1961. Comportamento delle temperature estreme in alcune località italiane. *Riv. Meteorol. Aeronaut.*, XXI(2): 28–48.

DEL TRONO, F., 1964. Analisi della distribuzione delle altezze medie stagionali della superficie isobarica di 500 mb sulla regione euro-atlantica. *Ric. Sci.*, 34(II–A): 121–130.

DEL TRONO, F., 1965. Tipi di circilazione media stagionale e loro fisionomia fenomenica sull' Italia. *Ric. Sci.*, 35(II–A): 331–334.

DEPIETRI, C., 1955. Analisi periodale della serie pluviometrica di Modena (1830–1954). *Geofis. Meteorol.*, III(6): 140–143.

EREDIA, E., 1905. Le precipitazioni atmosferiche in Italia dal 1880 al 1905. *Ann. Uff. Centr. Meteorol. Geodin.*, XXX, p. I.

EREDIA, F., 1909. La temperature in Italia. *Ann. Uff. Centr. Meteorol. Geodin.*, XXXI, p. I (data of the period 1866–1906).

EREDIA, F., 1912. La variazione diurna della temperatura in Italia, I. *Ann. R. Uff. Cent. Meteorol. Geodin.*, XXXIV: 1–44.

EREDIA, F., 1913. La nebulosità in Italia. *Ann. Uff. Centr. Meteorol. Geodin.*, XXXV, p. I (data of the period 1891–1910).

EREDIA, F., 1916. *La Nebbie in Val Padana.* Cecchini, Rome. (Reprinted from: *Riv. Meteorol. Agrar.*, XXXVII(10): 1–12.

EREDIA, F., 1942. *Distribuzione della Temperatura in Italia nel Decennio 1926–1935.* Ministero dei Lavori Pubblici, Servizio Idrografico, pubbl. 21.

FERRARA, G., 1917. *La Neve. Studio Sintetico Generale.* De Carolis, Teramo (Reprinted from: *Riv. Abruzzese.*)

FLOHN, H., 1968. *Clima e Tempo*. Il Saggiatore, Milano, 1968, 252 pp.

FROSINI, P., 1961. *La Carta Precipitazioni Medie Annue in Italia per il Trentennio 1921–1950*. Ministero dei Lavori Pubblici, Servizio Idrografico, Rome, Publ. 24, Fasc. XIII.

GAZZI, M., SIMONINI, G. and VINCENTINI, V., 1972. *Bibliografia "Nebbia" (1950–1970)*. CNR-IFA, Roma, 80 pp.

GAZZOLO, T. and PINNA, M., 1973. *La Nevosità in Italia nel Quarantennio 1921–1960*. (Gelo, neve e manto nevoso.) Ministerio dei Lavori Pubblici, Servizio Idrografico, Roma, 211 pp.

ISTITUTO CENTRALE DI STATISTICA, 1959ff. *Annuario di Statistiche Meteorologiche*. Roma.

LECCE, B., 1964. La ciclogenesi sul Golfo Ligure durante determinati periodi del decennio 1952–1961 e sua evoluzione media stagionale. *Attiv. Assoc. Geofis. Ital., Conv. Ann., XVII, Roma*.

LECCE, B. and DEL TRONO, F., 1957. Previsione a lunga scadenza mediante le singolarità di pressione e comportamento dell'onda portante a 500 mb. *Riv. Meteorol. Aeronaut.*, XVII(2): 3–19. (See also: XVIII(3): 35–52.)

LECCE, B. and DEL TRONO, F., 1960b. Flusso medio della circolazione atmosferica entro periodi statisticamente determinati e suo comportamento annuale. *Geofis. Meteorol.*, VIII (3–4): 53–56.

MELICCHIA, A., 1939. Variazioni climatiche nella pianura padana e loro rapporti col regime del Po. *CNR, Ricerche sulle Variazioni Storiche del Clima Italiano*, 7:

MELICCHIA, A., 1942. Variazioni climatiche nell'Italia centrale e loro rapporto col regime del Tevere. *CNR, Ricerche sulle Variazioni Storiche del Clima Italiano*, 9:

MENNELLA, C., 1956. L'andamento annuo della pioggia in Italia nelle osservazioni ultrasecolari. *CNR, Ricerche sulle Variazioni Storiche del Clima Italiano*, 13:

MENELLA, C., 1967. *Il Clima d'Italia nelle sue Caratteristiche e Varietà e Quale Fattore Dinamico del Paesaggio*. EDART, Napoli, 718 pp. (2nd volume 1972).

METEOROLOGICAL OFFICE, 1964. *Weather in the Mediterranean*. H.M. Stationery Office, London, II, 372 pp.

MONTALTO, M., PELLEGRINI, F. and RIVOSECCHI, I., 1967. Sulla climatologia delle ciclogenesi sottovento alle Alpi. *Riv. Meteorol. Aeronaut.*, XXVII(3): 3–29.

MONTERIN, U., 1937. Il clima sulle Alpi ha mutato in epoca storica? *CNR, Ricerche sulle Variazioni Storiche del Clima Italiano*, 2.

MORI, A., 1964. Climatologia e basi meteorologiche. In: *Un Sessantennio di Ricerche Geografica Italiana*. Soc. Geogr. Ital., vol. XXVI, 21 pp.

PALMIERI, S. and TODARO, P., 1958. *Frequenze in Giorni al mese di Visibilità Inferiore a 1000 m in Val Padana*. Ministero Difesa Aeronautica, Servizio Meteorologico, Roma.

PINNA, M., 1957. La carta dell'indice di aridità per l'Italia. In: *Atti del XVII Congresso Geografico Italiano*.

PINNA, M., 1970. Contributo alla classificazione del clima d'Italia. *Riv. Geogr. Ital.*, LXXVII(2): 24 pp.

POLLI, S., 1942. Analisi periodale di due serie climatiche centennali (Trieste 1841–1940). *Arch. Oceanogr. Sismol.*, II (2–3): 107–115. (Reprinted as: *Comitato Talassografico Italiano, Mem.*, CCXCVII.

POLLI, S., 1943. Analisi periodale di una serie pluviometrica bisecolare (Padova 1717–1940). *Riv. Meteorol. Aeronaut.*, VII (1): 19–23.

POLLI, S., 1945. Analisi periodale della serie pluviometrica di Mantova. *Geofis. Pura Appl.*, VII (1–4): 16–22.

POLLI, S., 1954. Criteri di realtà fisica per un ciclo climatico. *Geofis. Meteorol.*, II(3): 33–36.

POLLI, S., 1955. Analisi periodale di tre serie climatiche centennali (Milano 1851–1950). *Geofis. Meteorol.*, III(3): 64–69.

RIMA, A., 1962. Per una legge sulla periodicità delle serie cronologiche naturali. *Assoc. Geofis. Ital., Atti Conv. Ann.*, XII, Roma, 1962, 20 pp.

RIMA, A., 1968. *Periodi con Precipitazioni a sud delle Alpi e Caratteri Alluvionali*. Ass. Geofis. Ital., Napoli, 22 pp.

ROSINI, E., 1955. Alcuni risultati dell'esame statistico delle osservazioni meteorologiche in Italia dal 1946 al 1953. *Riv. Meteorol. Aeronaut.*, XV(1): 32–42.

ROSINI, E., 1963. A quantitative definition of the meaning of constancy and steadiness of climate. In: *Changes of Climate*. Unesco, Paris, pp.45–48.

ROSINI, E. and RISPOLI, F., 1962. *Elementi per la Climatologia della Troposfera sulla Regione Italiana*. C.N.R., CENFAM, Roma, 2, 8 pp.

ROSINI, E. and RISPOLI, F., 1966. *Elementi per la Climatologia della Troposfera sulla Regione Italiana* C.N.R., CENFAM, Min. Dif. Aeronaut., ITAV, Serv. Meteorol., CP, No. 24, Roma, 66 pp.

ROSTER, G., 1909. *Climatologia dell'Italia nelle sue Attinenze con l'Igiene e l'Agricoltura*. UTET, Torino, 1040 pp.

SERVIZIO IDROGRAFICO, 1918ff. *Annali Idrologici.* (Thermopluviometrical observations.) Minist. dei Lavori Pubblici, Roma.

SERVIZIO METEOROLOGICO DELL'AERONAUTICA, 1962. *Statistiche Meteorologiche Relative alle Principali Località Italiane.* (Upper-air data of the period 1956–1960 regarding 6 stations: surface data of the period 1946–1955 regarding more than 100 stations.) Minist. delle Difesa, Roma.

TODARO, C., 1964a. Sugli scambi di calore tra il mare e l'atmosfera nella regione mediterranea. In: *Associazione Geofisica Italiana, Atti Conv. Ann., XII, Roma, 1964*, pp. 343–348. (See also: *Riv. Meteorol. Aeronaut.*, XXIII(4): 17–31).

TODARO, C., 1964b. Sull'andamento delle temperature medie estive a Milano. *Riv. Meteorol. Aeronaut.*, XIV(2): 50–31. (See also: XXIII(2): 28–32; XXV(1): 36–42.)

URBANI, M., 1955. Su una classificazione di tipi di tempo in Europa. *Riv. Meteorol. Aeronaut.*, XV(3–4): 30–37.

URBANI, M., 1956a. Sulla ciclogenesi nel Mediterraneo durante l'inverno. *Riv. Meteorol. Aeronaut.*, XVI(1): 3–9.

URBANI, M., 1956b. Sui cicloni atlantici influenzanti il Mediterraneo durante l'inverno. *Riv. Meteorol. Aeronaut.*, XVI(4): 39–41.

URBANI, M., 1957. Sulla ciclogenesi nel Mediterraneo durante l'estate. *Riv. Meteorol. Aeronaut.*, XVII(3): 15–21.

URBANI, M., 1961. *Una classificazione di Tipi di Tempo sulla Europa e sul Mediterraneo.* Ministero Difesa Aeronautica, ITAV, Roma, 3° Reparto, 9° Divisione.

VERCELLI, F., 1915. Analisi armonica dei barogrammi e previsione della pressione barometrica. *Rend. Accad. Lincei.*

VERCELLI, F., 1916. Oscillazioni periodiche e previsione della pressione barometrica. *Mem. R. Ist. Lombardo Sci. Lett.*, XXXI: 1–31.

VERCELLI, F., 1930. Metodo generale per l'analisi della periodicità nei digrammi statistici e sperimentali. *Rend. Accad. Lincei.*

VERCELLI, F., 1940. Guida per l'analisi della periodicità nei diagrammi oscillanti. *Com. Talassogr. Ital., Mem.*, CCLXXV.

WALTHER, H. and LIETH, H., 1964. *Atlante dei Diagrammi Climatici.* VEB Fischer Verlag, Jena, Vol.2, 372 pp. (Klimadiagramm-Weltatlas).

ZANOLI, E., 1957. Sull'andamento diurno della pressione in Italia. *Riv. Meteorol. Aeronaut.*, XVII(3): 3–9.

ZENONE, E., 1959. *Klimatologie und Meteorologie des Mittelmeergebietes.* Meteorologische Zentralanstalt, Zürich, 21 pp.

TABLE III

CLIMATIC TABLE FOR DOMODOSSOLA (1901–1930)

Latitude 46°07′N, longitude 08°18′E, elevation 300 m

Month	Mean sta. press. (mbar)	Temperature (°C)					Precipitation		Mean daily cloud. (tenths)
		daily mean	mean max.	mean min.	extreme		mean (mm)	days with ⩾1 mm	
					max.	min.			
Jan.	1017.7	1.1	7.1	−2.6	22.5	−11.9	53.5	4.4	3.7
Feb.	1016.0	2.6	9.4	−1.6	25.0	−12.4	74.6	5.6	3.8
Mar.	1015.7	6.7	13.6	1.9	28.4	−8.3	139.5	8.9	4.4
Apr.	1014.3	10.6	17.4	5.5	29.0	−3.6	173.3	10.8	5.3
May	1015.1	15.1	22.0	9.6	32.8	−0.7	180.5	11.8	5.4
June	1015.3	18.6	25.4	13.1	36.6	4.5	122.1	10.4	5.0
July	1014.9	20.3	27.5	14.6	37.0	7.2	124.9	8.2	4.3
Aug.	1014.9	19.7	27.0	14.3	37.5	5.5	125.4	8.2	4.1
Sept.	1017.2	16.1	23.0	11.4	37.7	2.7	139.7	8.3	4.7
Oct.	1018.2	10.8	17.2	6.8	26.2	−4.1	227.2	9.2	4.8
Nov.	1016.5	5.3	11.3	1.6	23.2	−8.0	134.3	8.3	4.4
Dec.	1016.4	2.2	7.4	−1.2	22.6	−11.6	99.2	7.7	4.4
Annual	*	10.8	17.4	6.1	–	–	1594.1	101.8	4.6

Month	Mean daily sunshine (h)	Daily global radiation (cal. cm^{-2} day^{-1})	Max. sunshine hours	Min. sunshine hours
Jan.	4.3	183	189.1	92.7
Feb.	5.3	286	213.7	108.4
Mar.	5.8	418	259.5	124.0
Apr.	6.3	538	235.2	145.2
May	6.9	595	252.3	161.2
June	7.1	621	272.6	161.1
July	8.0	564	342.4	186.3
Aug.	7.0	500	281.8	145.1
Sept.	6.2	401	226.5	135.3
Oct.	5.6	305	226.6	102.9
Nov.	3.7	193	131.1	63.6
Dec.	3.6	156	144.4	65.4
Annual	**	**	**	**

*Milano Malpensa, 45°37′N 08°44′E; 211 m; 1951–1965.
**Pian Rosa, 45°56′N 07°42′E; 3,488 m; 1958–1969.

TABLE IV

CLIMATIC TABLE FOR TREVISO (1901–1930)

Latitude 45°40′N, longitude 12°17′E, elevation 20 m

Month	Mean sta. press. (mbar)	Temperature (°C)					Relative humid. (%)		Mean precip. (mm)
		daily mean	mean max.	mean min.	extreme		06h	12h	
					max.	min.			
Jan.	1017.2	2.5	5.7	−0.2	14.1	−9.4	87	75	58.0
Feb.	1015.7	4.1	7.9	1.2	16.3	−14.8	87	73	65.0
Mar.	1016.1	8.2	12.4	4.9	21.6	−4.8	87	69	96.0
Apr.	1015.0	12.5	17.0	8.6	26.9	−1.0	87	68	85.9
May	1014.5	17.5	22.5	13.2	33.5	4.1	86	68	106.2
June	1015.0	21.3	26.4	16.7	33.8	8.8	84	65	92.2
July	1014.4	23.4	28.7	18.6	39.2	11.7	83	63	98.2
Aug.	1014.2	23.0	28.1	18.2	36.1	10.6	85	63	81.3
Sept.	1016.6	18.8	23.4	14.8	34.0	6.5	87	64	93.1
Oct.	1018.0	13.5	17.5	10.3	26.5	−2.8	89	69	111.6
Nov.	1016.4	7.8	11.2	5.2	21.6	−5.8	88	75	85.0
Dec.	1016.0	4.3	7.3	2.0	14.5	−7.7	89	80	83.4
Annual	*	13.1	17.3	9.5	–	–	*	*	1055.9

Month	Number of days with						Mean daily cloud. precip.	Mean sunshine (h)	Mean global radiat. (cal. cm^{-2} per day)	Preval. wind direction		Monthly sunshine (h)	
	precip. ⩾1.0 mm	thunder- storm with (tenths)	fog							06h	15h	max.	min.
			03h	06h	12h	18h							
Jan.	5.4	0.1	4.2	4.3	3.9	3.2	5.3	2.9	95	NE	NE	124.0	52.7
Feb.	6.5	–	4.6	5.2	3.2	3.1	5.4	3.6	146	NE	NE	161.3	45.6
Mar.	9.6	0.1	2.1	3.3	0.4	0.7	5.7	5.0	244	NE	E	227.2	85.8
Apr.	9.3	1.3	0.2	1.2	0.2	0.2	5.9	6.1	295	NE	S	220.5	131.7
May	9.4	3.0	0.2	0.7	–	–	5.5	8.0	436	NE	S	291.4	211.4
June	8.6	5.4	0.2	0.1	–	–	5.1	8.4	457	NE	S	303.3	202.8
July	8.4	4.9	0.1	0.1	–	–	4.2	9.6	476	NE	SSE	347.2	256.1
Aug.	6.6	4.5	0.1	0.3	–	–	3.6	8.7	408	NE	SE	326.7	207.4
Sept.	7.3	2.5	0.8	1.9	0.3	0.1	4.6	6.9	306	NNE	SE	261.0	171.0
Oct.	8.1	1.7	1.1	2.6	0.5	0.4	5.3	5.3	206	NNE	E	243.4	108.2
Nov.	7.2	0.1	3.2	3.9	1.9	1.2	5.5	2.5	99	NNE	NE	101.1	47.4
Dec.	7.9	0.1	4.9	5.3	3.9	4.1	5.8	2.7	81	NNE	NE	110.7	58.3
Annual	94.2	*	*	*	*	*	5.2	**	**	*	*	**	**

*Venezia Lido, 45°26′N 12°23′E; 17 m; 1946–1960.
**Venezia Tessera, 45°30′N 12°20′E; 6 m; 1958–1969.

TABLE V

CLIMATIC TABLE FOR MILAN (1901–1930)

Latitude 45°28'N, longitude 09°11'E, elevation 147 m

Month	Mean sta. press. (mbar)	Temperature (°C) daily mean	mean max.	mean min.	extreme max.	min.	Relative humid. (%) 06h	12h	Mean precip. (mm)	Temperature (°C) mean max.	mean min.	extreme max.	min.	Mean precip. (mm)	Days with snow or ice 06h
Jan.	1017.8	2.0	5.0	−0.5	16.4	−10.2	92	83	55.4	4.2	−2.0	17.6	−15.0	53	17.9
Feb.	1016.5	4.0	7.7	0.8	22.0	−14.1	89	72	62.0	8.1	−0.8	20.8	−15.6	53	12.4
Mar.	1016.0	8.6	13.0	4.6	25.0	−3.3	89	65	93.0	13.2	3.4	24.3	−7.0	73	2.9
Apr.	1014.4	12.7	17.9	8.0	28.9	−0.5	88	58	93.9	18.4	7.7	29.8	−2.4	78	0.2
May	1014.5	17.9	23.6	12.7	35.0	3.8	87	58	83.3	22.7	11.9	31.7	0.8	86	−
June	1015.1	21.7	27.8	16.2	35.7	6.3	86	57	78.7	26.6	15.8	35.2	5.6	90	−
July	1014.7	24.1	30.4	18.4	38.3	11.3	86	55	62.8	29.0	17.6	36.4	10.0	56	−
Aug.	1014.5	23.5	29.3	17.9	36.8	10.3	90	57	67.3	28.1	16.9	35.8	9.2	71	−
Sept.	1017.0	19.1	24.1	14.6	35.1	6.3	93	61	75.2	24.6	13.8	32.2	4.2	68	−
Oct.	1018.1	13.1	17.3	9.6	26.6	−2.1	95	73	112.5	17.7	8.8	26.7	−1.3	107	−
Nov.	1016.1	7.1	10.3	4.4	19.7	−6.2	95	81	92.5	10.7	4.3	20.6	−6.2	118	1.5
Dec.	1016.3	3.8	6.3	1.3	17.4	−7.0	95	87	85.5	5.2	−0.2	16.6	−13.6	74	11.2
Annual	*	13.1	17.7	9.0	−	−	*	*	962.2	*	*	*	*	*	*

Month	Number of days with precip. ≥1.0 mm	thunder-storm with precip.	fog 03h	06h	12h	18h	Mean daily cloud. (tenths)	Mean sunshine (h)	Mean global radiat. (cal. cm⁻² per day)	Preval. wind direction 06h	15h	Monthly sunshine (h) max.	min.
Jan.	5.5	−	17.1	18.1	16.5	14.1	6.5	2.0	75	W	SW	105.7	29.1
Feb.	6.3	0.1	9.5	11.6	7.6	6.6	5.9	3.4	133	W	SW/W	177.8	30.0
Mar.	9.1	0.3	3.8	7.0	1.5	1.5	5.9	5.1	223	NE	E	269.4	66.7
Apr.	9.7	1.6	0.6	2.1	0.1	0.1	6.1	6.2	320	E	SW	227.1	161.1
May	8.7	1.7	0.2	1.0	−	0.1	5.7	7.2	384	E	S/SW	260.4	173.3
June	8.2	5.5	0.2	0.3	−	−	5.2	8.0	426	E	SW	274.8	183.9
July	6.0	5.2	0.3	0.9	−	−	4.3	9.1	433	ENE	E	310.0	217.0
Aug.	5.2	4.2	0.4	1.3	−	−	4.4	8.2	370	NNE	E	316.8	192.2
Sept.	6.2	2.5	1.8	6.1	0.1	0.1	5.1	6.0	279	E	SW	236.7	123.7
Oct.	9.0	0.7	8.9	13.7	3.1	2.9	6.1	3.9	166	ENE/E	SW	165.5	71.6
Nov.	8.6	0.5	10.7	12.7	18.2	7.2	6.8	1.7	79	E	SW	72.0	35.7
Dec.	8.6	−	16.2	18.2	17.9	15.7	7.1	1.5	60	WSW	SW	84.0	9.6
Annual	91.6	*	*	*	*	*	5.8	**	**	*		**	**

*Linate, 45°26'N 09°17'E; 103 m; 1946–1965..
**Linate, 1958–1969.

TABLE VI

CLIMATIC TABLE FOR BOLOGNA (1901–1930)

Latitude 44°30'N, longitude 11°21'E, elevation 84 m

Month	Mean sta. press. (mbar)	Temperature (°C) daily mean	mean max.	mean min.	extreme max.	min.	Precipitation mean (mm)	days with ≥1 mm	Mean daily cloud. (tenths)
Jan.		2.4	4.7	0.7	16.1	−8.6	40.7	5.3	5.4
Feb.		4.1	6.8	2.0	19.6	−11.5	35.4	4.5	5.0
Mar.		8.9	12.3	6.1	22.2	−3.7	47.1	6.7	5.0
Apr.		12.9	16.4	9.7	25.9	−0.3	54.4	7.4	4.9
May		17.9	21.6	14.3	31.4	3.0	51.2	6.5	4.2
June		21.9	26.0	18.0	34.6	8.8	48.0	5.5	3.7
July		24.6	28.8	20.6	37.0	12.8	31.5	3.4	2.6
Aug.		24.2	28.5	20.4	37.1	12.0	27.8	3.3	2.3
Sept.		19.8	23.4	16.7	33.0	8.1	59.8	6.7	3.7
Oct.		14.3	17.3	11.9	26.4	−1.6	73.9	7.4	4.9
Nov.		8.1	10.4	6.3	21.4	−7.1	70.2	7.3	5.7
Dec.		4.4	6.5	2.8	15.0	−5.9	49.1	6.2	6.2
Annual		13.6	16.9	10.8	−	−	589.0	70.3	4.5

Month	Mean daily sunshine (h)	Daily global radiation (cal. cm⁻² day⁻¹)	Max. sunshine hours	Min. sunshine hours
Jan.	2.8	111	135.2	57.7
Feb.	3.6	165	149.8	54.8
Mar.	4.7	247	228.5	84.6
Apr.	6.2	376	216.9	156.0
May	7.7	437	297.6	176.1
June	8.6	472	300.0	213.1
July	9.6	478	343.5	250.8
Aug.	8.6	416	324.0	220.1
Sept.	7.0	316	270.0	174.3
Oct.	4.8	205	215.8	99.8
Nov.	2.0	100	87.6	36.6
Dec.	2.0	83	97.3	35.0
Annual	*	*	*	*

*44°32'N 11°18'E; 49 m; 1958–1969.

TABLE VII

CLIMATIC TABLE FOR GENOA (1901–1930)

Latitude 44°25′N, longitude 08°55′E, elevation 54 m

Month	Mean sta. press. (mbar)	Temperature (°C)					Relative humid. (%)		Mean precip. (mm)	Temperature (°C)				Mean precip. (mm)	Days with snow or ice 06h
		daily mean	mean max.	mean min.	extreme max.	extreme min.	06h	12h		mean max.	mean min.	extreme max.	extreme min.		
Jan.	1015.4	7.9	10.3	5.9	18.7	−5.2	66	61	98.6	10.4	4.9	18.8	−6.0	93	2.6
Feb.	1014.1	8.3	11.1	6.1	19.9	−8.0	67	62	107.5	11.4	5.5	22.2	−6.0	87	1.8
Mar.	1014.6	11.0	13.7	8.7	23.7	−1.0	70	65	143.9	13.9	8.0	22.4	−1.0	94	0.2
Apr.	1013.8	13.7	16.4	11.3	26.7	3.1	72	66	93.8	17.3	11.1	27.2	3.4	99	−
May	1014.2	17.5	20.4	15.0	30.2	7.3	73	67	76.6	20.7	14.5	33.1	6.8	84	−
June	1015.1	21.3	24.1	18.7	32.6	11.0	72	68	71.9	24.3	18.0	32.7	11.2	54	−
July	1014.5	23.7	26.6	21.1	36.9	14.3	69	64	48.5	27.1	20.7	37.8	14.0	38	−
Aug.	1014.2	24.1	27.2	21.4	35.0	11.9	72	65	52.6	27.0	20.5	34.4	14.6	51	−
Sept.	1016.0	21.2	24.3	18.7	34.0	11.0	72	65	110.4	24.7	18.2	36.0	10.4	109	−
Oct.	1016.4	16.8	19.5	14.5	29.2	3.3	71	63	181.2	20.1	14.1	28.2	3.8	170	−
Nov.	1014.4	12.0	14.4	9.9	23.7	−1.0	71	65	173.8	15.2	9.8	23.4	1.4	190	−
Dec.	1014.0	9.2	11.5	7.4	20.4	−2.8	68	64	135.7	11.7	6.5	19.8	−1.8	114	0.4
Annual	*	15.6	18.3	13.2	−	−	*	*	1294.5	*	*	*	*	*	*

Month	Number of days with						Mean daily cloud. (tenths)	Mean sun-shine (h)	Mean global radiat. (cal. cm⁻² per day)	Preval. wind direction		Monthly sun-shine (h)	
	precip. ≥1.0 mm	thunder-storm with precip.	fog 03h	06h	12h	18h				06h	15h	max.	min.
Jan.	6.8	0.3	0.1	0.1	−	0.3	5.2	4.1	109	N	N	157.2	92.4
Feb.	6.9	0.3	0.1	0.1	0.2	0.2	5.0	4.5	156	N	N	185.4	75.1
Mar.	9.5	0.5	0.2	0.1	0.2	0.2	5.7	5.4	245	N	N	236.2	107.9
Apr.	9.1	1.1	0.2	−	−	−	6.0	6.3	329	N	S	236.4	155.7
May	7.2	1.4	−	−	0.1	−	5.6	7.6	406	N	S	269.1	191.9
June	5.9	2.7	−	−	−	−	5.0	8.4	445	N	S	302.4	199.2
July	3.2	2.3	−	−	−	−	4.0	9.6	460	N	S	317.1	273.7
Aug.	3.7	3.0	−	−	−	−	3.6	8.7	397	N	S	325.2	237.2
Sept.	6.1	2.4	−	−	−	−	4.5	6.8	296	N	S	244.8	170.1
Oct.	9.3	2.5	0.1	−	−	0.1	5.3	5.5	199	N	N	231.3	118.4
Nov.	8.9	1.6	0.1	−	0.1	−	5.6	3.5	113	N	N	172.2	77.4
Dec.	9.3	0.5	−	−	0.1	−	5.8	3.6	92	N	N	147.9	79.4
Annual	85.2	*	*	*	*	*	5.1	**	**	*	*	**	**

*44°25′N 08°51′E; 3 m; 1946–1965
**44°25′N 08°51′E; 3 m; 1946–1965

TABLE VIII

CLIMATIC TABLE FOR ALASSIO (1901–1930)

Latitude 44°00′N, longitude 08°10′E, elevation 33 m

Month	Mean sta. press. (mbar)	Temperature (°C)					Precipitation		Mean daily cloud. (tenths)
		daily mean	mean max.	mean min.	extreme max.	extreme min.	mean (mm)	days with ≥1 mm	
Jan.		9.3	11.7	7.1	21.5	−5.2	55.7	4.2	4.2
Feb.		9.7	12.0	7.3	26.0	−4.6	66.5	5.7	4.2
Mar.		11.9	14.2	9.4	29.5	0.2	80.7	7.3	4.6
Apr.		14.5	16.7	11.9	27.5	2.5	70.0	7.6	4.8
May		18.4	20.8	15.7	30.3	8.0	47.4	5.9	4.3
June		22.2	24.4	19.6	31.6	13.3	30.6	4.3	3.9
July		24.5	26.7	22.0	35.2	15.1	24.4	2.5	2.8
Aug.		25.0	27.1	22.5	34.6	14.2	24.8	2.6	2.8
Sept.		22.3	24.3	19.9	33.0	11.2	52.4	4.8	3.5
Oct.		17.9	20.0	15.6	29.9	6.8	100.8	8.0	4.4
Nov.		13.3	15.6	11.1	24.2	0.4	95.1	7.2	4.5
Dec.		10.5	13.0	8.4	22.5	−2.2	81.5	7.4	4.6
Annual		16.6	18.9	14.2	−	−	729.8	67.4	4.0

Month	Mean daily sunshine (h)	Daily global radiation (cal. cm⁻² day⁻¹)	Max. sun-shine hours	Min. sun-shine hours
Jan.	4.4	145	178.3	104.8
Feb.	5.0	198	198.2	72.8
Mar.	5.5	284	246.1	109.7
Apr.	6.7	375	268.8	150.9
May	7.7	443	278.7	209.3
June	8.7	489	307.2	224.4
July	9.8	502	323.6	280.9
Aug.	8.8	443	324.6	241.8
Sept.	6.9	340	249.6	156.1
Oct.	5.9	246	228.8	140.4
Nov.	4.0	151	147.6	70.4
Dec.	4.1	124	158.1	101.7
Annual	*	*	*	*

*Capo Mele, 43°57′N 08°10′E; 221 m; 1958–1969.

TABLE IX

CLIMATIC TABLE FOR FLORENCE (1901–1930)

Latitude 43°46′N, longitude 11°15′E, elevation 76 m

Month	Mean sta. press. (mbar)	Temperature (°C) daily mean	mean max.	mean min.	extreme max.	extreme min.	Relative humid. (%) 06h	12h	Mean precip. (mm)	Temperature (°C) mean max.	mean min.	extreme max.	extreme min.	Mean precip. (mm)	Days with snow or ice 06h
Jan.	1016.0	5.1	8.7	2.1	17.0	−6.6	84	70	60.5	9.4	1.7	18.7	−10.0	89	4.4
Feb.	1015.1	6.3	10.5	2.7	19.0	−10.3	84	65	57.9	11.4	2.2	21.4	−11.4	78	3.9
Mar.	1015.2	9.6	14.3	5.4	24.2	−3.8	85	60	72.3	14.9	4.8	24.4	−11.6	75	1.1
Apr.	1015.0	13.1	18.0	8.3	28.9	−1.5	88	57	66.5	19.4	7.9	30.6	−2.6	76	0.2
May	1014.7	17.7	23.3	12.1	33.0	3.3	88	54	62.2	23.4	11.6	32.8	1.0	64	–
June	1015.7	21.6	27.5	15.6	37.0	8.2	86	51	69.7	27.9	14.9	37.2	6.8	53	–
July	1015.1	24.2	30.3	18.0	38.2	10.4	84	45	25.5	31.0	17.1	40.2	8.0	32	–
Aug.	1015.0	24.1	30.2	18.0	38.8	9.8	87	47	35.5	30.7	16.7	40.6	9.2	34	–
Sept.	1016.9	20.2	25.6	15.0	34.5	5.2	88	53	68.4	26.9	14.5	38.2	5.0	83	–
Oct.	1017.1	15.1	19.6	11.0	28.6	−0.4	88	62	97.0	20.7	10.4	30.7	−0.6	113	–
Nov.	1015.3	9.8	13.6	6.5	23.2	−4.7	89	73	101.4	14.7	6.7	24.0	−5.6	123	0.8
Dec.	1014.7	5.7	10.0	3.8	18.6	−8.0	88	75	77.8	10.2	3.1	19.0	−7.0	98	3.0
Annual	*	14.4	19.3	9.8	–	–	–	–	794.6	*	*	*	*	*	*

Month	Number of days with precip. ≥ 1.0 mm	thunder-storm with precip.	fog 03h	06h	12h	18h	Mean daily cloud. (tenths)	Mean sun-shine (h)	Mean global radiat. (cal. cm⁻² per day)	Preval. wind direction 06h	15h
Jan.	7.0	0.1	6.6	7.2	6.2	5.3	5.7			N	N
Feb.	7.1	0.2	3.3	4.1	2.5	1.4	5.4			N	N
Mar.	9.0	0.2	0.9	3.1	0.7	0.4	5.8			NNE	N
Apr.	8.1	0.6	0.3	2.1	–	0.1	6.1			N	N,W
May	7.4	1.5	0.5	1.3	–	–	5.2			N	W
June	6.6	1.5	0.1	0.6	0.1	–	4.2			var	W
July	2.9	1.4	0.1	0.4	–	–	3.0			var	W
Aug.	3.3	1.5	0.2	0.5	–	–	2.6			var	W
Sept.	5.3	1.3	0.5	2.0	–	–	4.0			N	W
Oct.	8.5	0.9	2.7	6.8	0.3	0.3	5.3			N	N
Nov.	9.3	1.0	5.6	7.2	2.8	2.3	5.8			N	N
Dec.	8.8	0.2	7.7	8.7	6.6	5.4	6.4			N	N
Annual	83.1	*	*	*	*	*	4.9			*	*

*Peretola, 43°48′N 11°12′E; 38 m; 1946–1965.

TABLE X

CLIMATIC TABLE FOR PISA (1901–1930)

Latitude 43°43′N, longitude 10°24′E, elevation 9 m

Month	Mean sta. press. (mbar)	Temperature (°C) daily mean	mean max.	mean min.	extreme max.	extreme min.	Relative humid. (%) 06h	12h	Mean precip. (mm)	Temperature (°C) mean max.	mean min.	extreme max.	extreme min.	Mean precip. (mm)	Days with snow or ice 06h
Jan.	1015.3	6.2	11.6	2.2	19.0	−7.0	86	70	91.6	10.8	3.0	19.0	−6.8	86	5.7**
Feb.	1014.0	7.0	12.7	2.7	20.0	−10.7	85	66	78.0	11.7	2.9	19.8	−11.2	79	4.2
Mar.	1014.2	10.0	15.5	5.2	23.8	−3.8	89	65	109.8	14.6	5.4	22.9	−4.4	75	1.1
Apr.	1013.4	13.0	18.3	8.0	28.2	−1.2	90	60	83.4	18.3	8.0	26.9	−1.8	70	–
May	1014.6	16.8	22.5	11.3	31.6	2.4	89	59	71.4	22.1	11.3	31.6	1.2	57	–
June	1015.3	20.3	26.1	14.6	35.2	7.8	87	59	72.5	25.7	15.0	34.4	6.2	37	–
July	1014.8	22.6	28.7	16.5	38.0	9.5	86	54	31.1	28.6	17.3	37.0	11.0	19	–
Aug.	1014.6	22.8	29.1	16.5	37.3	8.2	86	55	40.9	28.3	17.2	36.0	10.0	38	–
Sept.	1016.3	19.7	26.0	14.1	36.2	4.4	86	58	87.1	25.6	15.1	32.6	6.2	90	–
Oct.	1016.4	15.4	21.3	10.4	29.3	−0.7	87	64	138.6	20.6	11.1	29.5	1.5	131	–
Nov.	1014.6	10.8	16.2	6.6	25.8	−4.3	88	72	138.9	15.5	7.7	22.6	−4.0	111	0.3
Dec.	1014.1	7.8	12.8	4.1	20.6	−7.6	89	77	105.1	12.1	4.6	19.4	−5.3	126	1.9
Annual	*	14.4	20.1	9.4	–	–	*	*	1048.4	*	*	*	*	*	*

Month	Number of days with precip. ≥ 1.0 mm	thunder-storm with precip.	fog 03h	06h	12h	18h	Mean daily cloud. (oktas)	Mean sun-shine (h)	Mean global radiat. (cal. cm⁻² per day)	Preval. wind direction 06h	15h	Monthly sun-shine (h) max.	min.
Jan.	8.6	0.7	0.8	1.0	1.0	0.5	5.1	4.0	121	E	E	176.1	80.9
Feb.	8.1	1.7	1.2	1.7	0.5	0.3	5.1	4.5	171	E	E	185.4	72.5
Mar.	10.5	1.0	1.8	2.7	0.1	0.1	5.3	5.2	246	E	WSW	229.4	101.4
Apr.	9.1	1.6	0.9	1.2	–	–	5.4	6.8	341	E	W	228.0	186.6
May	6.9	2.3	1.2	1.4	0.1	–	4.6	8.8	417	E	W	284.0	230.3
June	5.5	2.4	1.0	0.8	–	–	3.8	9.3	459	E	W	322.2	237.9
July	2.9	1.2	0.4	0.6	–	0.1	2.8	10.7	472	E	W	365.8	236.1
Aug.	3.5	2.3	0.4	0.6	–	–	2.5	9.4	402	E	W	338.3	235.9
Sept.	6.8	2.1	0.7	0.9	–	–	3.9	7.5	325	E	W	288.0	178.8
Oct.	10.3	3.2	0.6	1.4	0.6	0.1	4.9	6.0	220	E	E	227.5	132.4
Nov.	10.9	2.0	1.8	1.6	0.5	0.2	5.4	3.5	125	E	E	133.2	79.2
Dec.	11.0	1.4	2.0	1.9	1.2	1.2	5.8	3.0	93	E	E	114.7	78.7
Annual	94.3	*	*	*	*	*	4.6	***	***	*	*	***	***

*43°40′N 10°23′E; 1 m; 1951–1965.
**14 years.
***43°40′N 10°23′E; 1 m; 1958–1969.

TABLE XI

CLIMATIC TABLE FOR ANCONA (1901–1930)

Latitude 43°37′N, longitude 13°31′E, elevation 17 m

Month	Mean sta. press. (mbar)	Temperature (°C)					Relative humid. (%)		Mean precip. (mm)	Temperature (°C)				Mean precip. (mm)	Days with snow
		daily mean	mean max.	mean min.	extreme		06h	12h		mean max.	mean min.	extreme			
					max.	min.						max.	min.		
Jan.	1016.4	5.3	7.9	2.4	18.0	−7.6	80	75	65.3	8.2	3.2	19.8	−6.5	73	2.6
Feb.	1015.0	6.2	9.2	3.3	20.6	−7.5	77	71	43.4	9.4	4.0	19.6	−7.2	50	3.0
Mar.	1015.2	9.8	12.9	6.6	21.0	−2.5	77	70	40.4	12.1	6.6	23.6	−2.2	56	0.5
Apr.	1014.1	13.3	16.7	10.0	25.2	0.5	75	69	58.0	16.2	10.3	27.5	0.7	50	–
May	1014.2	17.3	21.1	13.9	31.0	4.2	75	70	53.6	20.0	14.1	28.6	5.5	50	–
June	1015.0	22.0	25.8	18.3	35.0	9.0	70	67	48.7	24.4	18.1	33.8	10.3	48	–
July	1014.4	24.4	28.3	20.1	36.0	8.3	67	63	37.1	27.3	20.9	35.0	13.4	36	–
Aug.	1014.2	24.3	28.1	20.1	39.0	9.2	69	63	38.4	27.3	20.9	36.5	14.5	37	–
Sept.	1016.6	20.5	24.0	16.8	34.2	6.5	75	67	89.1	24.0	18.1	34.5	11.6	68	–
Oct.	1017.1	16.0	19.2	12.7	27.0	4.3	80	72	95.2	18.7	13.5	26.6	5.4	80	–
Nov.	1015.2	11.6	14.4	8.3	26.0	0.8	80	75	63.8	14.0	9.2	27.0	1.0	78	–
Dec.	1014.9	7.9	10.5	5.0	19.0	−3.8	82	78	77.4	9.8	5.2	19.0	−3.3	79	0.8
Annual	*	14.9	18.2	11.5	–	–	*	*	710.5	*	*	*	*	*	*

Month	Number of days with							Mean daily cloud. (tenths)	Mean sun- shine (h)	Mean global radiat. (cal. cm⁻² per day)	Preval. wind direction		Monthly sun- shine (h)	
	precip. ≥1.0 mm	thunder- storm with precip.	fog								06h	15h	max.	min.
			03h	06h	12h	18h								
Jan.	8.0	–	6.4	5.7	6.3	5.1		7.1	2.4	100	W	NW	124.6	41.5
Feb.	6.4	0.1	4.0	4.6	4.7	4.6		7.0	3.7	171	W	W,NW	154.8	56.6
Mar.	7.5	–	2.9	3.6	3.1	2.3		6.2	4.7	268	W	N	208.6	87.4
Apr.	8.2	0.3	1.3	1.4	1.0	0.8		6.0	6.7	392	W	ESE	234.6	177.0
May	8.3	0.8	0.7	0.6	0.2	0.1		5.7	8.6	479	W	ESE	290.2	187.9
June	5.9	1.3	0.1	0.2	0.1	0.1		4.9	9.0	509	W	ESE	307.8	233.7
July	4.6	1.7	0.1	0.1	–	0.1		3.7	10.4	527	W	N	365.8	293.6
Aug.	4.4	2.0	–	0.1	–	0.1		3.4	9.5	466	W	E	349.4	260.4
Sept.	8.2	1.1	0.1	0.6	0.3	0.1		5.0	7.1	354	W	N	270.0	172.2
Oct.	8.9	0.9	0.6	1.8	1.4	0.8		6.2	5.2	232	W	N	207.7	83.7
Nov.	9.2	0.3	2.8	3.2	2.8	2.1		6.8	2.5	116	W	NW	108.9	51.0
Dec.	10.0	0.2	5.2	5.1	6.9	6.2		7.4	2.1	85	W	WNW	90.8	32.2
Annual	89.5	*	*	*	*	*		5.8	**	**	*	*	**	**

*Ancona Monte Cappuccini, 43°37′N 13°31′E; 105 m; 1958–1969.
**Ancona Monte Cappuccini, idem, 1958–1969

TABLE XII

CLIMATIC TABLE FOR CHIETI (1901–1930)

Latitude 42°21′N, longitude 14°10′E, elevation 341 m

Month	Mean sta. press. (mbar)	Temperature (°C)					Precipitation		Mean daily cloud. (tenths)
		daily mean	mean max.	mean min.	extreme		mean (mm)	days with ≥1 mm	
					max.	min.			
Jan.		5.1	8.6	2.0	21.0	−13.0	103.2	9.7	5.4
Feb.		5.4	9.3	2.1	20.3	−10.2	81.7	8.8	5.4
Mar.		8.2	12.5	4.8	24.0	−4.0	69.1	7.8	4.9
Apr.		11.5	15.7	7.8	25.0	−2.4	77.0	7.6	4.9
May		16.1	20.3	12.2	29.3	1.6	73.0	6.6	4.1
June		20.2	24.3	15.9	32.0	7.0	71.0	6.3	3.7
July		23.1	27.1	18.8	38.2	10.0	46.7	4.3	2.1
Aug.		23.1	27.0	19.1	40.8	8.8	52.6	4.2	2.1
Sept.		19.1	23.1	15.5	34.6	2.3	85.7	6.6	3.8
Oct.		14.4	18.3	11.2	28.8	2.0	109.3	9.1	4.9
Nov.		9.9	13.6	6.7	23.5	−4.4	119.9	9.9	5.9
Dec.		6.7	10.5	3.7	21.4	−5.6	106.8	10.4	5.7
Annual		13.6	17.5	10.0	–	–	995.9	91.4	4.4

Month	Mean daily sunshine (h)	Daily global radiation (cal. cm⁻² day⁻¹)	Max. sun- shine hours	Min. sun- shine hours
Jan.	3.2	112	128.7	59.5
Feb.	4.3	175	203.2	52.6
Mar.	4.8	251	243.0	91.5
Apr.	6.6	357	246.0	162.0
May	8.2	433	297.6	207.7
June	8.7	455	300.6	228.0
July	10.2	475	350.3	302.9
Aug.	9.5	416	348.8	261.3
Sept.	7.4	316	294.3	175.2
Oct.	5.7	218	233.4	116.6
Nov.	3.6	129	164.4	33.0
Dec.	2.7	96	129.0	43.4
Annual	*	*	*	*

*Pescara, 42°26′N 14°12′E; 16 m; 1958–1969.

TABLE XIII

CLIMATIC TABLE FOR ROME (Collegio Romano) (1901–1930)

Latitude 41°54′N, longitude 12°29′E, elevation 46 m

Month	Mean sta. press. (mbar)	Temperature (°C)					Relative humid. (%)		Mean precip. (mm)	Temperature (°C)				Mean precip. (mm)	Days with snow or ice 06h
		daily mean	mean max.	mean min.	extreme		06h	12h		mean max.	mean min.	extreme			
					max.	min.						max.	min.		
Jan.	1015.4	6.9	11.1	3.7	18.1	−5.0	85	68	75.9	12.3	3.4	20.8	−7.4	69	5.8
Feb.	1014.1	7.7	12.2	4.4	20.7	−5.4	84	62	88.3	13.3	3.5	23.0	−7.4	56	3.6
Mar.	1014.4	10.8	15.4	6.8	25.3	−1.2	84	58	77.2	16.1	5.5	25.0	−6.5	57	0.8
Apr.	1013.8	13.9	18.5	9.4	29.8	0.3	84	56	72.3	19.2	8.2	29.3	−2.4	57	−
May	1014.1	18.1	23.4	13.0	32.8	2.1	80	54	62.8	23.4	11.8	33.8	1.8	57	−
June	1015.7	22.1	27.4	16.6	34.9	9.2	76	48	48.0	27.8	15.3	36.7	5.6	30	−
July	1014.5	24.7	30.3	18.9	40.1	11.9	72	43	14.1	30.9	17.7	38.6	10.6	21	−
Aug.	1014.4	24.5	30.3	18.9	39.2	13.2	75	43	21.5	31.1	17.7	40.4	9.3	19	−
Sept.	1016.2	21.1	26.4	16.4	34.2	8.0	82	50	69.9	27.8	15.4	40.0	4.3	76	−
Oct.	1016.1	16.4	21.0	12.6	28.2	2.1	86	58	128.0	22.5	11.2	32.0	2.3	101	−
Nov.	1014.0	11.7	15.7	8.0	24.6	−2.4	86	67	116.3	17.2	8.1	24.4	−4.8	111	0.2
Dec.	1014.0	8.5	12.3	5.7	19.3	−5.0	87	70	106.4	13.5	4.7	21.2	−6.4	97	2.1
Annual	*	15.6	20.4	11.2	−	−	*	*	880.6	*	*	*	*	*	*

Month	Number of days with						Mean daily cloud. (tenths)	Mean sun-shine (h)	Mean global radiat. (cal. cm⁻² per day)	Preval. wind direction		Monthly sun-shine (h)	
	precip. ⩾ 1.0 mm	thunder-storm with precip.	fog							06h	15h	max.	min.
			03h	06h	12h	18h							
Jan.	6.3	0.9	0.5	0.6	0.2	0.1	5.3	4.3	143	NE	S	151.0	100.4
Feb.	6.2	1.2	1.0	0.9	0.2	0.1	5.4	4.7	198	NE	S	194.0	95.1
Mar.	7.6	0.9	0.8	1.2	0.1	−	5.4	6.6	286	NE	W	252.3	124.0
Apr.	8.3	2.1	1.3	1.6	−	−	5.6	7.0	387	S	WSW	238.5	172.2
May	6.1	1.9	0.7	1.0	0.1	−	4.8	8.6	477	NE	W	328.6	204.6
June	4.1	1.6	0.1	0.2	−	−	3.7	9.4	517	S	W	308.1	240.9
July	1.6	1.8	0.1	0.2	−	−	2.2	10.8	532	NE,S	W	356.5	312.5
Aug.	2.1	1.6	−	0.2	−	−	1.9	9.9	465	NE,S	W	352.5	213.6
Sept.	5.1	2.9	0.4	0.7	−	−	3.7	8.1	355	ENE	W	291.9	200.1
Oct.	5.6	3.6	0.9	0.8	0.1	0.1	4.9	6.4	252	NE	W	251.7	127.7
Nov.	7.6	2.6	0.5	0.4	−	−	5.7	4.1	149	NE	S	157.2	99.0
Dec.	7.9	2.2	0.9	0.8	0.2	0.1	5.8	3.3	114	NE	S	134.5	80.3
Annual	68.6	*	*	*	*	*	4.5	**	**	*	*	**	**

*Ciampino, 41°48′N 12°35′E; 101 m; 1946–1965.
**Ciampino, 41°48′N 12°35′E; 131 m; 1958–1969.

TABLE XIV

CLIMATIC TABLE FOR ISERNIA (1901–1930)

Latitude 41°35′N, longitude 14°14′E, elevation 402 m

Month	Mean sta. press. (mbar)	Temperature (°C)					Precipitation		Mean daily cloud (tenths)
		daily mean	mean max.	mean min.	extreme		mean (mm)	days with ⩾ 1 mm	
					max.	min.			
Jan.		5.4	9.2	2.3	17.0	−8.0	100.2	8.8	5.7
Feb.		5.8	10.0	2.5	21.0	−10.0	97.3	8.6	5.8
Mar.		8.4	13.4	4.5	23.4	−4.0	89.0	9.5	5.7
Apr.		11.3	16.3	7.0	26.0	−1.0	90.3	8.7	5.9
May		15.3	20.5	10.6	30.4	1.0	81.9	9.2	5.1
June		19.1	24.4	14.0	32.1	7.0	76.5	7.5	4.3
July		21.7	27.5	16.3	34.8	9.0	40.4	3.8	2.7
Aug.		22.1	28.1	16.5	37.0	10.0	36.9	3.1	2.1
Sept.		18.8	24.5	14.2	32.0	5.5	96.3	6.8	4.1
Oct.		14.3	18.7	10.3	28.4	−1.0	173.8	11.0	5.5
Nov.		9.8	13.9	6.4	22.4	−6.3	128.6	9.3	5.9
Dec.		6.8	10.3	3.7	17.8	−5.0	137.5	11.4	6.2
Annual		13.2	18.1	9.0	−	−	1148.7	97.5	4.9

TABLE XV

CLIMATIC TABLE FOR NAPLES (1901–1930)

Latitude 40°51′N, longitude 14°15′E, elevation 25 m

Month	Mean sta. press. (mbar)	Temperature (°C) daily mean	mean max.	mean min.	extreme max.	extreme min.	Relative humid. (%) 06h	12h	Mean precip. (mm)	Temperature (°C) mean max.	mean min.	extreme max.	extreme min.	Mean precip. (mm)	Days with snow or ice 06h
Jan.	1015.3	9.0	12.2	6.6	18.3	−3.9	80	66	93.0	12.2	5.2	20.3	−4.4	111	0.8
Feb.	1014.6	9.6	12.9	6.9	21.5	−3.2	80	65	81.8	13.0	5.1	20.2	−4.5	86	0.2
Mar.	1014.7	12.0	15.4	8.7	24.7	0.9	80	62	74.8	15.2	6.8	25.0	−4.0	71	0.3
Apr.	1014.3	14.6	18.3	11.3	27.5	3.9	82	61	66.9	18.7	9.3	27.8	1.3	59	–
May	1014.2	18.7	22.7	15.0	32.2	8.2	80	61	45.2	22.5	12.5	32.5	1.0	50	–
June	1015.4	22.2	26.3	18.4	35.2	11.5	77	57	45.7	26.6	16.2	34.8	7.1	24	–
July	1014.6	24.8	29.0	20.9	37.9	13.8	76	53	16.0	29.4	18.4	38.2	11.0	17	–
Aug.	1014.6	25.0	29.3	21.1	36.6	14.0	78	53	18.5	29.6	18.6	38.7	11.8	27	–
Sept.	1016.4	22.1	26.0	18.7	33.8	10.6	82	56	71.2	26.5	16.3	37.6	8.0	86	–
Oct.	1016.4	18.3	21.8	15.3	29.6	7.9	82	60	130.2	21.9	12.4	30.8	3.1	120	–
Nov.	1014.9	13.9	17.1	11.3	24.9	0.6	83	66	114.4	17.2	9.3	25.2	−1.8	136	0.2
Dec.	1014.3	10.9	14.0	8.6	20.8	−1.6	82	69	136.7	13.7	6.5	20.6	−3.6	132	0.4
Annual	*	16.8	20.4	13.6	–	–	*	*	894.5	*	*	*	*	*	*

Month	Number of days with precip. ⩾ 1.0 mm	thunder-storm with precip.	fog 03h	06h	12h	18h	Mean daily cloud. (tenths)	Mean sun-shine (h)	Mean global radiat. (cal. cm⁻² per day)	Preval. wind direction 06h	15h	Monthly sun-shine (h) max.	min.
Jan.	10.3	2.0	0.4	0.4	0.2	0.2	5.4	3.8	118	N	NE	156.6	87.4
Feb.	10.5	1.9	0.5	0.8	0.3	0.1	5.6	4.5	166	N	S	190.4	96.6
Mar.	10.3	1.6	0.6	1.2	0.1	–	5.3	5.2	226	NNE	SSW	238.1	111.9
Apr.	8.6	2.2	0.6	0.9	0.2	–	5.4	6.6	315	NNE	S	250.5	162.0
May	6.5	2.2	0.5	0.9	0.1	0.1	4.6	8.2	397	N	S	294.2	205.8
June	5.4	1.7	0.3	0.2	–	–	3.7	9.3	441	var.	S	307.2	260.4
July	1.9	1.5	0.2	0.4	–	–	2.4	10.4	447	N	S	344.1	287.4
Aug.	2.5	1.8	0.1	0.7	–	–	2.4	9.9	392	N	S	348.8	271.9
Sept.	6.8	2.9	0.5	0.4	0.2	0.1	3.8	8.1	302	N	S	297.3	203.1
Oct.	9.9	3.7	0.2	0.2	0.1	–	4.8	6.4	219	N	SSW	226.3	138.0
Nov.	10.6	3.5	0.3	0.6	0.2	0.2	5.7	4.1	131	N	S	155.7	96.0
Dec.	12.9	3.1	0.4	0.4	–	–	6.0	3.0	97	N	NE	116.6	67.9
Annual	96.0	*	*	*	*	*	4.6	**	**	*	*	**	**

*Capodichino, 40°53′N 14°18′E; 78 m; 1946–1965.
**Capodichino, 40°51′N 14°18′E; 72 m; 1958–1969.

TABLE XVI

CLIMATIC TABLE FOR POTENZA (1901–1930)

Latitude 40°38′N, longitude 15°48′E, elevation 826 m

Month	Mean sta. press. (mbar)	Temperature (°C) daily mean	mean max.	mean min.	extreme max.	extreme min.	Precipitation mean (mm)	days with ⩾ 1 mm	Mean daily cloud. (tenths)
Jan.		3.2	6.1	0.4	16.8	−9.6	89.1	8.6	7.0
Feb.		3.6	7.0	0.6	17.4	−16.0	84.5	7.8	6.9
Mar.		6.3	10.4	2.7	21.4	−6.0	66.6	9.3	6.6
Apr.		9.5	14.0	5.4	26.0	−3.6	82.3	9.4	6.6
May		14.0	18.8	9.4	30.2	1.1	71.7	8.7	5.8
June		17.9	22.8	13.0	34.0	5.7	57.9	5.8	4.8
July		20.7	26.2	15.5	36.0	7.7	29.2	3.4	2.9
Aug.		20.8	26.5	15.8	37.9	9.3	33.8	3.4	2.8
Sept.		17.4	22.3	13.1	33.0	2.6	64.0	6.8	4.8
Oct.		12.8	16.8	9.2	27.8	−1.0	90.5	8.6	5.9
Nov.		8.0	11.3	5.0	22.6	−6.2	115.0	10.4	7.1
Dec.		5.1	7.8	2.4	18.2	−10.0	106.0	10.0	7.2
Annual		11.6	15.8	7.7	–	–	890.7	92.2	5.7

TABLE XVII

CLIMATIC TABLE FOR TARANTO (1901–1930)

Latitude 40°28′N, longitude 17°13′E, elevation 16 m

Month	Mean sta. press. (mbar)	Temperature (°C)					Precipitation		Mean daily cloud. (tenths)
		daily mean	mean max.	mean min.	extreme		mean (mm)	days with ≥ 1 mm	
					max.	min.			
Jan.		8.8	12.6	5.6	19.5	−3.0	52.1	6.3	5.7
Feb.		9.2	13.0	5.8	20.4	−4.3	41.5	6.1	5.8
Mar.		11.7	15.5	7.9	23.3	0.1	36.6	5.4	5.4
Apr.		14.4	18.2	10.3	26.2	1.1	35.1	5.5	5.4
May		18.6	22.8	14.2	30.7	7.0	31.2	3.7	4.4
June		22.7	27.0	17.9	33.5	12.0	18.4	2.9	3.7
July		25.4	30.1	20.7	37.2	14.3	16.5	1.5	1.9
Aug.		25.7	30.1	20.9	38.3	13.6	16.1	1.7	1.7
Sept.		22.8	26.9	18.4	35.9	8.0	34.7	4.2	3.4
Oct.		18.3	22.3	14.6	30.7	3,2	71.6	6.5	4.8
Nov.		13.8	17.4	10.5	25.5	−3.3	54.5	6.3	5.8
Dec.		10.7	14.2	7.4	19.5	−2.8	67.7	7.4	5.9
Annual		16.8	20.8	12.8	–	–	476.0	57.5	4.5

TABLE XVIII

CLIMATIC TABLE FOR LECCE (1901–1930)

Latitude 40°21′N, longitude 18°10′E, elevation 72 m

Month	Mean sta. press. (mbar)	Temperature (°C)					Relative humid. (%)		Mean precip. (mm)	Temperature (°C)				Mean precip. (mm)	Days with snow or ice 06h
		daily mean	mean max.	mean min.	extreme		06h	12h		mean max.	mean min.	extreme			
					max.	min.						max.	min.		
Jan.	1014.6	8.9	13.0	5.1	27.0	−4.5	82	70	66.0	12.7	6.3	19.8	−3.4	68	0.5
Feb.	1013.7	9.1	13.4	5.3	20.4	−3.4	81	65	60.7	13.4	6.2	23.0	−1.6	52	0.2
Mar.	1014.0	11.7	16.1	7.3	24.6	−3.0	82	65	49.0	15.3	7.7	26.7	−4.2	47	0.4
Apr.	1013.4	14.3	19.3	9.4	30.2	0.0	83	64	45.8	18.5	10.0	28.2	2.4	30	–
May	1013.2	18.5	24.0	13.2	33.8	5.0	81	63	31.1	22.3	13.6	35.0	5.1	35	–
June	1013.8	22.7	28.3	17.2	41.5	10.2	77	62	30.8	26.3	17.5	37.2	10.4	14	–
July	1012.8	25.7	30.6	19.8	41.2	12.0	75	61	13.5	28.9	20.2	41.2	12.4	12	–
Aug.	1012.8	25.1	30.4	20.0	41.8	12.1	76	59	21.3	29.1	20.4	39.5	12.0	20	–
Sept.	1015.4	22.2	27.1	17.5	41.0	9.0	81	63	50.1	26.3	17.9	39.6	11.0	33	–
Oct.	1016.1	17.9	22.3	13.9	35.2	3.0	84	66	86.4	21.7	14.4	29.4	5.5	86	–
Nov.	1014.9	13.7	17.6	10.0	26.5	−0.2	85	70	90.0	17.9	11.1	25.6	0.4	77	–
Dec.	1013.9	10.7	14.6	7.2	26.5	−5.5	84	71	95.0	14.6	8.2	22.5	−2.5	74	0.4
Annual	*	16.6	21.4	12.2	–	–	*	*	639.7	*	*	*	*	*	*

Month	Number of days with						Mean daily cloud. (tenths)	Mean sun-shine (h)	Mean global radiat. (cal. cm⁻² per day)	Preval. wind direction		Monthly sun-shine (h)	
	precip. ≥ 1.0 mm	thunder-storm with precip.	fog							06h	15h	max.	min.
			03h	06h	12h	18h							
Jan.	8.6	0.5	0.4	0.4	0.1	–	5.4	4.2	124	S	S	164.3	83.4
Feb.	7.6	0.6	0.5	0.7	–	0.1	5.2	5.1	183	S	N	203.3	86.0
Mar.	7.0	1.0	0.5	0.4	0.1	0.1	5.1	5.5	250	S	N	244.9	119.4
Apr.	6.5	1.6	0.5	0.7	–	0.1	4.8	7.1	354	NW	NNW	259.8	170.1
May	3.9	1.6	0.4	0.2	–	–	3.8	9.1	432	NW	NNW	312.2	221.0
June	3.2	1.3	0.3	0.1	–	–	2.9	10.0	465	NW	NNW	322.5	252.3
July	1.9	1.3	0.2	0.1	–	0.1	1.5	11.2	470	NNW	NNW	365.8	318.7
Aug.	1.8	1.6	0.3	0.4	–	–	1.4	10.4	424	NW	NNW	356.5	282.1
Sept.	4.4	1.9	0.7	0.6	–	–	3.0	8.3	324	NW	N	300.3	210.6
Oct.	8.0	3.0	0.5	0.7	0.1	–	4.4	6.6	227	S	N	245.5	172.7
Nov.	8.7	2.1	0.8	0.8	0.1	0.1	5.5	4.4	140	S	S	168.6	93.0
Dec.	9.5	1.4	0.4	0.8	0.1	0.2	5.7	3.5	104	S	S	136.1	69.8
Annual	71.1	*–	*	*	*	*	4.1	**	**	*	*	**	**

*Brindisi, 40°40′N 17°57′E; 22 m; 1946–1965.
**Brindisi, 40°39′N 17°57′E; 10 m; 1958–1969.

TABLE XIX

CLIMATIC TABLE FOR CAGLIARI (1901–1930)

Latitude 39°13′N, longitude 09°06′E, elevation 75 m

Month	Mean sta. press. (mbar)	Temperature (°C) daily mean	Temperature (°C) mean max.	Temperature (°C) mean min.	extreme max.	extreme min.	Relative humid. (%) 06h	Relative humid. (%) 12h	Mean precip. (mm)	Temperature (°C) mean max.	Temperature (°C) mean min.	extreme max.	extreme min.	Mean precip. (mm)	Days with snow or ice 06h
Jan.	1015.5	9.4	13.1	6.3	18.8	−3.9	86	70	49.2	13.9	5.5	21.8	−3.5	49	–
Feb.	1015.2	9.9	13.8	6.7	21.4	−1.4	86	67	45.3	14.1	5.7	21.5	−3.2	43	0.2
Mar.	1014.8	11.8	16.1	8.2	24.6	0.3	86	64	48.3	16.3	7.4	27.3	−2.2	38	–
Apr.	1015.2	14.2	18.8	10.1	27.0	0.0	85	63	41.7	18.2	9.2	28.0	1.4	30	–
May	1014.5	17.8	22.9	13.1	32.9	4.9	84	61	34.9	22.1	12.5	29.9	3.5	30	–
June	1015.8	21.7	27.0	16.7	34.7	10.1	81	58	15.8	26.3	16.2	38.4	8.4	11	–
July	1015.3	24.5	30.1	19.2	38.7	12.3	80	56	3.8	29.3	18.6	38.0	12.5	3	–
Aug.	1014.9	24.7	29.9	19.8	38.5	13.3	81	58	4.4	29.3	19.1	38.2	12.4	12	–
Sept.	1016.2	22.1	26.8	18.0	35.4	10.3	85	60	33.9	27.0	17.4	39.0	10.2	28	–
Oct.	1015.8	18.3	22.4	14.7	29.1	6.0	88	64	67.0	22.7	13.8	31.0	3.8	76	–
Nov.	1014.8	14.0	17.8	10.8	24.4	0.4	86	68	53.5	18.3	10.0	27.2	0.3	52	–
Dec.	1014.4	11.0	14.4	8.1	22.5	−1.1	86	71	55.7	15.1	7.1	21.5	−0.6	62	–
Annual	*	16.6	21.1	12.7	–	–	*	*	453.4	*	*	*	*	*	*

Month	Number of days with precip. ≥1.0 mm	thunder-storm with precip.	fog 03h	fog 06h	fog 12h	fog 18h	Mean daily cloud. (tenths)	Mean sun-shine (h)	Mean global radiat. (cal. cm⁻² per day)	Preval. wind direction 06h	Preval. wind direction 15h	Monthly sun-shine (h) max.	Monthly sun-shine (h) min.
Jan.	7.1	0.8	0.5	0.8	0.1	0.1	5.4	4.5	156	NNW	NNW	178.6	70.4
Feb.	6.9	0.7	0.6	0.8	0.1	0.2	5.5	4.7	212	NW	NW	189.6	78.0
Mar.	7.1	0.9	0.6	1.0	0.1	0.1	5.2	6.2	302	NW	S	251.7	130.2
Apr.	6.4	1.1	0.4	0.9	–	–	4.9	7.2	384	NNW	S	253.2	177.3
May	4.4	1.2	0.7	0.8	–	0.1	4.0	9.0	467	NW	S	340.1	215.5
June	2.5	0.8	0.2	0.2	–	–	2.9	9.5	484	NW	S	339.0	210.9
July	0.4	0.9	0.3	0.5	–	–	1.3	10.7	518	NW	S	368.9	288.3
Aug.	0.4	1.0	0.1	0.5	0.1	–	1.1	10.2	458	NW	S	359.9	275.3
Sept.	3.5	1.9	0.2	0.3	–	0.1	3.3	8.3	356	NW/NNW	S	297.0	195.9
Oct.	6.4	2.5	0.1	0.2	–	–	4.4	6.3	253	NNW	S	259.5	151.6
Nov.	7.6	1.6	0.3	0.4	–	–	5.4	4.3	161	NNW	NNW	186.9	67.2
Dec.	7.4	0.9	0.4	0.7	0.1	–	5.5	3.6	122	NNW	NNW	148.5	61.1
Annual	60.2	*	*	*	*	*	4.1	**	**	*	*	**	**

*Cagliari Elmas, 39°15′N 09°03′E; 18 m; 1946–1965.
**Cagliari Elmas, 1958–1969.

TABLE XX

CLIMATIC TABLE FOR MESSINA (1901–1930)

Latitude 38°12′N, longitude 15°33′E, elevation 54 m

Month	Mean sta. press. (mbar)	Temperature (°C) daily mean	Temperature (°C) mean max.	Temperature (°C) mean min.	extreme max.	extreme min.	Relative humid. (%) 06h	Relative humid. (%) 12h	Mean precip. (mm)	Temperature (°C) mean max.	Temperature (°C) mean min.	extreme max.	extreme min.	Mean precip. (mm)	Days with snow or ice 06h
Jan.	1015.5	11.5	14.0	8.5	19.3	0.0	88	65	95.6	15.5	5.6	24.0	−3.2	97	–
Feb.	1015.6	11.4	14.4	8.4	22.9	−2.4	84	59	91.9	16.4	5.4	26.6	−4.0	39	0.1
Mar.	1015.3	12.8	16.0	9.6	23.4	1.2	86	60	81.0	17.8	6.7	30.4	−3.5	46	0.1
Apr.	1014.4	14.9	18.5	11.5	27.9	4.8	85	60	66.9	20.1	8.5	29.0	−2.2	38	–
May	1014.6	18.4	22.7	14.8	30.8	6.5	80	60	41.2	23.8	11.9	38.0	3.0	22	–
June	1015.4	22.6	26.9	18.6	35.5	11.7	73	58	25.5	28.0	16.0	41.5	10.2	6	–
July	1014.4	25.4	30.0	21.5	39.8	15.2	71	57	10.3	30.9	18.5	46.0	11.2	3	–
Aug.	1014.4	25.9	30.4	22.2	37.8	14.4	75	56	24.3	31.6	19.0	42.6	13.0	9	–
Sept.	1016.0	23.4	27.3	19.9	40.5	11.6	82	57	58.0	29.1	17.5	39.4	11.4	36	–
Oct.	1016.5	19.7	23.0	16.5	34.9	7.5	87	62	119.6	24.4	13.9	32.8	5.5	139	–
Nov.	1015.7	15.9	18.6	13.1	25.8	0.9	87	61	121.6	20.7	10.2	30.3	−2.6	92	–
Dec.	1015.0	13.0	15.8	9.8	23.1	−0.2	87	65	117.2	17.0	7.0	24.9	−1.6	75	–
Annual	*	17.9	21.5	14.5	–		*	*	853.0	*	*	*	*	*	*

Month	Number of days with precip. ≥1.0 mm	thunder-storm with precip.	fog 03h	fog 06h	fog 12h	fog 18h	Mean daily cloud. (tenths)	Mean sun-shine (h)	Mean global radiat. (cal. cm⁻² per day)	Preval. wind direction 06h	Preval. wind direction 15h	Monthly sun-shine (h) max.	Monthly sun-shine (h) min.
Jan.	11.4	0.5	0.3	0.1	0.2	–	6.2	3.7	132	W	W	149.1	79.1
Feb.	10.4	0.2	0.2	0.1	0.1	0.1	6.1	4.9	198	W	W	191.0	81.2
Mar.	9.1	0.6	0.1	0.3	0.2	0.1	5.6	5.5	261	W	E	229.4	107.1
Apr.	7.6	0.6	0.2	0.2	0.1	0.1	5.3	7.0	358	W	E	243.3	177.0
May	4.1	0.6	0.2	0.1	–	–	4.2	8.3	428	WSW,W	E	303.5	193.8
June	3.2	0.6	0.1	–	–	–	3.2	9.5	486	W	E	323.4	233.5
July	1.6	0.5	–	–	–	–	1.9	10.6	465	WSW	ENE,E	350.3	304.1
Aug.	2.2	1.0	–	–	–	–	2.0	10.0	415	W	E	345.7	260.8
Sept.	5.2	1.5	–	–	–	0.1	3.7	7.9	316	W	E	267.0	201.6
Oct.	10.0	2.7	0.3	0.4	0.1	0.2	5.1	6.1	225	W	E	243.0	131.1
Nov.	12.1	1.3	0.1	0.1	0.2	–	6.1	4.3	146	W	W	167.3	78.0
Dec.	13.8	0.5	0.1	–	–	–	6.3	3.3	114	W	W	131.4	80.6
Annual	90.9	*	*	*	*	*	4.6	**	**	*	*	**	**

*Catania Fontanarossa, 37°28′N 15°03′E; 14 m; 1946–1965.
**Messina, 38°12′N 15°33′E; 59 m; 1958–1969.

TABLE XXI

CLIMATIC TABLE FOR PALERMO (1901–1930)

Latitude 38°07′N, longitude 13°21′E, elevation 71 m

Month	Mean sta. press. (mbar)	Temperature (°C) daily mean	mean max.	mean min.	extreme max.	extreme min.	Relative humid. (%) 06h	12h	Mean precip. (mm)	Temperature (°C) mean max.	mean min.	extreme max.	extreme min.	Mean precip. (mm)	Days with snow
Jan.	1015.1	10.2	14.7	5.8	23.7	−1.7	72	63	97.4	14.4	8.6	27.2	−1.2	107	0.1
Feb.	1015.1	10.8	15.4	6.1	27.2	−0.9	71	61	87.1	14.9	8.5	26.2	0.0	75	−
Mar.	1014.7	12.8	17.6	7.4	31.4	−0.4	70	59	60.6	16.6	9.7	32.5	−0.3	60	0.1
Apr.	1014.4(*)	15.1	20.2	9.6	34.9	2.2	69(*)	58(*)	47.8	19.0	11.7	30.4	4.6	45(*)	−
May	1014.2(*)	18.3	23.5	12.3	37.8	6.0	69(*)	55(*)	28.3	22.9	14.9	36.2	8.4	18	−
June	1015.5(**)	22.2	27.4	15.9	40.7	9.0	66(**)	53(**)	14.4	27.0	18.8	39.6	11.0	7(**)	−
July	1014.7(*)	24.8	30.4	18.1	43.1	12.0	64(*)	52(*)	4.4	29.7	21.6	43.2	29.7	4(*)	−
Aug.	1014.6	25.1	30.6	18.6	41.3	12.0	63	52	15.2	29.9	22.1	45.0	29.9	19	−
Sept.	1016.2	23.1	28.4	17.4	42.2	8.0	67	54	51.9	27.5	20.2	41.0	27.5	33	−
Oct.	1015.7	19.1	24.1	14.1	36.4	5.0	71	60	95.5	23.3	16.5	34.0	23.3	93	−
Nov.	1015.2	15.3	19.9	10.6	31.8	1.1	70	61	105.4	19.3	13.2	29.8	19.3	93	−
Dec.	1014.4	11.9	16.4	7.5	24.0	0.1	72	64	113.8	15.9	10.2	26.0	15.9	94	−
Annual	*	17.4	22.4	11.9	−	−			721.8	*	*	*	*	*	*

Month	Number of days with precip. ≥ 1.0 mm	thunderstorm with precip.	fog 03h	06h	12h	18h	Mean daily cloud. (tenths)	Mean sunshine (h)	Mean global radiat. (cal. cm⁻² per day)	Preval. wind direction 06h	15h	Monthly sunshine (h) max.	min.
Jan.	11.4	1.3	−	−	−	0.1	6.5	4.5	169	WSW	W	169.3	106.6
Feb.	10.4	1.5	0.2	0.2	0.2	0.2	6.4	5.2	232	WSW	W	198.0	107.2
Mar.	7.9	0.7	0.1	0.2	0.3	0.1	5.8	6.1	309	WSW	E	246.8	135.2
Apr.	6.8	1.0	0.1	0.2	0.1	−	5.4	7.5	408	WSW	E	260.7	197.7
May	3.9	0.9	0.1	−	−	0.1	4.6	9.3	466	W	E	324.6	231.3
June	2.6	0.3(**)		−	−	−	3.3	10.1	474	W	E	337.5	284.7
July	0.8	0.6		−	−	−	1.6	11.3	498	SW,WSW	E	380.1	323.3
Aug.	1.2	1.0		−	−	−	1.7	10.4	496	WSW	E	347.2	262.9
Sept.	4.8	1.3		−	−	0.1	3.8	8.4	367	SW	E	285.3	201.6
Oct.	8.4	2.7		0.1	0.1	−	5.2	6.5	260	WSW	E	241.8	172.1
Nov.	10.4	1.7		−	−	−	6.1	5.2	188	WSW	W	186.9	106.5
Dec.	12.0	0.8		−	−	−	6.5	3.9	135	SW	W	143.4	92.4
Annual	80.6	*	*	*	*	*	4.7	**	**	*	*	**	**

`*Palermo Boccadifalco, 38°07′N 13°19′E; 117 m; 1946–1965; fog observations at 03h00 GMT not carried out in the 3 years' period June-Dec.1961–1963.
**Ustica, 38°42′N 13°11′E; 251 m; 1958–1969.
(*) 19 years; (**) 18 years.

TABLE XXII

CLIMATIC TABLE FOR SYRACUSE (1901–1930)

Latitude 37°03′N, longitude 15°18′E

Month	Mean sta. press. (mbar)	Temperature (°C) daily mean	mean max.	mean min.	extreme max.	extreme min.	Precipitation mean (mm)	days with ≥ 1 mm	Mean daily cloud. (tenths)
Jan.		10.5	14.7	6.8	20.6	0.0	90.2	9.6	6.0
Feb.		10.8	15.4	7.0	24.0	0.0	60.3	7.7	5.9
Mar.		12.3	17.3	8.3	26.0	1.0	43.0	6.5	5.5
Apr.		14.6	19.8	10.3	34.4	2.5	38.8	4.3	5.2
May		18.2	23.6	13.7	32.6	7.5	18.8	2.8	4.5
June		22.3	27.9	17.5	36.4	11.0	4.6	1.4	3.3
July		25.3	31.1	20.2	41.4	13.7	10.7	0.9	1.6
Aug.		25.7	31.5	20.8	40.4	14.2	10.1	0.9	1.7
Sept.		23.3	28.5	18.9	40.0	11.9	56.8	4.2	4.0
Oct.		19.5	24.2	15.5	35.0	8.3	97.5	6.6	5.4
Nov.		15.2	19.3	11.5	26.8	1.8	141.9	9.6	6.2
Dec.		11.9	16.4	8.2	25.0	0.0	93.1	8.7	5.8
Annual		17.5	22.5	13.2	−		665.7	63.2	4.6

Chapter 5

The Climate of Southeast Europe

D. FURLAN

Introduction

The eastern Mediterranean countries are regarded as the cradle of modern culture and civilization, and it is therefore not surprising that the earliest systematic meteorological observations were carried out in the region of southeast Europe, namely in Greece. The observations extend back to the 5th century B.C. and involve only those atmospheric elements for which no instruments were necessary.

Instrumental observations are much more recent. It is remarkable that the first observing stations were set up in various places in southeast Europe at about the same time, i.e., in the middle of the last century. These included Ljubljana and Split in the west, Uskudar and Athens (Athinai) in the south, and Sibiu and Bucharest (Bucuresti) in the east.

After the first phase in the organization of the meteorological network, when the observing stations were mainly in towns and consequently at no great height above sea-level, the second phase commenced at about the turn of the century. In order to provide data in the vertical dimension, observing stations were established at greater altitudes up to the limit of habitable territory and sometimes even higher.

The two Balkan Wars, and to an even greater extent World War I, not only hindered the development of the network but caused it to suffer setbacks. In particular, the network in the eastern half of southeast Europe was again limited to the largest towns, and no real progress was made in its re-establishment until the 1920s. World War II affected mainly the western half of southeast Europe, where the network broke down almost completely. In the extreme northwest, in the region adjacent to central Europe, less than 5% of the pre-war meteorological stations were still in operation in 1945.

This almost complete interruption of observations during World War II inevitably interfered seriously with the climatic records for the normal period 1931–1960. As a result of the necessary interpolations and reductions, sometimes only data for selected Clino stations were available; these were mainly low-level stations situated in the plains. There are no Clino data (normal values of pressure, temperature, amount of precipitation, and relative humidity) for mountainous regions, which in southeast Europe account for about 4/5 of the total area. Even fewer data are available for other parameters relating to these and other meteorological elements. The present climatological study is based on data for the period 1931–1960 when these were available; otherwise, it is based on data contained in the most recent climatological material for each country or larger region in question.

As already mentioned, about 4/5 of southeast Europe is mountainous. A large proportion of this area is uninhabited and is situated at altitudes for which hardly any data are available. It should be emphasized that the regions for which no data are available do not form a separate entity, but are scattered throughout the whole of southeast Europe. Under these circumstances and for technical reasons also, extrapolations can hardly be justified from an expert point of view.

Factors affecting the climate

Geographical position

Southeast Europe includes the following territories: the whole of the Balkan Peninsula and Archipelago, the eastern part of the Carpathians and the lowlands to the east and west of the Carpathians, i.e., the Pannonian Plain (only the southern part), and also Wallachia and Moldavia. From one end to the other, the region extends over about 13° of latitude and about 16° of longitude. The extent of the region, both in the meridional and zonal directions, precludes a uniform climate, although it falls mainly in the temperate zone (35°–48°N) (Fig.1).

Features of the relief

Southeast Europe is not a geographical entity. The Dinaric Alps embrace the greater part of the western half of the Balkan Peninsula, including Yugoslavia, as far as the Albanian frontier in the south. They then continue as the Pindus Mountains, which include the whole of Albania, the greater part of Greece, and Crete. The highest ridges and plateaus of the two mountain systems run parallel to the coast and usually in close proximity to it.

The eastern Carpathians form the backbone of Romania, and are continued in Bulgaria as the Balkan Mountains. Here the slopes are more gentle than in the west. The Rhodope Mountains lie between the Dinaric Alps and the Pindus Mountains in the southwest and between the Carpathians and the Balkan Mountains in the northeast. In addition to the less rugged land forms, this region is characterized by the broader basins.

Southeast Europe and especially the Balkan Peninsula can thus be regarded as a mountainous region. Really extensive plains are only to be found in the north. These are the Pannonian Plain to the west of the Carpathians and Wallachia and Moldavia to the east of the Carpathians.

In the east the height above sea-level increases almost imperceptibly and the Balkan Mountains as seen from the Danube do not appear to be real mountains. In contrast, the boundary between the northern spurs of the Dinaric Alps and the Pannonian Plain is clearly defined, even though these spurs are relatively low. The hills and uplands give way towards the south to mountain ranges with ridges and plateaus of over 2,000 m. The other mountain systems rise to similar heights.

These mountain systems are not separated by any broad basins (with the exception of the Maritsa Basin). Thus the lowlands of the north and the mountainous region form two geographical entities. The third entity consists of the coastal areas of the eastern Mediterranean. This includes the islands and coastal regions which, with the exception

Fig.1. Map of southeast Europe.

of the Black Sea coast, are sheltered from the severe continental winters by mountain ranges. The climate here is also more temperate owing to the relatively warm sea. This is borne out by the fact that in the coldest month the mean temperature is 5°C in the Bosporus, while in southern Istria it is as high as 10°C.

The three geographical entities exhibit very different climatic characteristics. The lowland in the north is cut off from the influence of the Mediterranean but lies open to central and eastern Europe. Consequently it has a continental climate. The mountainous region, rising to over 2,000 m, has an Alpine climate. The third entity, consisting of coastal regions and islands, enjoys a Mediterranean climate.

Radiation conditions

The very significant meridional extent of southeastern Europe is well reflected in the distribution of increasing global radiation.

The annual values of global radiation on a horizontal surface from a number of stations situated at various latitudes show a distinct increase towards the south; the increase per degree of latitude being around 1.3% (Table I). The differences in the direction from

TABLE I

AVERAGE MONTHLY (Ly/month) AND YEARLY (Ly/year) GLOBAL RADIATION TOTALS MEASURED ON A HORIZONTAL SURFACE, 1964–1969

Station	Jan.	Feb.	Mar.	Apr.	May	June	July	Aug.	Sept.	Oct.	Nov.	Dec.	Annual
Athens	4832.3	6339.3	9539.7	13244.3	16416.8	16805.8	18275.9	16125.3	12233.9	8894.7	5486.4	4350.9	132545.3
Skopje	4530.3	6186.1	9209.9	12385.4	16165.5	16570.3	17778.5	15935.3	12123.9	8945.0	4492.9	3101.6	127424.7
Bucharest	3849.9	5555.3	9060.3	12496.4	16100.5	16887.0	17288.0	15378.0	11340.6	8149.8	3516.4	2537.7	122159.9
Belgrade	3985.1	5404.1	8689.2	11807.4	15165.3	16062.2	17476.2	14564.0	11063.2	8404.2	4056.3	2535.2	119212.4
Ljubljana	2395.5	3940.1	6772.0	10149.5	13333.8	13951.3	14874.1	11955.4	8114.7	5392.3	2291.9	1470.2	94640.8

east to west, on the contrary, are small if one compares climatic areas, such as e.g. Wallachia and south Pannonia. In mountainous regions, however, especially in the basins and Karst-poljes of the northwestern parts of southeast Europe the conditions are considerably different; in places a decrease of ca. 20% may occur. Such a reduction in global radiation can be attributed to a great number of foggy days (in the Ljubljana Basin > 150 days per year), especially in the cold season, or to strong convective cloudiness in the hot months of the year.

In southeast Europe the monthly maximum in global radiation occurs in July and not in June (due to more favourable cloud conditions) and the minimum in December. If one excludes the Alpine Basin and the Karst-poljes, the mean value for January in the northern part of southeast Europe is ca. 4,000 Ly/month while it reaches ca. 5,000 Ly/ month in the south. The corresponding values for July are 17,000 and 18,500 Ly/month. The annual variation for the five stations presented in Fig. 2 shows great similarity in the average values. For individual years the differences may be considerably greater; this is connected with the great climatic differentiation in southeast Europe (Table II). The extreme values of the global radiation usually occur in the months of the winter and summer solstices or a month later. In the northern regions the hourly absolute maximum is nearly 45 Ly/h in December and 85 Ly/h in July. The corresponding values for a day are: 220 and 740 Ly/day.

No corresponding data are available for the southern part of southeast Europe. Because of the positive correlation between the amount of cloud and global radiation it is estimated that in southern regions the figure would be 58 or 100 Ly/h and 290 or 880 Ly/day.

Because the global-radiation data are so systematically distributed over southeast

Europe, it seems adequate to present data on direct radiation from the sun from only one centrally situated station, namely from Belgrade–Zeleno Brdo.

The maximum intensity of direct solar radiation on a horizontal surface, recorded at noon at the summer solstice under cloud-free conditions and with an average turbidity condition, amounts to 1.09 Ly/min while at the winter solstice the same value reaches only 0.40 Ly/min (Table III). At the autumn equinox this value amounts to 0.80 Ly/min.

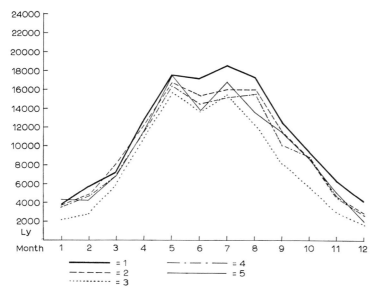

Fig.2. Annual variation of global radiation on a horizontal surface in 1969. *1* = Athens (Athinai); *2* = Bucharest (Bucuresti); *3* = Ljubljana; *4* = Skopje; *5* = Belgrade (Beograd).

TABLE II

AVERAGE TOTALS OF MONTHLY AND ANNUAL GLOBAL RADIATION ON A VERTICAL SURFACE AT BELGRADE–ZELENO BRDO (EXPRESSED AS PERCENTAGE OF THE GLOBAL RADIATION VALUES ON A HORIZONTAL SURFACE), 1968–1970

Orientation	Jan.	Feb.	Mar.	Apr.	May	June	July	Aug.	Sept.	Oct.	Nov.	Dec.	Annual
East	58.0	57.4	50.9	54.7	54.6	55.3	52.5	53.5	54.7	54.0	55.8	56.6	54.3
South	128.5	106.0	83.4	68.4	53.7	47.8	47.7	60.5	83.0	116.5	137.0	110.6	71.8
West	76.4	69.5	64.0	61.6	55.2	53.2	55.9	61.0	66.8	73.6	73.0	72.6	61.5
North	41.6	42.7	35.1	30.2	30.9	34.0	31.5	31.6	32.2	29.6	34.0	47.4	32.9

TABLE III

INTENSITIES OF THE DIRECT SOLAR RADIATION ON A HORIZONTAL SURFACE WITH CLOUD-FREE CONDITIONS AND AVERAGE DISTURBANCE FACTOR; BELGRADE–ZELENO BRDO, 1957–1966

Month	Day	12h	11h 13h	10h 14h	09h 15h	08h 16h	07h 17h	06h 18h	05h 19h
Mar.	21	0.84	0.81	0.69	0.52	0.32	0.11	0.00	0.03
June	22	1.09	1.05	0.94	0.79	0.59	0.37	0.17	
Sept.	23	0.80	0.77	0.66	0.51	0.30	0.10	0.00	
Dec.	22	0.40	0.37	0.27	0.13	0.01			

The maximum hourly value of direct solar radiation at Belgrade amounts to 64 Ly/h; whereas daily totals lie between 114 Ly/day in January and 538 Ly/day in June.

The general circulation over southeast Europe

The greater part of southeast Europe lies in the middle latitudes. Consequently the general circulation of the atmosphere is determined by waves on the polar front, particularly in winter. Throughout the year, troughs and ridges move eastwards across the region. At heights above about 700 mbar the sequences of southwesterly flow ahead of troughs, followed by northwesterly flow ahead of ridges, result in a mean westerly flow. This is characteristic not only of southeast Europe but of the whole temperate zone. However, the relative displacement of the Azores high towards the north in summer, results in the westerly flow having a more northerly component over the region.

The Aegean and Ionian islands and also the extreme south of the Balkan Peninsula are situated at such low latitudes that in the summer they are close to the centre of the subtropical belt of high pressure. Since these regions lie on the northern side of this high-pressure belt, westerly winds predominate. However, the strength and frequency of these westerlies have their annual minima at this time of year.

At levels below the 700-mbar surface, the circulation is more complex, particularly in connection with low-pressure systems, which are frequently established in the Gulf of Genoa and in the eastern Mediterranean. To the rear of these lows the northwesterly flow becomes northerly, and in the northwest of the Balkan Peninsula it may even become northeasterly as a result of orographic features. On the other hand, the southwesterly flow ahead of these lows, becomes southerly or southeasterly. On average, it is flow from the southern quadrant and not from the southwestern quadrant which prevails in such synoptic situations.

An anticyclone develops regularly over middle and eastern Europe in winter. As a result of the difference in temperature between the cold anticyclone over the heart of Europe and the warm air over the Mediterranean, an easterly flow develops over southeast Europe. Since this anticyclone is usually cold and shallow, the easterly flow aloft may be reversed and the westerly flow of the general circulation remains undisturbed.

The main features of the weather

The above mentioned centres of action are most active in winter, and often give rise to diametrically opposite types of weather in different parts of southeast Europe. The above indicated weather is mainly influenced by: (*1*) the Eurasian anticyclone, in the circulation of which, polar continental air flows from the interior of eastern Europe as a European winter monsoon, and (*2*) the lows in the eastern Mediterranean.

On account of its easterly direction, the winter monsoon blocks the entering of the polar front waves over Europe and often steers the latter over the Mediterranean. The nearer an area is situated to the source region of the winter monsoon (i.e., to the anticyclone-forming part of the Eurasian belt of high pressure), the lower becomes the temperature and the smaller becomes the precipitation. In southeast Europe this applies particularly to Moldavia and Wallachia.

At the same time, the greatest activity of the polar front and of depressions associated with it occurs over the Mediterranean. This is particularly the case in the southeastern half of the Mediterranean and over the southern part of the Balkan Peninsula. Accordingly, the frequency and the amount of precipitation are maximal in this area.

Differences of latitude and of the physical characteristics between water, land and snow cover inland, are responsible for enormous contrasts in the distribution of temperatures in winter. Another factor, which is of even greater significance for these temperature contrasts, is the advection of different air masses. In the north of this part of Europe, continental air masses from the north predominate, in connection with the above mentioned anticyclone. In the south, on the other hand, the flow has a southerly component due to cyclonic activity, bringing in tropical or subtropical air masses. Thus large temperature contrasts occur in southeast Europe, greater than in many other parts of the world.

In summer, as in other continental regions of the temperate zone, the pressure is generally low over Europe and westerly winds prevail. The polar-front waves affecting western and central Europe may sometimes reach the northern part of southeast Europe. Convection due to heating accentuates the summer maximum of precipitation, occurring in that region.

In the south the conditions are quite different in summer also. Polar-front depressions reach the western Mediterranean but towards the east they become less frequent and only exceptionally penetrate into the eastern Mediterranean. Furthermore, even when they do, there is little precipitation. Although the sea is warm, it is still cooler than the land and there is no convective activity. Moreover, polar air is strongly modified after traversing the whole of Europe. The result of this modification is that the processes causing precipitation at frontal surfaces between air masses are appreciably weakened. This is particularly the case along the coasts and over the Aegean Islands. In this region the steady etesian winds blow from the northern quadrant. These winds are partly due to the displacement of the Azores high towards northwest Europe, and to the low which develops over the Iranian Plateau. In spite of the low pressure over the eastern Mediterranean, the weather in the region of the etesian winds is fine and clear.

During the two seasons of transition, polar-front waves penetrate into all parts of southeast Europe. The interruption of the normal movement of waves in a zonal flow is usually due to: (*1*) extraordinarily deep depressions over northern Italy, which usually develop over the Gulf of Genoa, the Plain of Lombardy or over the northern Adriatic; and (*2*) sub-tropical anticyclones over the continent of Europe.

The above mentioned area, from the Gulf of Genoa to the northern Adriatic, is one of the regions of cyclogenesis in Europe, known to be most active. The most intense cyclogenesis occurs when the originally secondary depressions are cut off and become independent deep depressions. The deepening process slows down the advance of the depression and increases the circulation associated with it. This, in turn, results in an increase in the duration, amount, and area of precipitation.

Cut-off depressions are steered by the upper winds and consequently the coastlines have little influence on their path. The behaviour of the ordinary, more shallow polar-front waves is rather different. Owing to their small vertical extent, they cannot cross the Apennines or the Dinaric Alps, and are consequently obliged to follow the coasts southward. Since these are shallow waves, their radius of influence is small.

The tracks of the depressions, steered by the upper winds (the polar "jet" stream), correspond broadly to the position of the polar front. With the displacement of the latter towards the southeast in the autumn, the tracks of depressions are also displaced towards the southeast, and in winter the eastern Mediterranean becomes a principal region of cyclonic activity.

The building up of anticyclones is most frequent over the Alps in midwinter. In central and eastern Europe they quite often occur in the autumn, particularly during September and October. The greater part of the continent of Europe, including the southeastern regions, then has fine clear weather.

Air masses

The same air masses are found in southeast Europe as in central and western Europe. Polar air masses prevail because the territory is situated in the temperate zone, but tropical air is drawn in, in connection with the depressions on the polar front. Since the depressions move from west to east, i.e., from the Atlantic, maritime air masses are advected. In winter, continental polar air (cP) usually covers the continent. The mP advected in connection with the cyclones are rapidly transformed into cP as slow moving ridges of high pressure build up in the rear of the cold fronts. The cP remains for long periods, particularly in the basins.

The cP air masses which in winter enter the area from the northeast to southeast are very cold, and the humidity is very low. The physical characteristics of these air masses are very similar to those of cA air masses. Indeed, the occurrence of true cA in southeast Europe is not at all exceptional. cA is found more frequently in the eastern half of southeast Europe, especially when an anticyclone builds up over the Carpathians. In this latter case, there is a flow of cA direct from arctic latitudes on the eastern side of the anticyclone.

In winter, the cP, mP, and cA air masses having a predominantly northerly component encounter tropical air, cT, and mT, which, under the influence of the Azores high, flows towards western Europe from the southeast. However, due to the situation of the high pressure system, this air reaches southeast Europe from the northwest. The frequency of such situations decreases, however, as one proceeds towards eastern Europe. More frequently, mT air masses moving in an air current of a marked southerly component occur in southeast Europe when upper troughs move eastwards into France, or when very deep depressions are moving eastwards across the British Isles.

In winter, cT air reaches southeast Europe only when to the west there is an outbreak of polar air extending as far as the north coast of Africa or even further south. In connection with Mediterranean depressions, cT air masses from the Sahara are drawn into southeast Europe, in the region of the upper trough. Over the Mediterranean this air becomes humid, and frequently spreads into central Europe and the greater part of eastern Europe as the "sirocco". It brings with it large quantities of Sahara dust and causes the well known "blood rain". During the remaining three seasons cA and cP air masses are seldom found, and only mP and mT air masses, mainly from a northwesterly direction, reach southeast Europe in association with polar-front depressions.

In anticyclonic situations, and particularly in summer, cT air masses spread over southeast Europe when the centre of the anticyclone over Europe forms part of the ridge of high pressure centred over North Africa. Owing to subsidence, the relative humidity is low even though the air crosses the Mediterranean. In such situations the west of the Balkan Peninsula comes under the influence of *cT* air masses. The eastern half of southeast Europe experiences *cT* air masses when ever there is a centre of high pressure over northern Europe, bringing air from the Ukraine or Asia Minor.

Owing to the large area included in southeast Europe, the occurrence of the various air masses (in widely separated areas) and the flow direction, are seldom uniform.

In the central region of southeast Europe, between the estuary of the Sava (Belgrade) and the central Adriatic (Split), 45% of days during the period 1955–1957 had a homogeneous air flow, i.e., a single air mass and a uniform direction of flow (disregarding the influence of relief). On closer examination of this period of 45% of all days having homogeneous flow, it is found that 48% had *mP* air masses coming from the northwest, 34% had *mT* air masses coming from the southern quadrant, and 18% had *cP* air masses coming from the eastern quadrant.

The maximum frequency of occurrence of all three air streams was during the winter. This shows that the durations of the various airstreams are greatest in winter, and that frontal passages and slack pressure gradients are rare during that season.

The temperature differences between the various air masses are large, and vary owing to the effects of seasonal changes, height above sea-level, and geographical location. Observations extending over three years in Belgrade showed that in the lowest layers, the annual average temperature was about 8°C higher in southerly than in northwesterly air streams. At the 500-mbar level this difference decreased to one-half. In this short series of observations air masses from the northwest and from the east had practically the same temperatures in the lowest layers.

The main characteristics of the most important climatic elements

Temperature

In the first section it was mentioned that there is hardly any other region in the world where the temperature contrasts are so great as in southeast Europe. The reasons for this are the great meridional extent of southeast Europe, differences in height, and especially the high mountains along the southwest coast. The highest plateaus of the Karst and ridges of the Pindus Mountains run parallel to the Adriatic and Ionian coasts, and, what is particularly important, they lie in close proximity to these coasts. At the coldest time of year these mountain ranges provide effective shelter for the narrow coastal region against invasions of polar air, and not infrequently even arctic air. The picture for the north Aegean coast is similar; there, shelter is provided by the eastern flank of the Rhodope Massif. By contrast, the coast of the Black Sea has little shelter and the coldest air masses spread almost unhindered. At other times of the year, including summer, these contrasts disappear, and the conditions in the coastal belt become broadly similar to those inland owing to the Mediterranean climate, at least during the warmest six months of the year. Accordingly, the temperatures over the whole region are mainly determined by latitude and height.

Mean temperature in January

From a mean January temperature of about 12°C on the north coast of Crete, the mean temperature decreases steadily along the Ionian and Adriatic coasts, i.e., in a north-westerly direction. Along the southern Adriatic coast there are temperatures of about 9°C, in the central part of the Adriatic coast temperatures of about 7°C, whilst temperatures in the Gulf of Trieste decrease by a further 2°C. The unsheltered coast of the Black Sea is appreciably colder. In the Danube delta (Sulina), the difference from the corresponding position in the northern Adriatic (Opatija), amounts to about 5°C. Towards the south the difference becomes steadily smaller (Pula–Constanta about 4°C; Kotor–Burgas about 3°C), until along the coasts of the southern Aegean, the temperatures become equal. In the direction from Crete towards the northwestern Adriatic, the mean temperature for January decreases on average by 0.8°C for each degree of latitude, whilst the corresponding decrease towards the Danube delta amounts to about 1.3°C (Fig.3).

The temperature differences just described are not due solely to the varying degree of shelter. There is a warm current from the south, coming from the centre of eastern Mediterranean along the coast of Yugoslavia. On the other hand, along the Black Sea coast there is a cold current coming from the north. The water of this current not only becomes colder while in areas situated to the north, but mixes with large quantities of markedly cold water from the rivers which flow into the Black Sea. The situation is thus similar to that along the west coast of the Adriatic, which, for the same reason, is about 2°C colder than the east coast.

The January temperatures in the regions of the interior are about the same as on the western coast of the Black Sea, especially if we consider the broad low-lying basins and broad valleys and take height into account (Belgrade (Beograd) and Constanta about 0.7°C).

Of all the regions in southeast Europe, Bukovina is situated in the immediate neighbourhood of the coldest regions of Europe and accordingly has the lowest mean January temperatures (Iasi −4.1°C). The cold outbreaks towards the south have a long unobstructed path. After passing over Wallachia, such outbreaks spread into northern Bulgaria and over the Balkan Mountains to the valley of the Maritsa and even further south. The Balkan Mountains form in fact the first major obstacle to further movement of large masses of the coldest air towards the south. The absence of mountains in Moldavia (Moldova) and Wallachia is the reason why the temperature gradient is only 0.9°C per degree of latitude from Sofia to Iasi, whilst from Sofia to Sparta it has almost doubled this value. The Carpathians provide similar effective shelter from cold outbreaks, for the Pannonian Plain. At the same latitude, mean January temperatures are higher by more than 2°C than those in Wallachia.

South of the Balkan Mountains the temperatures are higher by 2–3°C; it only becomes markedly warmer on the southern slopes of the Rhodope Mountains, i.e., in the region directly influenced by the Aegean. At the foot of the Rhodope Mountains the increase in temperature amounts to a further 5°C.

In the interior of the Balkan Peninsula mountain barriers often present an obstacle to the advance of cold air in the lower layers. In these southern latitudes this results in a gradual increase of temperature inland in the Balkan Peninsula, but this increase does not become marked until we reach Thessaly. Here the Khasia Mountains provide

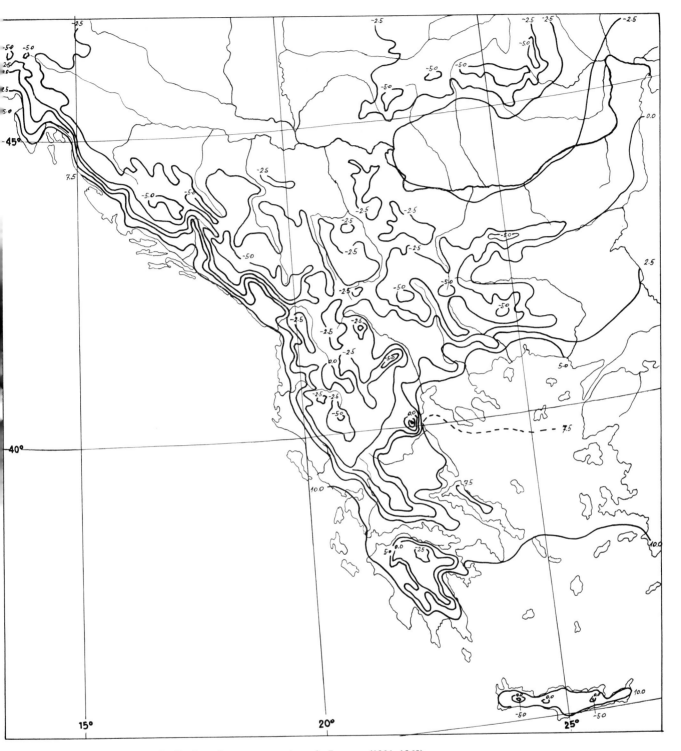

Fig.3. Distribution of mean temperatures in January (1931–1960).

effective shelter against northerly outbreaks. Compared with southern Macedonia, Thessaly is warmer in January by about 3°C. However, it must be remembered that the Balkan Peninsula becomes steadily narrower towards the south, so that the central regions are more accessible to *mT* air. This explains why the temperature between

195

Thessaly and the interior of the southern Peloponnese plateau increases by more than 2°C per degree of latitude, while between Thessaly and the northern slopes of the Rhodope Mountains (Trikkala–Belgrade) the corresponding figure is only 0.9°C.

The meridional temperature gradient of 0.9°C per degree of latitude leads to the conclusion that differences of temperature in inland regions of the Balkan Peninsula and in the remaining part of southeast Europe north of the Sava and the Danube, are mainly due to differences of latitude. The doubling of the gradient shows however that in the south, the Mediterranean influence is effective well inland.

In southeast Europe peaks and ridges are relatively warmer than the lowlands, as can be seen by comparing temperatures at pairs of stations in all regions of southeast Europe. In January, valleys and basins, including the poljes of the Karst are 2–3°C colder than mountain slopes and peaks at the same height. Since the lapse rate of temperature is about 0.4°C/100 m in the north and 0.5°C/100 m in the south, the temperatures in basins occur on peaks 400–600 m higher up (Fig.4). There are therefore distinct inversions.

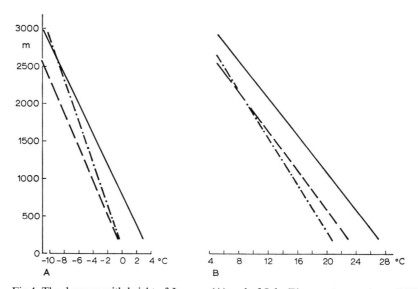

Fig.4. The decrease with height of January (A) and of July (B) mean temperatures (1931–1960). Rhodopes (———); S.E.Alps (— · — ·); Carpathians (– – –).

Both the relatively high temperatures on the peaks and mountain slopes and the inversions in valleys in winter, are the main characteristics of an Alpine climate, and so are not confined to southeast Europe.

The Carpathians are separated from the massifs of the Balkan Peninsula. At heights of 2,500 m the January mean temperatures (see Fig.4) are about 1.5°C lower than in the Julian Alps (Mount Onul, 2,509 m, −10.5°C; Kredarica, 2,515 m, −8.7°C). The nearest stations to these in the plains show almost the same difference, both in the Julian Alps and in the Carpathians. These temperature differences are quite small, and show that inland in southeast Europe, in winter, there is a uniform distribution of temperature. Differences of temperature are mainly due to differences of height and latitude.

Mean temperature in July

The contrasts between the archipelago and the southern and western coasts on the one hand and the interior of southeast Europe (together with the Black Sea coast) on the other, which are so marked in January, are not only greatly reduced but reversed in July, as in other regions of the temperate zone. On account of the high temperatures in the Mediterranean region these differences are small, 1°–2°C (Fig.5).

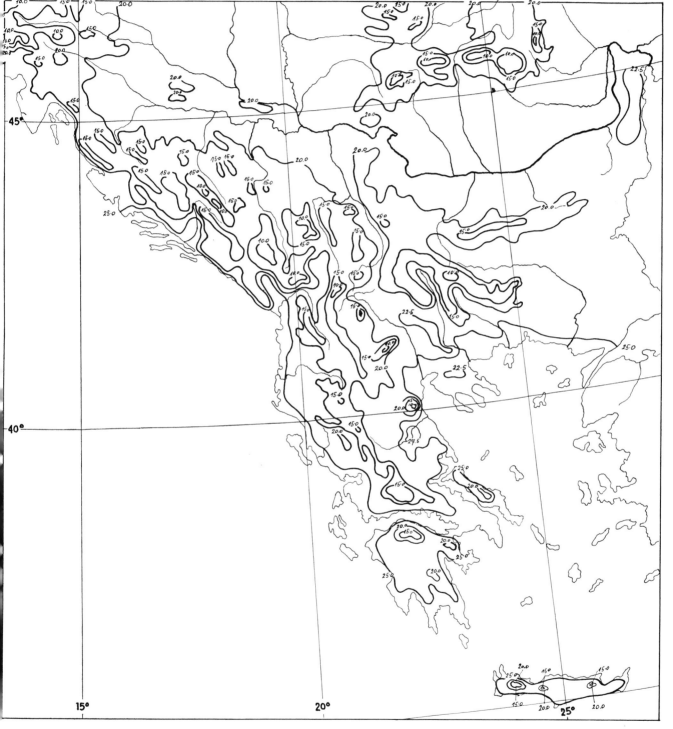

Fig.5. Distribution of mean temperatures in July (1931–1960).

The main differences in summer are due to differences in height, since the temperature lapse-rate increases from about 0.4°C/100 m in January to about 0.7°C/100 m in July. Such conditions occur in the northern regions; in the south the lapse-rate in summer is even greater—over 0.8°C/100 m, and consequently the effect of height is even greater (Fig.4).

Along the Ionian and Adriatic coasts the temperatures steadily decrease, and differences are very small. The temperature difference between the northern coast of Crete and the Gulf of Trieste is barely 3°C (about 26° and 23°C). Along the Black Sea coast, on the opposite side, conditions are practically isothermal, with the temperature between 22° and 23°C. The Aegean coast is 2°–3°C warmer.

On account of differences of height, conditions in the interior are somewhat more erratic. In the extreme north in Bukovina, at a height of about 100 m, the mean temperature in July is between 21° and 22°C, while on the Peloponnese Plain, at a height of about 200 m it is about 28°C. If we consider a few stations lying in between, we find a uniform increase in temperature in a north–south direction. The gradient per degree of latitude is about 0.7°C.

In July, the basins and valleys are on average more than 1°C cooler than the peaks. In the early morning the average differences may be over 3°C.

The characteristics of the temperature regime in July are thus as follows: the effect of continentality disappears almost completely, and the temperature change per degree of latitude decreases from 0.9°C in January to 0.7°C in July.

Annual variation of temperature (Fig.6)

The mean values for the extreme months, January and July, are very little different from those for February and August. This is particularly true of August, which only in exceptional cases, is as much as 1°C cooler than the warmest month of the year. Sometimes August is even the warmest month; this occurs on the coast, in the Ionian Archipelago, and at some localities in the Aegean Archipelago. Similar exceptions also occur in winter; at a height of about 2,500 m, the February temperatures are similar to those in January or even lower. The smallest differences between the monthly means (Table IV) occur in the periods following the summer and winter solstices, whilst the greatest of these differences occur between March and April (3°–7°C). However, on the coast, in the archipelagos, and also in the higher regions, the time of greatest warming is retarded by a month. In the extreme south the time of greatest warming is delayed for another month and occurs between May and June. In autumn the maximum cooling (4°–6°C) occurs between September and October. In the Aegean Sea and at heights above 2,500 m the greatest cooling occurs between October and November.

TABLE IV

MONTH TO MONTH DIFFERENCES BETWEEN MEAN MONTHLY TEMPERATURES (°C)

Station	Jan.	Feb.	Mar.	Apr.	May	June	July	Aug.	Sept.	Oct.	Nov.	Dec.
Kredarica	1.0	1.8	2.1	4.3	3.3	2.5	0.1	1.9	4.1	4.5	3.0	1.6
Athens	0.6	0.7	3.1	3.8	4.3	1.8	0.3	2.9	3.4	3.3	3.1	1.6
Bucharest	2.1	5.2	7.1	5.3	3.9	2.4	0.6	4.4	6.3	6.5	5.1	3.1

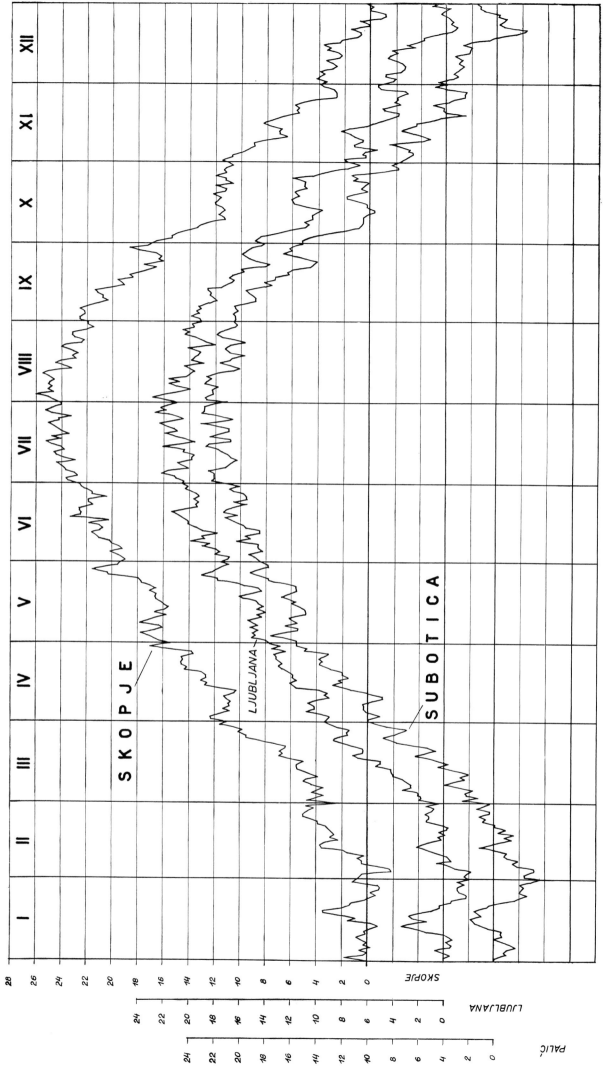

Fig.6. Annual variation of daily mean temperature at three stations in the central region and in the western half of southeast Europe (1947–1956).

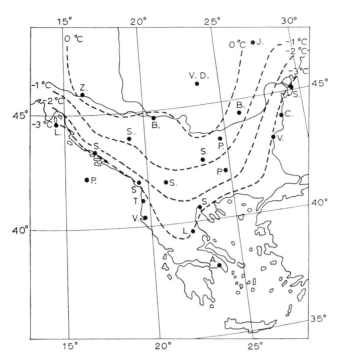

Fig.7. Differences between the mean temperatures for April and October (1931–1960).

The variation of continentality is demonstrated in Fig.7, which shows the distribution of the differences between mean temperatures in April and October (1931–1960).
The distribution of hourly mean temperatures over the year at Bucharest is shown in Fig.8 as an example of how the daily variation of temperature changes over the year.

Extremes of temperature

The contrast between the southwest and east coasts (and also the interior) of the Balkan Peninsula is especially noticeable in the distribution of the mean monthly minimum (average of lowest temperature each month) during winter. In January, in the Gulf of Trieste, this parameter is more than 10°C higher than in the Danube delta (Koper, −2.9°C; Sulina, −14.4°C). The difference decreases rapidly towards the south, and at the latitude of the Bosporus amounts to only 5°C (Durak, about 2°C; Istanbul, about −3°C).

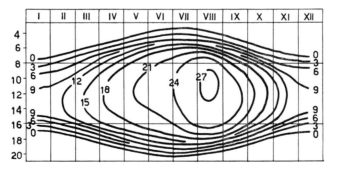

Fig.8. Annual variation of hourly mean temperatures at Bucharest-Filaret (1928–1943, 1948–1955).

In the interior, north of the Sava and the Danube, the mean monthly minima remain above −15°C, but in Moldavia (Moldova) and Wallachia they even approach the value of −20°C. In the interior of the Balkan Peninsula too, absolute minima in January below −15°C are far from rare, although mean absolute minima between −15°C and −10°C are more usual, as far as southern Macedonia. In Thessaly the mean monthly minimum temperature is still below −5°C. In the extreme south of the interior of the Peloponnese, this parameter lies below the freezing point (Sparta, −1.7°C), even though it is 5°C warmer. However, Naxos, situated in almost the same latitude, has a value of nearly 4°C and the same value is found on the northern coast of Crete.

In mountainous regions, at over 2,000 m, the mean monthly minimum temperature in January lies between −20 and −25°C, i.e., lower than in the coldest regions of southeast Europe—Bukovina and Wallachia.

In the whole of southeast Europe, the absolute minimum temperatures are negative; even in Crete the absolute minimum at Candia is −0.1°C. No other type of temperature data brings out the effect of climatic factors as clearly as the absolute minimum temperature. Some examples are given below.

(*1*) Difference of latitude: Larisa, 39°37′N, −12.8°C; Belgrade, 44°48′N, −25.5°C.

(*2*) Different degrees of continentality: Shkoder, −9.4°C; Plovdiv, −31.7°C. At this point we can take into consideration the difference between eastern and western coasts: Koper, −12.8°C, Sulina, −25.5°C.

(*3*) Difference of height: Zagreb–Grič, 168 m, −21.6°C; Mount Onul, 2,509 m, −38.0°C.

(*4*) Difference of relief: Bjelašnica—summit, 2,067 m, −29.6°C; Gospić—in a Karst-polje, 565 m, −32.4°C.

The lowest temperatures, below −30°C, were observed in shallow basins and in Karst-poljes having gentle slopes. Absolute minima below −30°C occur at the lowest points of certain basins in the lower Alps and the Carpathians, and of course in the lowlands of Romania and Bulgaria, where there is nothing to prevent the advection of the coldest air masses from eastern Europe.

The Khasia Mountains form an effective barrier, although the absolute minimum at Larisa is still about −13°C.

In regions having an Alpine climate, temperatures are low due to great altitude. Above 2,000 m the absolute minima approach those in the coldest basins, i.e., about −30°C, and in the Carpathians at a height of 2,500 m, they even approach −40°C (Mount Onul, −38°C).

The mean monthly maxima in July reach their highest values in the most southerly regions, but only inland at places sheltered from cool sea-breezes. This is particularly the case in the interior of the Peloponnese and in Thessaly, where the mean monthly maximum temperatures are nearly 40°C. The temperatures decrease only slowly northwards, and values of over 35°C still occur in the Pannonian Plain. On cloudless days the heat in the afternoon is quite unbearable. An exception to this is the narrow coastal belt of the peninsula and the archipelago. Near the coast, even if the afternoon temperatures rise well above 35°C, the heat is relieved at the time of the afternoon maximum by refreshing sea-breezes. In places not reached by these breezes the heat is unbearable, especially as the humidity is very high in the basins and valleys in the interior. In the mountains, above 2,000 m, there are values of about 15°C in the extreme north in the Carpathians.

Another feature of the temperature regime of southeast Europe are the high values on the Black Sea coast (Sulina, about 32°C), which are higher than those at the same latitude in the northern Adriatic (Trieste, about 30°C). This difference is due to the warm easterly winds which in summer bring warm *cT* air from Asia Minor and southern Ukraine.

This displacement of the highest temperatures in the eastern half of the Balkan Peninsula and on the coast of the Black Sea is even more marked in the absolute maxima. In these regions there are few stations, at heights up to 500 m, where the absolute maxima do not exceed 38°C. This value may also be exceeded in the west, but the frequency of occurrence is lower.

The highest values are of course found at places far from the coast. The temperature maxima of Larisa and Bucharest approach closely the value of 45°C. The difference between the Adriatic and the Black Sea coasts is shown by the following: north of 41° latitude, the absolute maxima on the west coast are below 40°C, while on the east coast they are over 40°C.

In mountainous regions, the absolute maxima have low values. At heights above 2,500 m they barely reach 20°C.

The greatest differences between the absolute maxima and the absolute minima exceed 60°C in the most northeasterly regions, i.e., Romania and northern Bulgaria. In the remainder of Bulgaria, including the Black Sea coast and also in Yugoslavia, such differences decrease to 50°–60°C. In the interior of Greece they decrease by a further 5°C, and the coastal region has differences even of less than 45°C. Similar differences between temperature extremes are also found at high altitudes in the mountains.

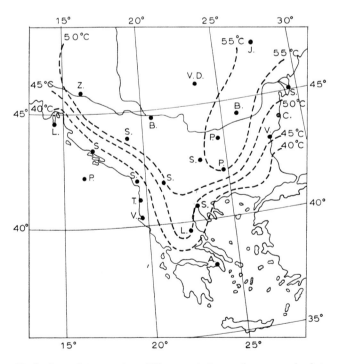

Fig.9. Annual temperature differences between the mean absolute maximum and the mean absolute minimum (after BIEL, 1944).

Precipitation

As in the case of temperature, precipitation shows considerable differences between the various regions in southeast Europe.

Annual precipitation

The mountain ranges form the backbone of the pattern of the annual totals of precipitation. In spite of the great uniformity of the highest mountain ridges and plateaus of all the mountain systems, the annual amounts of precipitation exhibit strong differences. In the southernmost part of the Dinaric Alps, the Crkvice station, at a height of 1,050 m in the hinterland of Boka Kotorska, has an annual precipitation of over 4,600 mm. This figure is based on actual measurements and is the maximum for Europe. Since the amounts of precipitation increase with height, it is likely that the highest plateaus in the hinterland of Boka Kotorska would have over 5,000 mm, whilst others of the higher plateaus and ridges of the Dinaric Alps would have about 4,000 mm. The orientations of the mountain barriers, the steep slopes which rise directly from the coast, and the high humidity of the southwesterly flow of air are the reasons why the Dinaric Alps have the highest precipitation in southeast Europe, and indeed in the whole of Europe (Fig.10). Precipitation decreases with increasing distance from the highest plateaus and ridges of the Dinaric Alps. This is particularly the case in the direction of the Adriatic, along the shores of which, annual totals of more than 1,000 mm are seldom recorded; some islands have only half this amount (Palagruža, about 400 mm). Towards the northeast, precipitation only decreases rapidly at first: in the Niška kotlina there is about 600 mm, and towards the Black Sea coast (Dobrudža) the precipitation decreases by a further 200 mm (Sulina, Constanta, 350–400 mm). At the same time, the distance between Dobrudža and Crkvice is nearly six times that between Crkvice and Palagruža. The Dobrudža coast, with less than 400 mm, has the lowest precipitation in southeast Europe and in Europe generally. Certain islands in the Aegean (Naxos) also have less than 40 mm, but against this the coastal regions of the Aegean have an annual precipitation of 500–600 mm. Comparable amounts are also found in Bukovina and Wallachia. The southern part of the Pannonian Plain has 600–700 mm; towards the west the amounts gradually increase, and in the spurs of the southeastern Alps there is nearly 1,000 mm (Fig.11). The characteristics of annual precipitation are the contrasts between mountains and lowlands on the one hand (Table V), and between western and eastern coasts on the other. In the first case the ratio is over 2:1, and in the second also over 2:1 (Fig. 10).

TABLE V

THE RATIO OF THE AMOUNTS OF PRECIPITATION AT MOUNTAIN STATIONS TO AMOUNTS AT LOW-LEVEL AND COASTAL STATIONS

Station	Jan.	Feb.	Mar.	Apr.	May	June	July	Aug.	Sept.	Oct.	Nov.	Dec.	Year
Komna/Bled	1.8	2.1	2.3	3.4	2.2	1.7	1.8	1.8	1.9	2.1	2.3	2.1	2.0
Mt. Onul/													
Bucharest	2.8	5.3	4.0	2.2	1.5	1.9	2.5	2.1	1.5	2.1	2.5	2.3	2.3
Crkvice/Kotor	2.6	2.3	2.9	2.6	2.0	1.2	2.3	1.3	1.8	2.4	3.1	2.4	2.4

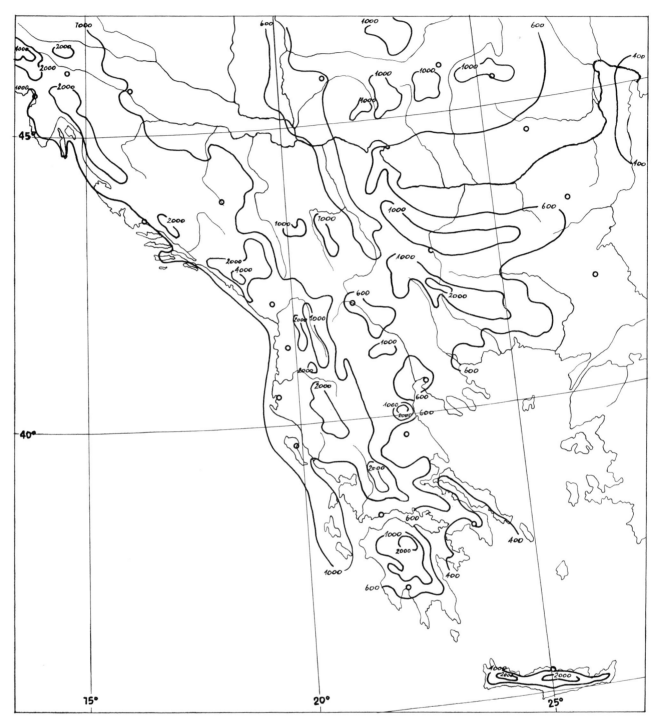

Fig.10. Distribution of total annual precipitation.

Since the mountains in the southwest rob the air of much of its humidity, precipitation in the three other systems—the Rhodope Mountains, the Balkan Mountains, and the Carpathians—is less plentiful, especially as the moist southwesterly winds do not blow at right angles to these ranges. However, the southwest winds are not the only source of humidity, and consequently these mountains have far more precipitation than the surrounding lowlands. The former estimates for the highest regions of the Rhodope,

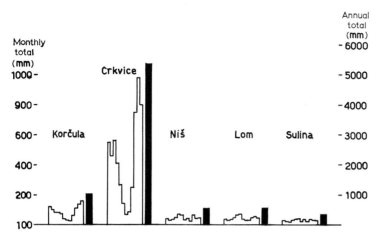

Fig.11. Profile of the total amounts and annual distribution of precipitation from west-southwest to east-northeast.

Balkan, and Carpathian Mountains were about 1,500 mm. However, on the basis of investigations carried out in the Julian Alps, the annual precipitation in the above mountains can be estimated as at least 2,000 mm.

A contrast between north and south is observed, similar to that observed for temperatures in the coldest months, but in this case, the difference appears in the summer. The July or June maximum precipitation occurs on the northern windward slopes of the Carpathians and the Balkan Mountains, where there is well over 150 mm; on the other hand, Crkvice receives only 70 mm. The main areas of precipitation are transferred from the southwest, to the north and northeast. This displacement corresponds to the pattern of air flow over central Europe in summer, which has a northerly component when there are outbreaks of polar air. Winds having a southerly component predominate during the other three seasons, and the greatest amounts of precipitation occur in the Dinaric Alps.

The tendency for a decrease from north to south during summer can also be seen in low-lying regions; thus Wallachia and the southern part of the Pannonian Plain have 80–60 mm in June, the Skopje Basin, about 35 mm, and the interior of the Peloponnese, only about 15 mm. The same tendency is found along the Adriatic and Ionian coasts; for example, the Gulf of Trieste has about 100 mm, the region of Skadarsko Jezero about 30 mm, and the coast of Crete about 1 mm, whilst the Aegean Islands receive no precipitation at all (July and August). A similar situation obtains in the east. The northern coasts of the Aegean have about 20 mm, and the western coasts of the Black Sea over 40 mm.

Annual variation of precipitation

Since the continental regions of the temperate zone are, in winter, predominantly under the influence of anticyclones, winter is the season of minimum precipitation. During the period 1931–1960, the minimum did not usually occur until March in continental regions of southeast Europe. However, it cannot be assumed that this is always the case (Tables VI, VII). It is interesting to note that in Yugoslavia, the dialect word "Sušec" (i.e. "dry") is often used for March. Coastal regions, together with the interior of Greece, but

TABLE VI

THE RELATIVE EXCESS OF PRECIPITATION (%)

Station	Jan.	Feb.	Mar.	Apr.	May	June	July	Aug.	Sept.	Oct.	Nov.	Dec.
Split	+12	+ 9	−22	− 9	−12	−22	−41	−53	−19	+ 4	+62	+ 91
Athens	85	11	13	−31	−31	−58	−83	−80	−56	52	67	112
Bucharest	−11	−26	−24	− 2	43	80	14	1	−38	− 9	−11	− 15

TABLE VII

THE AVERAGE DAILY INTENSITY OF PRECIPITATION (mm)

Station	Jan.	Feb.	Mar.	Apr.	May	June	July	Aug.	Sept.	Oct.	Nov.	Dec.	Year
Crkvice	40.1	39.7	41.2	28.4	17.5	14.0	12.9	13.5	28.9	54.3	62.3	45.7	36.4
Constanta	3.0	2.8	2.0	4.4	4.3	6.3	6.2	7.7	7.6	6.4	4.7	3.7	4.6

excluding the Black Sea coast, have their minimum in July and August, in agreement with the annual variation of a Mediterranean climate. Thus, the minimum occurs in totally different seasons. In the north, the minimum occurs at the end of winter; in the south, in the middle and the end of summer (Table VI).

We find a similar contrast in the times of occurrence of maximum precipitation. The time of maximum cyclonic activity in winter, especially December, corresponds to the time of maximum precipitation in Greece in the extreme south of southeast Europe. The northern coasts of the Aegean, the greater part of Macedonia, and the greater part of the Dinaric Alps, including the Adriatic coast, have their maximum in November.

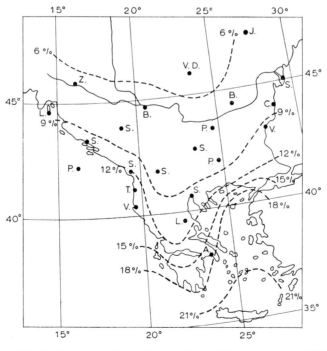

Fig.12. Percentage of annual precipitation falling in January (1931–1960).

206

In the uplands and hills in the extreme northwest, the maximum precipitation occurs in October. The autumn maximum is characteristic of a modified Mediterranean climate. As has already been mentioned, the regions in the north experience the maximum precipitation in June, i.e., in summer (Fig.13). The boundary between Mediterranean and continental regions (so far as precipitation is concerned) coincides in general with the limit of the March precipitation minimum. The weather regimes characteristic of each climatic region quite often extend beyond the boundaries of the region, so that these boundaries are not clearly defined. The characteristic features of each climate extend into the most contrasting regions, to such an extent, that not even the most representative places, Iasi and Heraklion, have an idealized precipitation regime with a single maximum and minimum, though they do approximate to such a regime. The fact that the seasons in which the characteristic months for precipitation occur, are reversed, shows that differences in the precipitation regime of southeast Europe are even greater than in the case of temperature.

Frequency and annual variation of the number of days with precipitation

The distribution of the number of days with precipitation ($\geqslant 0.1$ mm) reflects only partially the distribution of amounts of precipitation in southeast Europe. The figures for the number of days with precipitation are also lower in the east. Some localities on the coast of the southern Aegean Islands frequently have only about 50 days with precipitation, but in general, the figure for the Aegean Islands is 75. Much the same situation obtains in the northern half of the western Black Sea coast. These two regions have already been mentioned, since the annual precipitation does not even reach 400 mm. The western coasts of the peninsula have a greater number of days with precipitation

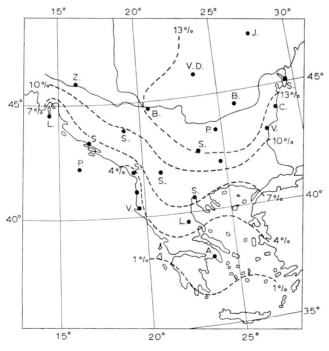

Fig.13. Percentage of annual precipitation falling in June (1931–1960).

than the eastern coasts. In the Gulf of Durrës there are nearly 100 days with precipitation, but in the Gulf of Kavala, barely 80. Between the delta of the Danube and the Gulf of Trieste, the difference increases to 50 days (about 125 and 70). These figures agree fairly well with the annual precipitation. The number of days with precipitation is greater inland than on the western coast, and this is in agreement with the preponderance of summer precipitation in the interior of the peninsula. In southern Thessaly there are over 100 days, and this figure increases to over 150 (Belgrade) on the southern edge of the Pannonian Plain. In the mountains the number of days with precipitation is very high, e.g., nearly 180 at Bjelašnica.

In spite of the contrasting precipitation regime, the minimum of the number of days with precipitation occurs in the second half of summer throughout southeast Europe. Only in Romania does the minimum occur in September. The southern edge of the Pannonian Plain has the greatest number of days with precipitation (about 10) during the month of minimum frequency. The Aegean Islands are another extreme case, having practically no days with precipitation in either July or August. A marked difference between east and west coasts occurs only in the extreme north. In the Gulf of Trieste there are twice as many days with precipitation (8–9) as in the Danube delta (about 4). In the interior, the month with the greatest number of days with precipitation is June, or quite often May or even April. The coasts and the archipelago have the maximum number of days with precipitation in winter: in the west the maximum occurs mainly in December (Table VIII), but in some places (extreme northwest) it occurs as early as November; on the Aegean coast the maximum may occur in January or February as well as in December. As in the case of the annual variation of the amount of precipitation, the Black Sea coast forms part of continental southeast Europe.

In general, in the month of maximum frequency there are 13–15 days with precipitation in the west, and 10–13 in the east. The only area in southeast Europe with less than 10 days is the Black Sea coast.

Large differences in the annual number of days with precipitation can usually be ascribed to differences in the summer. The ratio of the amount of precipitation in regions where it is most abundant to that in regions where it is poorest is 12:1; the ratio of the numbers of days with precipitation is for the same regions 3.5:1.

Maximum daily precipitation

Daily precipitation amounts of over 100 mm can occur over the whole of southeast Europe. In some localities the falls greatly exceed 100 mm: Shkoder 204 mm, Heraklion 194 mm, Bucharest 135 mm, Zakinthos 177 mm, Kerkira (Corfu) 150 mm. The situations of these stations do not bear out the possibility that these particularly high values might

TABLE VIII

THE NUMBER OF DAYS WITH PRECIPITATION $\geqslant$ 10 mm

Station	Jan.	Feb.	Mar.	Apr.	May	June	July	Aug.	Sept.	Oct.	Nov.	Dec.	Year
Crkvice	8.4	7.7	8.6	7.7	6.3	3.7	2.0	2.3	5.1	9.3	11.4	12.0	84.5
Skopje	0.9	0.7	0.6	1.6	1.7	1.2	0.8	0.8	1.1	2.2	1.1	1.3	14.0

be due to relief. However, a completely different picture emerges if we examine the data not only for regions at low altitudes but also for high-level and mountain stations. The relationship between relief and daily rainfall is just as evident as that for annual and monthly amounts of precipitation. Maximum falls of over 300 mm in 24 hours occur in the extreme northwest of the Dinaric Alps and in the Julian Alps (Snežnik, Tržaški kras, Trnovski gozd, Bohinjski greben; the maximum on Mount Matajur is 356 mm). Similar values are found in the Velebit Mountains. It is certainly no coincidence that, of the stations mentioned above, only Shkoder, situated at the foot of a mountain, has a daily maximum of over 200 mm. Not far from this station is Crkvice with a daily maximum of over 550 mm, i.e., more than the average annual precipitation for a fairly large part of southeast Europe. It is highly probable that in the other mountains too, the daily maxima may exceed 300 mm, particularly as a value of 349 mm was reached in eastern Romania (Ciupercenii Vechii). In general, the month for maximum daily precipitation is the same as that for maximum monthly precipitation.

Snow and snow cover

Although in winter the region of the greatest amount of precipitation is around the Aegean, the greatest numbers of days with snowfall occur in the continental part of southeast Europe. The reason for this lies in the distribution of temperature. In Wallachia about 75% of all days with precipitation in winter are also days with snowfall, and in the interior of the Peloponnese this proportion decreases to 40%. Temperatures of 0°C seldom occur on the coast and in the archipelago, and consequently snowfall is very rare. On Kithera, for example, there is only 1% of such days. The number of days with snowfall increases from the coast towards the interior, in the same way as the number also increases towards the north. This is particularly the case on the western coast of the Balkan Peninsula; in the east the decrease is not so marked. There is seldom any snow on the coast of Crete—perhaps once every 10 years. In the interior of the Peloponnese there are 12 days with snowfall; on the Pannonian Plain and in Wallachia this number is doubled (20–30 days). Towards the north differences in height above sea level greatly modify the pattern of the number of days with snowfall. In the central part of the Balkan Peninsula, and likewise in the Carpathians, there are about 35–40 days with snowfall at heights between 500 and 600 m. Bjelašnica (2,067 m) has about 100 days, and Kredarica in the Julian Alps (2,515 m) only 74. It is presumed that the lower number of Kredarica is due to the fact that in the Julian Alps the maximum precipitation occurs in summer.

The difference between the west and east coasts, which has been mentioned so often, applies also to snowfall. On the Black Sea coast there are hardly any localities with less than 10 days with snowfall. In the west it is exceptional for a station to have more than 5 days with snowfall in January. The same applies to the north. The contrast becomes even more marked at the latitude of Constanta or Zadar, where the ratio of days with snowfall is about 10:1 (12.4 and 1.3).

The number of days with snowfall increases from 5 to 25 on the northern Adriatic coast to the southwestern edge of the Pannonian Plain. Over a similar distance from the Black Sea in the direction towards Wallachia the number increases from 15 to 20 days.

In the highest regions in the north of southeast Europe it is almost impossible to distinguish between the past and the coming seasons of snowfall, since even in July and August snow is not exceptional. There is a high probability for the first snowfall to occur by mid-September, even far to the south, including the mountains of northern Greece. A month later, in mid-October, the first snow falls on the highest peaks of the Peloponnese. Again a month later, in November, the first snowfall occurs in low-lying regions in the interior. The first snowfall is delayed until even later on the coast and in the archipelago. The southern coasts of the peninsula and the Aegean Islands do not have their first snowfall until December.

The last snowfall in the Aegean Islands is at the beginning of March, but in the lower lying regions of the interior of Greece this occurs about the middle of the last decade of March; the same is true of the Pannonian Plain and Wallachia. In regions at higher altitudes snowfall does not cease until April or even May. The peaks of the Rhodope Mountains, the Dinaric Alps, and the northern half of the Pindus Mountains may have a snow cover until the end of June. In the Julian Alps and in the Carpathians snowfall persists even in the middle of summer.

On the southern and western coasts the newly fallen snow melts immediately, and even in the extreme north of the Adriatic, in the Gulf of Trieste, the snow cover does not last for more than a day. With increasing distance from the west coast towards the east, and with increasing height above sea level, the snow cover lasts much longer. There is also a big difference between the southwesterly and the northeasterly slopes of the Dinaric Alps and the Pindus Mountains. A snow cover remains on the high plateaus for over 100 days of the year, but in low-lying areas in the interior for less than 50 days. Dobrudza on the Black Sea coast has more than 20 days with snow cover; in the Pannonian Plain there may be as many as 30 days. In between, in Wallachia, the snow cover lasts for 40 days, while in the extreme northeast it may last for more than 50 days.

The contrast between the Adriatic and Black Sea coasts is understandable. The large difference between the number of days with snow cover in the Pannonian Plain on the one hand (less than 30) and in Wallachia and Bukovina on the other (40–55) shows that relatively small temperature differences may have a considerable effect, for the mean temperature in January in the Pannonian Plain is only 2°C higher than in Wallachia.

Clouds and sunshine

As regards cloud and sunshine, southeast Europe cannot be said to form a uniform region. Along the coast and in the archipelago, the annual average amount of cloud is less than 4/10. In the north, in the Pannonian Plain and in Wallachia, the annual average cloud amount exceeds 5/10, and the same applies to the western Black Sea coast. In the highest regions of all five mountain systems, the annual average is greater than 6/10 (Fig.14).

The annual variation of cloud amount does not follow the variation of precipitation—a closer relationship can be traced with the temperature (Table IX). However, the relationship is inverse and there is a slight displacement in time. The maximum for cloud amount occurs in winter, usually in December or January. Along the Adriatic, Ionian, and Aegean coasts, the cloud amounts in the three winter months are very uniform. This winter maximum for cloud amount is understandable, since this is the season of maximum cyclonic activity in the Mediterranean region. However, the fact that the maximum

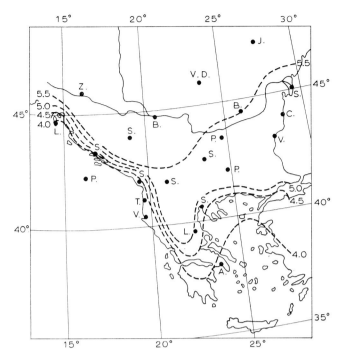

Fig.14. Mean annual amount of cloud (after BIEL, 1944).

TABLE IX

PERCENTAGE FREQUENCIES OF CLOUD AMOUNTS

Station	Cloudiness (tenths):										
	0	1	2	3	4	5	6	7	8	9	10
Split	17	12	11	7	8	3	5	5	8	5	19
Zagreb	22	4	6	6	4	4	3	4	5	4	38

also occurs in winter further inland is explained by the frequency of days with low stratus. This type of cloud occurs only very occasionally over the sea and in the coastal regions, especially in the south. In fact, at the time of the winter precipitation maximum and even during spells with precipitation, there are also clearances lasting for several hours. In the whole of the interior, in the month of maximum cloud amount, this cloud amount is over 7/10.

In the centre of the eastern Mediterranean the cloud amount in the middle of winter is less than 6/10, and this figure is not exceeded either on the Aegean coast, or on the Adreatic coast up to the Gulf of Trieste.

The winter maximum does not occur in all the mountainous regions, but mainly in the south. Towards the north, the cloud amounts decrease in these mountainous regions, and the maximum becomes so displaced that in the Julian Alps and Carpathians it is the minimum instead of the maximum which occurs in the winter. In the Julian Alps this is due to the influence of the Azores high, which is very marked in January. Localities at heights above about 1,500 m under the regime of this high enjoy ideal, clear weather. The average amount of cloud in January is about 5/10, and this is the annual minimum. The winter minimum in the higher regions of the Carpathians is due to the summer maximum, which in turn is due to convection by day. Thus on Mount Onul the average amount of cloud in December is over 7/10, i.e., as much as in Wallachia and Moldavia, but in summer it is over 8/10, which is more than in the surrounding region.

There is a gradual transition from the maximum in winter to the minimum in summer. In most inland localities this minimum occurs in August. In the Carpathians (predominantly) and on the highest plateaus of the Dinaric Alps, the minimum is delayed until September. The coastal regions and also the interior of Greece have the minimum in July (Fig.15). There is not much difference between the interior and the coastal region in the time at which the annual minimum cloud amount occurs. The average amount of cloud at the time of the annual minimum is less than 3/10 in the whole coastal region as far as the Gulf of Quarnero. The same applies partly to the Black Sea coast. Inland, the minimum average is 3/10–5/10. In the highest regions there is a marked increase due to heavy convection clouds in the afternoon. As already mentioned, the monthly mean cloud amount in the Carpathians is 8/10, and this is sufficient to make it the annual maximum. The same is true of the Julian Alps.

In spite of the neighbouring eastern Mediterranean, where there is practically no cyclonic activity during the summer, the influence of thermal convection is also evident in the south (Bjelašnica, 5/10). Differences in the amount of cloud between the lowlands in the north and the archipelago of the southern Aegean are thus most marked in summer, particularly in July and August.

With regard to the duration of sunshine (Table X), certain regions of southeast Europe rank first in the "old" continent. The western coasts of the Peloponnese together with the Ionian coasts and the archipelago, and likewise the western coasts of Asia Minor and the islands in the Aegean, have more than 3,000 h of sunshine per year. The southern half of the Adriatic, where as far as Split there are about 2,650 hours of sunshine, is very favoured. The remaining coasts of Greece have about 2,500 h, and inland the figure

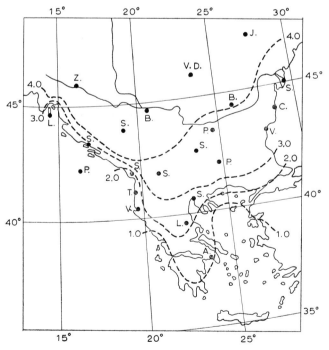

Fig.15. Mean amount of cloud in July (after BIEL, 1944).

TABLE X

PERCENTAGE OF THE ASTRONOMICALLY POSSIBLE DURATION OF SUNSHINE

Station	Jan.	Feb.	Mar.	Apr.	May	June	July	Aug.	Sept.	Oct.	Nov.	Dec.	Year
Sulina	21	30	27	44	63	66	71	75	58	46	30	29	56
Sofia	25	28	35	44	50	58	67	68	55	44	29	18	43
Athens	49	50	52	55	53	66	81	81	73	61	43	37	60
Lošin Mali	40	46	48	51	59	62	71	73	61	49	40	33	54
Bjelašnica	31	29	30	25	28	33	47	54	46	35	29	26	34

runs at about 2,300 h. The Black Sea coasts and the coasts of the northern Adriatic receive more than 2,200 h. Wallachia has just as much sunshine, but elsewhere inland the number of hours is only about 2,000 h.

The duration of sunshine decreases rapidly in the mountains. A good example of this is Bjelašnica, with barely 1,550 h, while Sarajevo (to the north of Bjelašnica) has about 1,700 h and Split (on the Adriatic coast) even 2,650 h. The same applies to the peaks of the Carpathians; Iasi in the lowland has over 2,000 h, and Mount Onul only about 1,400 h.

Relative humidity

In contrast to other meteorological elements, the distribution of the relative humidity is comparatively uniform.

Inland, the annual relative humidity varies between 70 and 80%, but tends to be nearer to the lower of these two figures. In the mountains, from data for Bjelašnica and Sonnblick, the values run to over 80%. In contrast to the mountains, the relative humidity in the coastal regions is below 70%, but nowhere less than 60%. This is only the broad picture, but it should be pointed out that the Adriatic coast up to the Gulf of Kvarner has the same relative humidity as Crete (over 60%), and that values on the Black Sea coast are similar to those in Wallachia and in the Pannonian Plain (over 70%). It should further be pointed out that inland in Greece, but excluding the mountains, the relative humidity is the same as on the coast, i.e., below 70%.

The annual variation of the relative humidity is similar to that of the amount of cloud and is very simple. In most cases the maximum occurs in December. However, in Greece, all three winter months have the same percentage relative humidity. In December the differences are very small. In the coastal region from the Gulf of Trieste, through Crete and the Aegean, and further northwards including the southern half of the western coast of the Black Sea, the relative humidity is below 80%. The whole of the interior, including the Dinaric poljes and the basins of the Rhodope Mountains, the Pannonian Plain and Wallachia, and probably all mountains, have a relative humidity of over 80%. In regions which are often in fog, the relative humidity can be estimated at over 90% in the month of maximum humidity.

The minimum occurs in the middle of summer, in July and August. In the western part of southeast Europe, i.e., in the southwest of the Pannonian Plain and also in the eastern Alps, the minimum relative humidity occurs as early as April, but the differences between April and September are small over the whole of Europe. The distribution of the minimum relative humidity is very simple. The greater part of southeast Europe has a relative humidity of over 60% in the month of minimum humidity. From the central Adriatic coast, through Macedonia and south of the eastern Rhodope Mountains, the relative humidity is below 55%. The eastern coasts of Greece, including Thessaly and the western Aegean Islands, have a relative humidity below 50%.

The annual variation of the relative humidity is also simple in the mountains, exhibiting one maximum and one minimum. The rule for the temperate zone is that in mountain regions the maximum occurs in summer and the minimum in winter. This is the opposite to the situation in the lowlands. From data for Sonnblick, which is outside southeast Europe, it would appear that the same annual variation applies in the mountains in the north.

The differences in average humidity deficit during fine weather conditions and bad weather conditions in Ljubljana are shown in Fig.16.

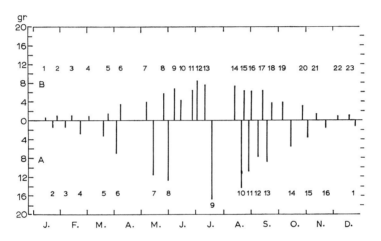

Fig.16. The average humidity deficit (in g) at 14h00 at Ljubljana during fine weather conditions (A) and bad weather conditions (B) (1947–1956).

Evapotranspiration

The exceptionally high proportion of mountainous country in southeast Europe precludes all hopes that it will ever be possible to give a true representation of the actual evapotranspiration in this region, and the outlook for potential evapotranspiration is hardly any more promising.

The empirically derived formula of Thornthwaite to calculate potential evapotranspiration, which is often the most practical, depends only on temperature. This is a great advantage from the practical point of view, but at the same time gives rise to certain disadvantages. In a region whose relief is as irregular as that of southeast Europe, the fact that other elements, particularly wind, are neglected means that the results are not very accurate.

Even applying the assumption that potential evapotranspiration is equal to evaporation from a water surface does not easily enable us to obtain data for a region with complex relief. The evaporation is usually determined by using so-called Class-A-pan evaporimeters. Attempts to determine the potential evapotranspiration using this instrument

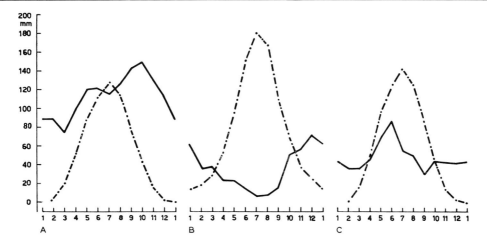

Fig.17. Annual variation of potential evapotranspiration (according to Thornthwaite) compared with annual variation of precipitation at three stations in southeast Europe. A. Ljubljana; B. Athens; C. Bucharest.

(in the extreme northwest of southeast Europe) showed that it could not be used under the following circumstances:

(*1*) In regions having a high number of days with precipitation. The values on such days had to be interpolated. Where the wind was not constant or was light, or where there was a preponderance of calms, the interpolated values were very doubtful. This means that in regions where the annual number of days with precipitation is greater than about 150, nearly one half of the data are questionable.

(*2*) In mountainous regions.

There is no sound basis for determining a coefficient by means of which the effects of the isolated position of the evaporimeter could be eliminated.

In the region investigated, the valley of the upper Sava, the ratio of precipitation to run-off led to a coefficient of 0.9. This value is obviously exceptionally high, and it would be very risky to apply it in areas where it is impossible to carry out water-balance measurements.

It is therefore quite possible that the use of a doubtful coefficient for evaporation from a water surface (and thus also for potential evapotranspiration) has led to doubtful estimates for the extreme south for which the following values have been given: Athens > 1,600 mm; and for Thessaly > 1,300 mm per year.

In the northern regions of southeast Europe, potential evapotranspiration as determined by Thornthwaite's method, amounts to over 700 mm in the Pannonian Plain as well as in Wallachia, up to about 150 m altitude. In the higher situated (approx. 150 m) pre-Alpine basins and plateaus in the extreme northwest, the evapotranspiration diminishes but still remains over 600 mm. Moving towards the southeast, the annual potential evapotranspiration (P.E.) slowly increases until, in Macedonia, at a height of ca. 400 m, it again exceeds 700 mm.

In the Julian Alps, Carpathians, and Rhodope Mountains, the potential evapotranspiration is some 300 mm at an altitude of about 2,500 m.

The southwestern and southern coasts of the Balkan Peninsula show a greater P.E. as a result of their higher temperatures. In the Gulf of Trieste it amounts to over 750 mm, along the Albanian coast and in the Gulf of Salonica to over 850 mm, while on the coast of Attica, to a value some 100 mm higher.

The differences in P.E. value may appear to be relatively small compared to the large variations in temperature in several regions of southern Europe. However, this is only to be expected when one realizes that the summer temperatures, which are almost the same over this region, are the significant factors in the determination of the annual P.E. The climatological pattern of southern Europe is better presented in terms of the actual evapotranspiration calculated according to Thornthwaite's method. On the coast of Attica this comes to over 350 mm, but in the southern part of Wallachia, to over 500 mm. The proportion is thus reversed in comparison with the P.E. This contrast becomes still more marked when the actual evapotranspiration is expressed in percentages of the P.E. On the coast of Attica the actual evapotranspiration amounts to only 40% of the P.E., whilst in Wallachia it amounts to almost 80%, being therefore twice as large. The approximation of the actual evapotranspiration to the P.E. in areas in the north only occurs in the middle high, or high, mountains, due to the relatively high precipitation during summer at the lower temperatures of mountainous areas.

It should be firmly emphasized that for the extreme northwest of southeast Europe the values calculated with Thornthwaite's formula seem to agree with those obtained from measuring the water balance.

Pressure distribution and winds

Pressure distribution

Being small and surrounded by three continents, the Mediterranean Sea has little effect on the annual variation of pressure. Even in Crete, in the centre of the eastern Mediterranean, the maximum occurs during the cool half of the year and the minimum during the warm half of the year. This is a continental type of annual pressure variation (Fig.18). The pressure gradient has a predominantly northerly component, both in summer and winter (Table XI). In the cool half of the year the belt of high pressure extending from the Pacific coast to the Atlantic coast dominates the situation. An additional influence

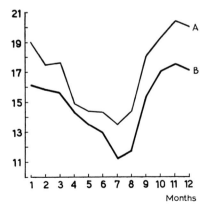

Fig.18. Annual variation of pressure: *A* = Athens; *B* = Bucharest.

TABLE XI

THE PRESSURE GRADIENT (mbar) BETWEEN LJUBLJANA AND TRIESTE AND THE FREQUENCY OF NORTHEAST WINDS IN TRIESTE

	Jan.	Feb.	Mar.	Apr.	May	June	July	Aug.	Sept.	Oct.	Nov.	Dec.
Gradient (mbar)	3.2	2.1	1.3	0.5	0.3	0.3	0.7	0.9	1.5	1.7	2.7	3.5
Frequency	47	42	38	34	30	28	30	34	35	39	46	47

is that of the Aegean low to the southeast of Cyprus. During the summer the direction of the gradient is determined by an extension of the Azores high in northwest Europe and particularly by the low over the Iranian Plateau.

From the maximum in the coldest months of the year the pressure gradually falls until mid-summer (Table XI). The increase in pressure in the autumn is not steady. The greatest increase occurs between August and September and amounts to more than 3 mm, i.e., about 50% of the annual amplitude, which in the extreme south and the extreme north is about 7 mbar (Iasi, 1021.7 mbar, 1014.8 mbar; Heraklion, 1017.3 mbar, 1010.7 mbar). During the period 1931–1960 the pressure maximum in most regions was reached by November and not in January. Such displacements in time are assumed to be due to climatic irregularities and mainly to the "peninsula" character of Europe. Averaged over a longer period, the pressure maximum occurs in January, the coldest month of the year.

Winds

According to the pressure gradient, northerly winds should prevail in southeast Europe during both the warm and the cold times of the year; however, this is prevented by the complexity of the relief. The northernmost ridges of the Dinaric Alps and the whole of the Balkan Mountains run in a zonal direction, and as a result of this the northern part of southeast Europe has mainly easterly or westerly winds (Fig.19). Wind directions which are not due to relief are of little importance. On the other hand, calms are more

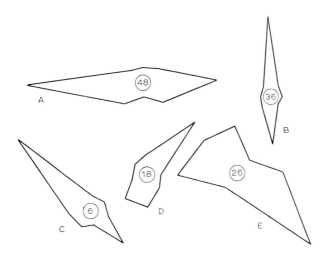

Fig.19. Typical wind roses due to: relief (A. Plovdiv; B. Cvrljivica, annual); etesians (C. Korčula, July); bora (D. Postojna, January); kosava (E. Belgrade, January).

frequent. Since the prevailing wind direction in each locality is due to the relief, the seasons have very little effect.

Inland there is a predominance of winds of Beaufort force 2–3. Exceptionally, however, wind strengths up to force 12 may occur anywhere.

The following local winds occur either partly or only in southeast Europe: etesian winds, bora, vardarac, košava and föhn.

Etesians

Northerly components, corresponding to the pressure distribution, are found mainly along the coasts. The annual average direction is largely determined by the flow during the warm half of the year, known as the "etesians". Including some interruptions, these last from mid-May to mid-September. The duration and frequency of interruptions decrease towards the southeast Mediterranean. The chief characteristic of the etesians is their constancy in direction and strength. In the region of southeast Europe they are light winds, blowing from the north in the east, and from the northwest in the west. The etesians are dry and invigorating, since the air comes from the north. Before the advent of steam ships these winds were of great importance for navigation because of their constancy in direction and strength. On the Adriatic coast the etesians are known as the mistral. Although they are directly due to the low over the Iranian Plateau, their onset is a sign of settled, fine weather.

Bora

The bora occurs in southeast Europe from the Gulf of Trieste to the Gulf of Medovani, but the classical region for its occurrence is mainly the northern coast of the eastern Adriatic. Immediately adjacent to this, in the transition zone from the Dinaric Alps and the Julian Alps, is the deepest and narrowest pass between central Europe and the Mediterranean Basin. When the pressure gradient is sufficiently large, polar air from the northwest or northeast most frequently breaks through to the Mediterranean via this pass. The speed of the descending air is greatest at the southwestern foot of the Karst Plateau, which faces the sea. Speeds of over 130 km/h are a frequent occurrence in the Gulf of Trieste and in the Senj Channel. The main characteristic of the bora is its gustiness; some gusts reach a speed of 60 m/sec. Near Ajdovščina, below Pass Col in the Vipava Valley, the bora also reaches a strength comparable with that in the Gulf of Trieste. The temperature difference between the old air mass and the new one flowing in, is too great for it to be neutralized by adiabatic warming of the descending air. On the island of Hvar, where a weak bora regularly occurs, the temperature on days with bora, is, in winter, more than 3°C below normal, but the relative humidity is also 20% lower. The bora is a healthy wind, but it does give rise to some difficulties. Inland it causes snow drifts, while on the Adriatic, waves reach a height of up to 2.5 m. This wind may thus temporarily bring all shipping to a standstill. This happens particularly at Kraljevica, Senj, Šibenik, and Dubrovnik.

The main season for the bora is the cold half of the year; nevertheless, it may also occur in mid-summer, provided that polar air penetrates to the eastern coast of the northern Adriatic.

It is only in exceptional cases that the occurrence of bora is not connected with cyclonic activity (especially the Adriatic cyclone) above the Mediterranean.

The polar air, which penetrates as bora into the Mediterranean Basin, forms the cold sector of the cyclones, whilst the warm sector is represented by tropical air currents from the southeast. This is the jugo or juzina, the southern wind which normally precedes the bora and whose annual path, therefore, is bound to that of the bora. Since the jugo is neither a squall nor is it gusty, but rather blows hard and continuously, the waves created as a result, are even higher than those of the bora (over 3 m).

The tropical air of the jugo is of the same North African origin as the air brought to Europe by the sirocco. The distinction between these kinds of winds lies mainly in the difference in thickness of the air layer in which they occur.

The sirocco is bound to the large waves of polar air which reach to the border of the stratosphere, while the jugo is generated by smaller waves and therefore remains restricted to the lower air layers. Both names, however, are used.

Vardarac

This wind blows on the northern coast of the Gulf of Salonica and in the hinterland along the valley of the River Vardar. Its physical characteristics are similar to those of the bora, but it is much less gusty.

The vardarac only occurs when a high extends over the Balkan Peninsula while there is a low over the Aegean Sea.

Košava

The usual season for this wind is from October to April. It, too, is a gusty wind, with gusts up to 30 m/sec. It blows from the east, and the region where it occurs is the extreme southeastern part of the Pannonian Plain together with the northern slopes of the Carpathians. The košava most frequently lasts for 2–3 days, but it may also last for weeks, perhaps with some interruptions. The extreme limits for its occurrence are the western Mačva in the west and the basin of Kragujevac in the south; in the north it occurs over the whole Vojvodina. The košava brings an increase rather than a decrease of temperature, but because of the low relative humidity, and especially its strength (over 12 m/sec in Belgrade), it is regarded as a cold wind. The most frequent synoptic situation with which it is associated is an anticyclone over eastern Europe with a depression over the western Mediterranean. Polar-continental or arctic air flows from the higher ridges of the Balkan Mountains.

Föhn

The föhn occurs with a cyclonic southwesterly circulation to the lee of the Julian Alps and the Dinaric Alps. It blows in the valleys of the right-bank tributaries of the Sava as far as the Pannonian Plain; towards the east it reaches Belgrade. As in the valley of the Inn, it is most frequent in spring. Zagreb has 70 days per year with föhn, and Sarajevo, about 100 days. In winter, the temperature on days with föhn is about 6°C higher than on other days, while in the summer the difference is only about 2°C. On days with föhn the relative humidity decreases by about 10% in valleys and basins. The strongest föhn is experienced in the Bosna Valley, whilst towards the southeast and northwest its influence gradually decreases.

The valleys and basins of the southeastern Alps also experience northwesterly föhn, in the same way as the Plain of Lombardy. This occurs most frequently during summer and usually does not cause any changes of temperature. Northwesterly föhn is a sign of an improvement in the weather, while southwesterly föhn is a sign of a deterioration. Both types of föhn are caused by an incursion of polar air.

Remarks on the climatic variation in southeast Europe

A general increase in temperature over the Northern Hemisphere has been traced from the beginning of this century until the 1940's; since then temperature has again decreased. From the series of observations, extending over nearly 80 years, for the city of Belgrade, situated in the heart of southeast Europe, it can be seen that the most recent decades (1931–1960) are warmer than preceding ones (1891–1920) by about 0.7°C. In addition to the general increase in temperature in the Northern Hemisphere over most of this period, this large increase is undoubtedly partly due to the increase in the size of the city. The temperature increase amounts to 0.3°C for the winter months, but is about 1.0°C between April and November. If, however, we assume that the temperature increase due to the growth of the city is almost neutralized by the košava winds, we have a temperature increase of 0.3°–0.5°C, which is in agreement with the tendency for the whole of the Northern Hemisphere. The reverse is true of the tendency during the warmer months, for this tendency remains positive even after allowance has been made for the influence of the city (about 0.5°C). The tendency during summer in southeast Europe is thus not in agreement with that for the remainder of the Northern Hemisphere. The Belgrade station is certainly not representative of all three climatic zones of southeastern Europe. It has the advantage, however, that during the whole of the 80 years observation period, all measurements have been made at the same place, which is obligatory for a study of possible temperature shifts. Even a slight displacement of the temperature cabin may cause an inhomogeneity in the records. In the case of displacement of the pluviometer the consequences are even worse, especially when the relative height of the instrument is altered.

Due to the latter fact, one can rely only on the data of a series of observations in the turbulent region of southeast Europe. Although the variability of precipitation is much larger than the variability in temperature, analysis of the continuous series of observations covering 80–115 years, shows the following result—the Belgrade station, lying in the centre of southeast Europe, has, during the last decade recorded over 10% more annual precipitation than in former decades, this excess occurring mainly in wintertime. In the east and south (Bucharest, Athens) the annual precipitation shows no differences between the last decade and previous decades. Neither are there notable seasonal discrepancies. The same is also true for the extreme western part of southeast Europe (Ljubljana). Here during the last 50 years, some unusually large shifts took place and the highest precipitation values of the whole 115 year observation period were recorded in the two decades 1920–1940. In the following two decades, however, only the average minimum of the total observation series was reached.

References

AULITZKY, H., 1967. Lage und Ausmass der "warmen Hangzone" in einem Quertal der Inneralpen. *Ann. Meteorol.*, 3: 154–165.

BAUR, F., 1958. *Physikalisch-statistische Regeln als Grundlagen für Wetter- und Witterungsvorhersagen.* Akademische Verlagsgesellschaft MBH, Frankfurt, II.B, 152 pp.

BERGERON, T., 1960. Problems and methods of rainfall investigation. "Project Pluvius". In: H. WEICKMAN, *Physics of Precipitation: Geophys. Monograph*, 5: 5–30; 152–157.

BERMAN, S., 1965. Estimating the longitudinal wind spectrum near the ground. *Q. J. R. Meteorol. Soc.*, 91: 302–317.

BIEL, E., 1927. Klimatographie des ehemahligen österreichischen Küstenlandes. *Öst. Akad. Wiss., Wien, Kl.*101: 137–195.

BIEL, E., 1944. *Climatology of the Mediterranean.* University of Chicago Press, Chicago, Ill., 180 pp.

BJERKNES, J., 1963. Climatic change as an ocean–atmosphere problem. *Proc. W.M.O.–Unesco Symp. Climatic Changes, Rome, 1961, Arid Zone Res.*, 20: 297–321.

BLÜTHGEN, J., 1964. *Allgemeine Klimatographie.* De Gruyter, Berlin, 599 pp.

BUETTNER, K. J. K., 1968. Valley wind, sea-breeze and mass fire: three cases of quasi-stationary airflow. *Colo. State Univ., Dept. Atmos. Sci. Pap.*, 122: 103–129.

DAGNINO, I., 1952. Sull andamento annuo della pressione atmosferica del Mediterraneo. *Geofis. Pura Appl.*, 21: 43–48.

DEFANT, F. and DE BOOGAARD, H. M. E., 1963. The global circulation features of the troposphere between the Equator and 40°N based on a single day's data. *Tellus*, 15: 251–261.

DJUKANOVIĆ, D., 1958. Osunčavanje vertikalnih površina u Beogradu. *Vjesn. Hidrometeorol. službe FNR Jugoslavije*, 7: 34 pp.

DOBRILOVIĆ, B., 1960. *Visinsko strujenje iznad Jugoslavije i prizemni karakteristični vetrovi.* Prirodosl.-mat. Fak. Beogr. Meteorol. zavod, Beogr., 3: 146 pp.

DUBIEF, J., 1959. *Le Climat du Sahara.* L'Institut de Météorologie et de Physique du Globe de l'Algérie, Algiers, 1: 312 pp.

EKMAN, V. W., 1932. Über die Beeinflussung von Windbahnen durch Gebirge. *Beitr. Phys. Atmos.*, 19: 272–274.

FLOHN, H., 1948. Zur Kenntnis des jährlichen Ablaufes der Witterung im Mittelmeergebiete. *Geofis. Pura Appl.*, 13: 167–188.

FLOHN, H., 1953. Studien über die atmosphärische Zirkulation in der letzten Eiszeit. *Erdkunde*, 7: 266–275.

FLOHN, H., 1954. *Witterung und Klima in Mitteleuropa.* Hirzel, Stuttgart, 214 pp.

FLOHN, H. and HUTTARY, J., 1950. Über die Bedeutung der Vb-Lagen für das Niederschlagsregime Mitteleuropas. *Meteorol. Rundsch.*, 7(8): 167–170.

FURLAN, D., 1954. Nekaj novejših podatkov o padavinah v Julijskih Alpah. *Letno Poročilo Meteorološke službe Slovenije za Leto 1954, Ljubljana*, 1: 93–97.

FURLAN, D., 1959. O uveljavljanju srednjeevropskih singularitet na prodročju Jugoslavije. *Geog. Vestn., Ljubljana*, 17: 121–130.

FURLAN, D., 1960. L'influence du relief sur la répartition des températures. *Congr. Int. Meteorol. Alpine, 6me, Bled*, pp.263–273.

FURLAN, D., 1961. Padavine v Sloveniji. *Acad. Sci. Artium Sloven., Acta Geog., Ljubljana*, 6: 159 pp.

FURLAN, D., 1963. Determinazione delle precipitazioni mediante le portate unitarie. *Geofis. Meteorol.*, 11: 175–176.

FURLAN, D., 1964. *Ugotavljanje Evapotranspiracije s Pomočjo Normalnih Klimatskih Pokazateljev.* Hidrometeorološki zavod SR Slovenije, Ljubljana, 74 pp.

FURLAN, O., 1965. Temperature v Sloveniji. *Opera Acad. Sci. Artium Sloven.*, 15: 166 pp.

GAMSER, F. and GAMSER, K., 1967. Zamučenost atmosfere i direktno sunčevo zračenje u Beogradu. *Zb. rad. Savezni hidrometeorološki zavod, Beograd*, pp.21–55.

GRUNOW, J., 1952. Beiträge zum Hangklima. *Ber. Deut. Wetterdienstes U.S.-Zone*, 5: 293–298.

HOČEVAR, A., 1967. Globalno sevanje v Ljubljani. *Razpr. Pap., D.M.S., Ljubljana*, 8: 3–30.

HUTTARY, R., 1950. Die Verteilung der Niederschläge auf die Jahreszeiten im Mittelmeergebiet. *Meteorol. Rundsch.*, 5(6): 111–119.

ISIDOROV, C. C., 1955. *Klima e Shoiperise.* Tirana, 159 pp.

KARAPIPERIS, L. N., 1953. Influence of the etesian winds on the summer temperatures in Athens. *Meteorol. Mag.*, 82: 238–239.

KARAPIPERIS, L. N., 1954. Über eine Klassifizierung der Etesien auf Grund der herrschenden isobarischen Zustände. *Meteorol. Rundsch.*, 7: 6–9.

KIRIGIN, B., 1963. *Prikaz klimatskih prilika planine Medvednice.* Hidrometeorloški zavod SR Hrvatske, Zagreb, 72 pp.

KOHLER, M. A., NORDENSON, T. J. and FOX, W. E., 1955. Evaporation from pans and lakes. *U.S. Weather Bur., Res. Pap.*, 38: 340 pp.

KOVAČEVIĆ, M., 1942. Temperature vazduha. In: S. ŠKREB et al., *Klima Hrvatske*. Geofiz. zavod Zagrebu, Zagreb, 35–68.

LAMB, H. H., 1965. Frequency of weather types. *Weather*, 20: 9–12.

LANDSBERG, H., 1960. *Physical Climatology*, Gray Printing Co., Du Bois, Pa., 446 pp.

LANDSBERG, H., 1967. Two centuries of New England climate. *Weatherwise*, 20: 52–57.

LAUSCHER, F., 1960. Lufttemperatur. *Klimatograph. Österreich, Wien*, 3: 137–206.

LEE, R. and BAUMGARTNER, A., 1966. The topography and insolation climate of a mountainous forest area. *Forest Sci.*, 12: 258–267.

LINSLEY JR, R. K., KOHLER, M. A. and PAULUS, J. L. H., 1958. *Hydrology for Engineers*. McGraw-Hill New York, N.Y., 340 pp.

LORENZ, E. N., 1967. *The Nature and the Theory of the Atmosphere*. W.M.O., Geneva, 161 pp.

MANOHIN, V., 1939. Ein Beitrag zum Studium der Singularitäten. *Meteorol. Z.*, 11: 485–487.

MARGETIĆ, F., 1942. Oborine. In: S. ŠKREB et al. *Klima Hrvatske*, Geofiz. Zavod Zagrebu, Zagreb, pp.104–123.

MARIOLOPOULOS, E. G., 1961. An outline of the climate of Greece. *Meteorol. Inst., Univ. Athens*, Publ., 6: 51 pp.

MELIK, A., 1935. *Slovenija I.1.* Slovenska Matica, Ljubljana, 192 pp.

MELIK, A., 1948. *Jugoslavija*. Slovenska Matica, Ljubljana, 393 pp.

MILOSAVLJEVIĆ, M., 1963. *Klimatologija*. Naučna knjiga, Beograd, 259 pp.

MILOSAVLJEVIĆ, K., 1963. *Rezultati osmatranja meteorološke observatorije u Beogradu u periodu 1888–1962.* Hidrometeorološki zavod SR Srbije, Beograd, 107 pp.

MITCHELL, J. M., 1963. On the world-wide pattern of secular temperature change. In: Proc. W.M.O.-Unesco, Symp. Clim. Change, Rome 1961—Arid Zone Res., 20: 161–181.

PARADIŽ, B., 1957. Burja v Sloveniji. *10 let Hidrometeorološke Službe, Hidrometeorološki zavod Slovenije, Ljubljana*, pp.147–172.

PASQUILL, F., 1962. *Atmospheric Diffusion*. Van Nostrand, London, 297 pp.

PENZAR, I., 1959. Globalna radiacija u Zagrebu na temelju 10-godišnjega merjenja. *Radovi Geofiz. Inst.*, Ser. 3, 12: 35 pp.

PHILIPPSON, A., 1947. Griechenlands zwei Seiten. *Erdkunde*, 1(2): 144–162.

POPOVIĆ, M., 1974. Sume sunčanog zračenja izmerene u Beogradu na vertikalno orijentiranim površi nama. *X. Savetovanje klimatologa Jugoslavije, Pale-Sarajevo 1973*. Hidrometeorološki zavod SFLR Jugoslavije. In press.

RADINOVIĆ, D. and LALIĆ, D., 1959. *Ciklonska Aktivnost u Zapadnom Sredozemlje*. Izdanje Saveznog hidrometeorološkog zavoda, Beograd, 57 pp.

REYA, O., 1933. Odnošaji med padavinami in cikloni v Jugoslaviji. *Geogr. Vestn., Ljubljana*, 9: 165–180.

ROSSBY, C. G., 1949. On the nature of the general circulation of the lower atmosphere. In: G. P. KUIPER (Editor), *The Atmospheres of the Earth and Planets*. Univ. Chicago Press, Chicago, Ill., pp.16–48.

SCHAMP, H., 1954. Die Etesien, ein Luftkörper? *Meteorol. Rundsch.*, 7: 174–177.

SCHERHAG, R., 1948. *Neue Methoden der Wetteranalyse und Wetterprognose*. Springer Verlag, Berlin, Göttingen, Heidelberg, 424 pp.

SCHERHAG, R., 1950. Betrachtungen zur allgemeinen Zirkulation, I. Die Schwankungen der allgemeinen Zirkulation in den letzten Jahrzehnten. *Ber. Deut. Wetterdienstes U.S.-Zone*, 12: 40–44.

SCHINZE, G., 1932. Die Erkennung der troposphärischen Luftmassen aus ihren Einzugsfeldern. *Meteorol. Z.*, 5: 169–179.

SCHNEIDER-CARIUS, K., 1947/1948. Die Etesien. *Meteorol. Rundsch.*, 1: 464–470.

SCHNELLE, F., 1965. *Frostschutz im Pflanzenbau*. Bayrischer Landwirtschaftsverlag, München, 604 pp.

SEIDL, F., 1935–1936. Dinarskogorski fen. *Geogr. Vestn., Ljubljana*, 11–12:

SELLERS, W. D., 1965. *Physical Climatology*. Univ. Chicago Press, Chicago, Ill., 272 pp.

STONESCU, ST. M. and TISTEA, D., 1962. *Clima Republicii Populare Romine*. Institut Meteorologic, Bucuresti, 164 pp.

SUTCLIFFE, R. C., 1960a. Depressions, fronts and airmass modification in the Mediterranean. *Meteorol. Abhandl.*, 1: 135–144.

SUTCLIFFE, R. C., 1960b. The Mediterranean in relation to the general circulation, *Meteorol. Abhandl.*, 1: 125–133.

TANNER, C. B. and PELTON, W. L., 1960. Potential evapotranspiration estimated by the approximate energy balance method of Penman. *J. Geophys. Res.*, 65: 391–413.

URFER-HENNENBERGER, C., 1967. Zeitliche Gesetzmässigkeiten des Berg und Talwindes. *Veröff. Schweiz. Meteorol. Zentralanst.*, 4: 245–252.

UTTINGER, H., 1951. Zur Höhenabhängigkeit der Niederschlagsmengen in den Alpen. *Arch. Meteorol., Geophys. Bioklimatol., Ser. B*, 4: 360–382.

VUJEVIĆ, P., 1953. Podnebje Jugoslavije. *Arhiv Poljoprivredne nauke*, 6: 3–46.

VUJEVIĆ, P., 1948. *Meteorologija. Beograd,*

WILLIS, E. H., 1961. The validity of radiocarbon dating. *Times Sci. Rev.*, Winter, pp.6–9.

Climatological data sources

Clima Republicii Populare Romine, 1961. Vol. II, Date Climatologice, Institut Meteorologic, Bucuresti 283 pp.

Climatological Normals (Clino) for Climate and Climate Ship Stations for the Period 1931–1960. WMO/OMM, No. 117, TP.52.

Padavine u Jugoslaviji, 1957. Rezultati osmatranja za period 1925–1940. Hidrometeorološka služba FNR Jugoslavije, Beograd, 571 pp.

Rezultati osmatranja Meteorloške observatorije u Beogradu u periodu 1888–1962, 1963. Hidrometeorološki zavod S.R. Srbije, Beograd, 107 pp.

Solar Radiation Data. Published monthly by A. I. Voeikov Observatory, Leningrad.

Temperature, vetar i oblačnost u Jugoslaviji, 1952. Rezultati osmatranja za period 1925–1940. Hidrometeorološka služba FNR Jugoslavije, Beograd, 104 pp.

TABLE XII

CLIMATIC TABLE FOR IASI

Latitude 47°10'N, longitude 27°36'E, elevation 101 m

Month	Mean sta. press. (mbar)	Temperature (°C)				Mean vap. press. (mbar)	Precipitation (mm)		Days with snow (mean)[*1]
		mean daily	daily range	extreme			mean	max. in 24 h	
				max.	min.				
Jan.	1020.6	−4.1		15.4	−29.0	3.2	31.9	34.5	7.3
Feb.	1018.8	−2.3		18.2	−30.0	3.4	29.6	23.5	6.2
Mar.	1019.7	2.5		25.5	−22.7	4.4	21.3	28.0	3.9
Apr.	1016.7	10.0		31.5	−7.0	6.2	37.4	38.1	0.9
May	1016.2	16.0		36.4	−2.1	9.1	47.9	48.5	0.0
June	1015.9	19.5		36.9	3.5	11.0	73.3	83.0	–
July	1014.8	21.6		39.0	8.5	12.6	61.6	110.9	–
Aug.	1015.7	20.7		39.6	6.0	10.3	64.0	51.2	–
Sept.	1019.0	16.2		38.0	−1.0	9.8	35.8	33.0	0.0
Oct.	1020.7	10.0		33.9	−6.5	7.2	34.8	55.0	0.5
Nov.	1021.7	4.0		21.1	−18.0	5.4	39.0	47.3	2.8
Dec.	1020.8	−1.0		18.2	−29.5	4.0	29.1	20.6	4.8
Annual	1018.3	9.4		39.6	−30.0	7.4	505.7	110.9	26.4

Month	Mean evap. (mm)	Number of days with			Mean cloud-iness (tenths)	Mean sun-shine (h)	Wind		Days with snow cover		Relat. humid. (%)
		precip. ⩾0.1 mm	thunder-storm	fog			preval. direct. [*3]	mean speed (m/sec)	mean[*2]	max.	
Jan.		10.0	–	8.8	7.3		NW (20.0)		20.2		83
Feb.		10.8	–	5.8	7.2		NW (24.4)		17.2		80
Mar.		8.5	0.2	3.8	6.3		NW (23.3)		5.6		72
Apr.		9.1	0.4	0.8	6.1		NW (20.1)		0.3		62
May		10.7	3.5	0.6	5.8		NW (23.3)		–		61
June		12.1	4.0	0.4	5.7		NW (28.5)		–		62
July		9.5	4.1	0.6	4.5		NW (31.7)		–		60
Aug.		9.0	3.0	1.1	4.4		NW (28.2)		–		63
Sept.		7.3	0.9	2.0	4.4		NW (20.8)		–		66
Oct.		8.4	0.3	4.0	5.6		NW (16.4)		0.1		73
Nov.		10.0	0.0	5.5	7.2		SE (18.1)		2.0		81
Dec.		10.2	–	8.3	7.6		NW (15.2)		11.4		85
Annual		115.7	16.4	41.7	6.0		NW (22.3)		56.8		71

[*1] 1896–1915, 1921–1935, 1938–1939, 1941–1943, 1945–1955; [*2] 1926–1955; [*3] 1946–1960.

TABLE XIII

CLIMATIC TABLE FOR CLUJ

Latitude 46°47'N, longitude 23°42'E, elevation 313.4 m

Month	Mean sta. press. (mbar)	Temperature (°C)				Mean vap. press. (mbar)	Precipitation (mm)		Days with snow (mean)[*1]
		mean daily	daily range	extreme			mean	max. in 24 h	
				max.	min.				
Jan.	1012.7	−3.4		12.2	−29.8	3.4	26.8	21.4	7.2
Feb.	1014.0	−2.7		16.6	−27.7	3.5	36.0	22.1	5.6
Mar.	1016.0	3.2		24.6	−21.0	4.2	20.0	22.6	3.5
Apr.	1016.3	8.9		25.1	−5.8	6.0	49.1	27.0	1.2
May	1016.7	14.0		31.6	−2.9	8.7	72.1	26.2	0.1
June	1015.7	18.1		31.7	0.4	11.4	72.5	65.4	0.0
July	1014.0	19.9		36.2	5.6	11.9	64.0	40.4	–
Aug.	1013.2	19.3		38.0	4.7	11.7	71.3	51.1	–
Sept.	1015.4	14.6		31.6	−2.7	9.2	33.9	45.5	0.0
Oct.	1015.8	8.7		32.6	−7.2	6.8	29.9	24.6	0.2
Nov.	1015.5	3.5		18.6	−12.7	5.2	30.8	25.7	2.0
Dec.	1011.7	−0.3		16.3	−22.4	4.2	31.4	16.2	4.5
Annual	1014.8	8.6		38.0	−29.8	7.2	537.8	65.4	24.3

Month	Mean evap. (mm)	Number of days with			Mean cloud-iness (tenths)	Mean sun shine (h)	Wind		Days with snow cover		Relat. humid. (%)
		precip. ⩾0.1 mm	thunder-storm	fog			preval. direct. [*2]	mean speed (m/sec)	mean	max.	
Jan.		13.4	–	7.8	7.3		W (9.9)				86
Feb.		13.2	0.1	7.2	7.1		W (10.6)				84
Mar.		11.8	0.2	2.0	5.8		NW (13.4)				70
Apr.		14.0	1.1	0.2	6.7		NW (15.1)				66
May		15.6	5.7	1.0	6.7		NW (11.8)				67
June		13.7	8.0	1.1	6.3		NW (18.6)				68
July		11.4	6.7	0.8	5.2		NW (18.0)				63
Aug.		11.2	5.2	2.3	4.8		NW (15.4)				66
Sept.		9.0	2.1	4.4	4.7		NW (12.6)				68
Oct.		8.6	0.1	7.4	5.4		SE (8.9)				75
Nov.		10.0	0.1	7.5	7.0		SE (11.2)				83
Dec.		14.5	–	10.3	7.7		E (8.0)				88
Annual		147.2	29.3	52.0	6.2		NW (12.0)				74

[*1] 1896–1915, 1923–1939, 1949–1955; [*2] 1947–1960.

TABLE XIV

CLIMATIC TABLE FOR ZAGREB

Latitude 45°49′N, longitude 15°58′E, elevation 163 m

Month	Mean sta. press. (mbar)	Temperature (°C)				Mean vap. press. (mbar)	Precipitation (mm)		Days with snow (mean)*[1]
		mean daily	daily range	extreme			mean	max. in 24 h*[1]	
				max.*[1]	min.*[1]				
Jan.	1019	0.2		17.9	−24.6		56	26.8	9.0
Feb.	1018	2.2		19.5	−30.5		54	25.2	6.2
Mar.	1017	6.8		23.5	−17.6		47	32.8	5.0
Apr.	1015	12.0		27.4	−5.0		59	42.1	1.1
May	1015	16.4		30.3	−2.2		86	53.2	0.1
June	1016	19.9		37.0	0.2		95	53.6	−
July	1015	22.0		37.1	5.2		79	84.2	−
Aug.	1016	21.3		34.7	4.4		74	116.9	−
Sept.	1018	17.7		32.1	−1.1		70	46.1	−
Oct.	1019	11.8		28.4	−4.3		88	82.8	0.6
Nov.	1019	6.6		23.0	−15.6		89	63.6	1.8
Dec.	1019	2.4		18.3	−26.3		67	29.5	9.9
Annual	1017	11.6		37.1	−30.5		864	116.9	33.7

Month	Mean evap. (mm)	Number of days with			Mean cloud-iness (tenths)*[1]	Mean sun-shine (h)	Wind		Days with snow cover		Relat. humid. (%)
		precip.*[1] ⩾ 0.1 mm	thunder-storm*[1]	fog			preval. direct.*[1]	mean speed (m/sec)	mean	max.	
Jan.		13.2	0.1		8.2		E (16)		15.9		80
Feb.		9.8	0.1		6.6		NE (17)		11.2		75
Mar.		11.8	0.5		6.8		NE (19)		5.5		67
Apr.		13.3	1.9		6.8		NE (19)		0.4		64
May		16.5	6.4		6.6		NE (17)		−		67
June		13.8	7.9		5.8		NE (16)		−		6/
July		10.6	6.2		4.5		NE (16)		−		66
Aug.		11.5	5.6		4.7		NE (18)		−		67
Sept.		11.1	3.1		5.5		NE (18)		−		71
Oct.		13.9	1.9		6.8		NE (17)		0.1		78
Nov.		14.2	0.6		7.9		NE (16)		0.4		82
Dec.		16.0	0.1		8.3		NE (19)		12.1		83
Annual		155.7	34.4		6.5				45.6		72

*[1] 1925–1940.

TABLE XV

CLIMATIC TABLE FOR TIMISOARA

Latitude 45°47′N, longitude 21°13′E, elevation 89.6 m

Month	Mean sta. press. (mbar)	Temperature (°C)				Mean vap. press. (mbar)	Precipitation (mm)		Days with snow (mean)*[1]
		mean daily	daily range	extreme			mean	max. in 24 h	
				max.	min.				
Jan.	1018.5	−1.6		16.7	−27.0	3.8	44.5	32.5	6.5
Feb.	1017.0	0.4		18.6	−29.2	4.3	43.4	30.3	5.0
Mar.	1016.7	5.5		27.0	−20.0	5.4	40.1	24.8	2.9
Apr.	1014.6	11.4		32.0	−5.2	7.6	42.6	29.1	0.4
May	1014.9	16.4		34.5	−5.0	10.6	71.0	54.8	−
June	1014.5	19.7		38.4	3.6	12.8	75.8	83.1	−
July	1014.0	21.7		39.6	7.3	14.0	56.1	73.2	−
Aug.	1014.5	21.0		40.0	5.8	13.2	50.3	38.3	−
Sept.	1017.6	17.2		39.7	−1.0	11.0	39.8	54.0	0.0
Oct.	1018.3	11.3		33.8	−5.1	8.4	53.0	38.9	0.2
Nov.	1018.8	6.0		23.0	−10.6	6.3	58.5	40.2	1.6
Dec.	1018.8	1.5		18.5	−24.5	4.8	49.3	25.4	4.2
Annual	1016.4	10.9		40.0	−29.2	8.5	625.4	83.1	20.8

Month	Mean evap. (mm)	Number of days with			Mean cloud-iness (tenths)	Mean sun-shine (h)	Wind		Days with snow cover		Relat. humid. (%)
		precip. ⩾ 0.1 mm	thunder-storm	fog			preval. direct.*[3]	mean speed (m/sec)*[1]	mean*[2]	max.	
Jan.		11.1	−	6.9	7.1		E (20.1)	2.0	11.8		85
Feb.		11.2	0.1	5.4	6.7		N (16.2)	2.2	9.6		81
Mar.		10.4	0.3	2.2	5.7		N (17.8)	2.4	2.9		75
Apr.		10.7	1.0	0.7	5.4		N (20.3)	2.6	−		65
May		12.8	3.3	0.5	5.3		N (16.8)	2.3	−		66
June		12.1	4.2	0.4	4.7		N (18.9)	2.2	−		65
July		9.0	4.0	0.2	3.7		N (18.2)	2.2	−		62
Aug.		8.5	2.4	0.4	3.5		E (15.8)	2.2	−		63
Sept.		7.4	0.8	0.8	3.6		E (17.2)	2.8	−		67
Oct.		9.7	0.3	2.5	4.9		E (17.7)	2.8	−		75
Nov.		12.7	0.1	4.3	·6.9		E (19.4)	1.8	0.3		83
Dec.		11.9	0.0	5.2	7.1		E (21.1)	2.0	5.2		86
Annual		127.5	16.5	29.5	6.3		N (16.8)	2.0	29.8		73

*[1] 1896–1915, 1922–1955; *[2] 1926–1955; *[3] 1941–1960.

TABLE XVI

CLIMATIC TABLE FOR PULA

Latitude 44°52′N, longitude 13°51′E, elevation 36 m

Month	Mean sta. press. (mbar)	Temperature (°C)				Mean vap. press. (mbar)*¹	Precipitation (mm)		Days with snow (mean)*¹
		mean daily	daily range	extreme			mean*²	max. in 24 h*²	
				max.	min.				
Jan.						6	50	24.6	1.6
Feb.						7	43	32.8	1.2
Mar.						8	56	42.8	0.7
Apr.						10	39	29.2	0.2
May						13	54	36.4	–
June						17	40	48.4	–
July						18	47	65.4	–
Aug.						18	37	56.6	–
Sept.						16	89	52.4	–
Oct.						13	80	58.0	–
Nov.						10	106	73.0	0.4
Dec.						8	58	53.0	0.9
Annual						12	699	73.0	5.0

Month	Mean evap. (mm)	Number of days with			Mean cloud-iness (tenths)*¹	Mean sun-shine (h)*¹	Wind		Days with snow cover		Relat. humid. (%)*¹
		precip.*¹	thunder-storm*¹	fog			preval. direct.	mean speed (m/sec)	mean	max.	
Jan.		10	0.5		5.4	123					75
Feb.		8	0.8		4.9	115					75
Mar.		10	1.7		4.8	161					75
Apr.		11	1.9		4.7	203					74
May		10	4.7		4.4	271					74
June		9	6.3		4.0	288					71
July		8	8.9		3.0	335					69
Aug.		6	6.0		2.7	319					69
Sept.		9	6.3		3.7	210					73
Oct.		12	4.3		5.1	158					78
Nov.		11	1.9		5.2	120					77
Dec.		12	1.1		5.6	82					78
Annual		116	44.4		4.4	2334					74

*¹ After Biel, 1944; *² 1925–1940.

TABLE XVII

CLIMATIC TABLE FOR BELGRADE

Latitude 44°48′N, longitude 20°27′E, elevation 132 m

Month	Mean sta. press. (mbar)	Temperature (°C)				Mean vap. press. (mbar)	Precipitation (mm)		Days with snow (mean)*¹
		mean daily	daily range	extreme*¹			mean	max. in 24 h*¹	
				max.	min.				
Jan.	1019	−0.2		19.8	−19.5		48	23.4	6.6
Feb.	1018	1.6		20.1	−25.5		46	26.8	5.2
Mar.	1017	6.2		26.3	−14.4		46	40.5	4.2
Apr.	1015	12.2		30.9	−6.1		54	31.0	0.7
May	1014	17.1		33.2	−1.4		75	68.7	0.0
June	1015	20.5		36.7	4.8		96	77.9	–
July	1014	22.6		39.4	9.3		60	32.3	–
Aug.	1015	22.0		39.2	8.3		55	87.5	–
Sept.	1018	18.3		35.4	1.7		50	53.9	–
Oct.	1019	12.5		34.7	−1.0		55	48.3	–
Nov.	1019	6.8		29.3	−5.0		61	29.8	0.8
Dec.	1019	2.5		20.3	−19.3		55	28.9	7.7
Annual	1017	11.8		39.4	−25.5		701	87.5	25.2

Month	Mean evap. (mm)	Number of days with			Mean cloud-iness (tenths)*¹	Mean sun-shine (h)	Wind		Days with snow cover		Relat. humid. (%)
		precip. ≥0.1 mm*¹	thunder-storm*¹	fog			preval. direct.*¹	mean speed (m/sec)	mean*¹	max.	
Jan.		13.1	0.0		7.1		SE (20)		13.1		81
Feb.		10.6	0.0		6.4		SE (16)		8.5		77
Mar.		12.3	0.5		6.0		SE (21)		4.6		68
Apr.		12.6	1.9		5.7		SE (17)		0.1		62
May		15.0	5.5		5.6		SE (15)		–		65
June		13.0	5.8		4.8		W (13)		–		65
July		9.4	4.5		3.7		W (15)		–		62
Aug.		10.4	4.7		3.8		W (13)		–		62
Sept.		9.6	1.5		4.2		SE (14)		–		64
Oct.		12.3	0.8		5.3		SE (19)		–		72
Nov.		12.4	0.0		6.5		SE (26)		0.2		80
Dec.		15.2	0.0		7.7		SE (20)		10.2		82
Annual		145.9	25.2		5.6				36.7		70

*¹ 1925–1940.

TABLE XVIII

CLIMATIC TABLE FOR BUCHAREST-FILARET

Latitude 44°25'N, longitude 26°06'E, elevation 82 m

Month	Mean sta. press. (mbar)	Temperature (°C)				Mean vap. press. (mbar)	Precipitation (mm)		Days with snow (mean)*1
		mean daily	daily range	extreme			mean	max. in 24 h	
				max.	min.				
Jan.	1019.0	−2.7		16.6	−30.0	3.5	42.6	49.8	6.9
Feb.	1017.5	−0.6		20.8	−23.6	3.9	35.7	34.1	5.7
Mar.	1017.6	4.6		28.8	−13.3	4.8	34.9	36.2	4.0
Apr.	1014.9	11.7		34.4	−4.0	6.8	46.9	45.1	0.8
May	1014.4	17.0		36.6	0.8	9.6	68.9	61.6	0.0
June	1014.3	20.9		37.2	7.1	11.9	87.3	88.2	−
July	1013.5	23.3		39.3	8.6	13.1	55.0	46.8	−
Aug.	1014.4	22.7		41.1	7.1	12.4	48.9	135.4	−
Sept.	1018.1	18.3		39.6	0.0	10.1	30.0	57.3	0.0
Oct.	1019.3	12.0		35.5	−4.0	7.9	44.1	42.1	0.1
Nov.	1020.4	5.5		23.5	−11.3	6.0	43.0	34.7	2.2
Dec.	1020.0	0.4		19.7	−19.9	4.4	40.9	34.3	5.0
Annual	1017.0	11.1		41.1	−30.0	7.9	578.2	135.4	24.7

Month	Mean evap. (mm)	Number of days with			Mean cloudiness (tenths)	Mean sunshine (h)	Wind		Days with snow cover		Relat. humid. (%)
		precip. ≥0.1 mm	thunderstorm	fog			preval. direct.*3	mean speed (m/sec)*4	mean*2	max.	
Jan.		10.3	−	14.8	7.4		W (21.8)	2.2	20.0		86
Feb.		9.5	−	12.6	7.0		W (22.1)	2.3	15.0		82
Mar.		9.1	0.3	6.8	6.2		E (24.4)	2.4	6.9		71
Apr.		10.1	1.0	1.5	6.0		E (27.7)	2.4	0.2		63
May		12.7	5.2	0.7	6.1		E (25.7)	2.0	−		62
June		12.2	6.4	0.7	5.5		E (19.1)	1.5	−		61
July		8.8	5.3	0.6	4.1		E (19.1)	1.4	−		58
Aug.		6.1	3.6	1.4	3.8		E (22.5)	1.5	−		57
Sept.		5.8	1.2	2.5	4.1		E (25.1)	1.5	−		61
Oct.		7.9	0.3	7.6	5.3		E (23.5)	1.8	0.1		73
Nov.		10.4	0.1	12.8	7.2		E (22.6)	2.0	1.7		84
Dec.		10.9	−	16.8	7.5		W (21.1)	2.0	8.7		87
Annual		113.8	23.4	78.7	5.8			2.0	52.6		70

*1 1896–1915, 1921–1955; *2 1926–1955; *3 1941–1960; *4 1896–1955.

TABLE XIX

CLIMATIC TABLE FOR CONSTANTA

Latitude 44°10'N, longitude 28°37'E, elevation 52 m

Month	Mean sta. press. (mbar)	Temperature (°C)				Mean vap. press. (mbar)	Precipitation (mm)		Days with snow (mean)*1	Snow depth (cm)	
		mean daily	daily range	extreme			mean	max. in 24 h		mean	max.*2
				max.	min.						
Jan.	1014.0	−0.8		16.7	−24.7	4.2	29.6	36.5	3.4		25
Feb.	1012.7	0.5		19.6	−20.2	4.4	25.8	24.0	3.4		54
Mar.	1013.1	3.8		30.4	−11.6	5.2	23.9	30.2	2.0		28
Apr.	1010.9	9.1		29.4	−4.5	7.2	32.3	33.0	2.1		5
May	1010.0	14.9		36.6	1.0	10.5	37.3	35.5	−		−
June	1009.3	19.5		36.4	3.8	13.4	49.0	98.0	−		−
July	1008.3	22.2		35.6	7.6	14.7	34.6	112.3	−		−
Aug.	1009.3	22.0		36.2	7.0	14.8	32.2	111.6	−		−
Sept.	1013.2	17.8		35.0	1.0	12.3	27.2	54.3	0.0		−
Oct.	1014.4	12.8		29.4	−5.4	9.4	39.7	56.9	0.0		−
Nov.	1015.7	7.2		24.4	−12.8	7.0	43.8	32.2	1.0		22
Dec.	1015.3	2.1		20.4	−18.6	5.2	37.3	24.5	2.6		14
Annual	1012.2	10.9		36.6	−24.7	9.0	412.7	112.3	15.5		54

Month	Mean evap. (mm)	Number of days with			Mean cloudiness (tenths)	Mean sunshine (h)	Wind		Days with snow cover		Relat. humid.*1 (%)
		precip. ≥0.1 mm	thunderstorm	fog			preval. direct.*3	mean speed (m/sec)*1	mean*2	max.	
Jan.		9.9	−	7.4	7.1		W (24.4)	4.7	8.8		88
Feb.		9.2	0.1	6.6	7.1		N (22.9)	4.3	7.4		86
Mar.		7.9	0.2	5.3	6.5		N (21.8)	4.0	1.3		83
Apr.		7.4	0.3	5.3	5.9		SE (17.4)	3.6	−		80
May		8.6	1.9	4.6	5.6		NE (17.1)	3.4	−		81
June		7.8	2.6	2.0	4.5		SE (15.0)	3.4	−		77
July		5.6	1.8	1.3	3.1		N (13.7)	3.4	−		74
Aug.		4.2	1.8	1.8	2.9		N (13.7)	3.4	−		74
Sept.		3.6	0.7	1.4	3.6		N (17.4)	3.8	−		78
Oct.		6.2	0.3	3.4	5.5		N (20.5)	4.0	−		82
Nov.		9.3	0.1	5.2	7.2		N (22.0)	4.3	0.9		86
Dec.		9.9	0.0	7.2	7.4		W (22.7)	4.5	5.3		88
Annual		89.6	9.8	51.5	5.5		N (18.0)	4.0	23.7		80

*1 1896–1915, 1921–1955; *2 1926–1955; *3 1941–1960.

TABLE XX

CLIMATIC TABLE FOR SARAJEVO

Latitude 43°52'N, longitude 18°26'E, elevation 537 m

Month	Mean sta. press. (mbar)	Temperature (°C)				Mean vap. press. (mbar)	Precipitation (mm)		Days with snow (mean)[*1]
		mean daily	daily range	extreme[*1]			mean	max in 24 h	
				max.	min.				
Jan.	1019	−1.4		16.5	−20.9		71	36.3	9.6
Feb.	1018	0.7		18.0	−23.4		69	49.8	6.8
Mar.	1018	4.9		23.6	−15.9		50	38.0	6.6
Apr.	1016	9.8		30.0	−6.4		69	35.1	2.0
May	1016	14.3		30.8	−2.3		84	53.8	0.2
June	1016	17.4		35.3	2.8		86	61.4	−
July	1015	19.5		36.9	5.5		68	26.5	−
Aug.	1015	19.7		38.1	4.6		62	41.6	−
Sept.	1018	16.0		33.7	−2.2		71	86.8	0.2
Oct.	1020	10.2		32.2	−3.1		84	49.4	0.8
Nov.	1019	5.4		24.7	−7.9		98	40.7	2.2
Dec.	1019	1.7		16.9	−22.4		87	32.6	9.4
Annual	1017	9.8		38.1	−23.4		889	86.8	38.4

Month	Mean evap. (mm)	Number of days with			Mean cloud-iness (tenths)[*1]	Mean sun-shine (h)[*1]	Wind		Days with snow cover		Relat. humid. (%)
		precip. ≥0.1 mm[*1]	thunder-storm[*1]	fog			preval. direct.[*1]	mean speed (m/sec)[*1]	mean[*1]	max.	
Jan.		12.4	0.1		7.1	63	E (12)	1.0	20.7		80
Feb.		10.2	0.0		6.3	89	E (15)	1.6	10.1		75
Mar.		13.0	0.1		6.5	125	E (17)	1.8	10.2		68
Apr.		13.3	0.2		6.3	136	E (14)	1.8	1.7		66
May		16.2	0.9		6.3	168	E (11)	1.5	−		68
June		13.4	1.7		5.3	207	W (14)	1.2	−		68
July		9.3	3.5		4.0	240	E,W (12)	1.3	−		66
Aug.		10.1	1.7		3.9	243	E (17)	1.6	−		63
Sept.		10.7	1.3		4.7	180	E (12)	1.5	0.4		69
Oct.		12.9	1.4		5.7	110	E (12)	1.5	1.1		76
Nov.		13.4	0.2		6.8	69	E (15)	1.2	4.1		80
Dec.		15.4	0.2		8.1	55	E (11)	0.9	16.8		81
Annual		150.3	11.3		5.9	1686		1.4	65.1		72

[*1] 1925–1940.

TABLE XXI

CLIMATIC TABLE FOR PLEVEN

Latitude 43°36'N, longitude 24°35'E, elevation 110 m

Month	Mean sta. press. (mbar)	Temperature (°C)				Mean vap. press. (mbar)	Precipitation (mm)		Days with snow (mean)	Snow depth (cm)	
		mean daily	daily range	extreme			mean	max in 24 h		mean	max.
				max.	min.						
Jan.	1000.7	−2.0		19.4	−25.5	4.8	36.8	24.0		11.2	56
Feb.	999.1	0.4		21.6	−28.3	5.2	29.5	20.0		9.7	60
Mar.	999.0	5.4		34.2	−15.9	6.8	30.3	22.4		4.8	38
Apr.	996.6	12.4		32.0	−3.8	9.2	50.0	37.5		1.0	11
May	996.0	17.4		37.0	0.6	13.2	64.6	47.9		−	−
June	996.4	21.0		38.6	2.6	16.4	83.4	57.6		−	−
July	995.7	23.5		40.6	8.3	17.6	63.1	60.2		−	−
Aug.	996.4	23.4		41.8	8.9	16.4	32.9	73.6		−	−
Sept.	999.7	18.8		40.8	2.5	13.6	33.1	59.5		−	1
Oct.	1000.7	12.4		37.3	−3.4	10.8	47.7	38.5		0.3	1
Nov.	1001.6	5.9		23.7	−11.1	8.4	42.6	32.5		1.7	20
Dec.	1001.3	0.8		22.4	−23.1	6.0	40.7	35.0		7.6	48
Annual	998.6	11.6		41.8	−28.3	10.7	554.7	73.6			60

Month	Mean evap. (mm)	Number of days with			Mean cloud-iness (tenths)	Mean sun-shine (h)[*1]	Wind		Days with snow cover	
		precip. ≥1.0 mm	thunder-storm	fog			preval. direct.	mean speed (m/sec)	mean	max.
Jan.		10.6	0.0	10.8	7.1	74.6	W (28.0)	1.5	18.4	
Feb.		9.1	0.03	7.1	6.7	102.5	W (34.2)	2.0	16.3	
Mar.		9.2	0.2	5.5	6.3	143.5	W (27.5)	2.3	4.8	
Apr.		9.7	0.9	1.7	5.6	208.8	W (26.1)	2.2	0.2	
May		11.6	2.9	1.4	5.6	247.1	W (25.6)	1.9	−	
June		10.4	4.5	0.9	6.0	287.0	W (29.0)	1.6	−	
July		7.5	3.2	0.2	3.6	331.7	W (28.0)	1.5	−	
Aug.		5.3	2.1	0.3	3.2	319.7	W (24.9)	1.6	−	
Sept.		5.3	1.1	1.6	3.8	240.2	W (20.5)	1.4	−	
Oct.		8.3	0.5	6.1	5.3	154.1	W (21.9)	1.5	0.3	
Nov.		10.3	0.1	9.9	7.3	72.1	W (23.3)	1.6	1.5	
Dec.		10.2	0.0	12.6	7.3	63.3	W (24.7)	1.4	10.0	
Annual		107.5	15.5	58.1	5.6	2,244.6		1.7	48.8	

[*1] 1943–1962.

TABLE XXII

CLIMATIC TABLE FOR SPLIT—MARJAN

Latitude 43°31'N, longitude 16°26'E, elevation 128 m

Month	Mean sta. press. (mbar)	Temperature (°C) mean daily	daily range	extreme max.	min.	Mean vap. press. (mbar)[1]	Precipitation (mm) mean	max. in 24 h[2]	Days with snow (mean)[1]
Jan.	1015	7.8				6	76	39.1	0.4
Feb.	1014	8.1				7	74	44.4	0.5
Mar.	1015	10.3				9	53	53.4	0.1
Apr.	1014	14.0				10	62	55.3	0.1
May	1014	18.6				13	60	53.7	–
June	1015	22.9				17	53	88.5	–
July	1014	25.6				18	40	44.2	–
Aug.	1014	25.4				17	32	50.9	–
Sept.	1016	21.6				15	55	103.2	–
Oct.	1017	16.8				13	71	75.7	–
Nov.	1016	12.3				9	110	86.3	–
Dec.	1016	10.1				8	130	78.2	0.1
Annual	1015	16.1				12	816	103.2	1.2

Month	Mean evap. (mm)	Number of days with precip. ≥0.1 mm[2]	thunder-storm	fog	Mean cloud-iness (tenths)[2]	Mean sun-shine (h)[1]	Wind preval. direct.[2]	mean speed (m/sec)[1]	Days with snow cover mean	max.	Relat. humid. (%)
Jan.		12.7			5.4	148	NE (42)	3.5			62
Feb.		8.6			5.2	158	NE (32)	3.6			63
Mar.		11.4			5.7	195	NE (29)	3.2			63
Apr.		11.0			5.4	206	NE (20)	3.3			60
May		13.8			5.2	255	SE (20)	3.3			60
June		9.1			3.8	323	NE,SW (19)	2.8			57
July		5.4			2.2	354	NE (24)	2.6			52
Aug.		5.4			2.5	345	NE (27)	2.7			50
Sept.		8.0			3.7	246	NE (26)	2.7			58
Oct.		11.3			4.9	185	NE (24)	3.0			63
Nov.		12.7			5.6	128	NE (29)	3.7			66
Dec.		13.4			6.0	115	NE (38)	3.1			67
Annual		122.8			4.6	2658	NE (326)	3.1			60

[1] After Biel, 1944; [2] 1925–1940.

TABLE XXIII

CLIMATIC TABLE FOR VARNA

Latitude 43°12'N, longitude 27°55'E, elevation 3 m

Month	Mean sta. press. (mbar)	Temperature (°C) mean daily	daily range	extreme max.	min.	Mean vap. press. (mbar)	Precipitation (mm) mean	max. in 24 h	Days with snow (mean)	Snow depth (cm) mean	max.
Jan.	1017.1	1.2		20.4	−23.5	6.0	35.5	58.7		6.9	55
Feb.	1015.4	2.4		21.4	−15.8	6.0	30.9	38.0		6.8	40
Mar.	1015.7	5.0		27.5	−9.7	6.8	26.4	38.7		1.4	16
Apr.	1013.3	10.0		29.5	−2.3	9.6	35.3	30.5		–	–
May	1012.6	15.5		34.7	2.4	13.6	39.6	37.4		–	–
June	1011.9	20.2		35.4	7.2	17.6	56.4	64.0		–	–
July	1010.9	22.9		38.7	10.1	19.6	38.9	93.3		–	–
Aug.	1011.8	22.6		39.4	9.8	19.2	37.5	257.8		–	–
Sept.	1013.3	18.9		35.4	0.0	16.4	25.1	80.2		–	–
Oct.	1016.8	14.0		32.4	−1.6	12.4	43.4	65.6		–	–
Nov.	1017.3	8.6		24.2	−8.2	9.6	49.1	35.5		0.8	15
Dec.	1017.5	4.1		21.0	−12.8	7.2	55.6	38.1		4.8	40
Annual	1014.5	12.1		39.4	−23.5	12.0	473.7	257.8			55

Month	Mean evap. (mm)[1]	Number of days with precip. ≥0.1 mm	thunder-storm	fog	Mean cloud-iness (tenths)	Mean sun-shine (h)	Wind preval. direct.	mean speed (m/sec)	Days with snow cover mean	max.
Jan.	–	9.5	0.1	3.5	7.2	82.8	NW (30.6)	3.6	7.4	
Feb.	–	8.5	0.2	3.5	7.0	95.2	NW (28.6)	3.6	4.8	
Mar.	–	7.9	0.3	3.2	6.8	126.8	NW (20.3)	3.5	1.4	
Apr.	3.2	8.9	0.6	3.7	6.2	177.3	E (24.3)	3.0	–	
May	4.1	9.7	2.9	3.4	5.8	231.6	E (24.0)	2.8	–	
June	5.2	8.9	3.8	1.1	4.8	272.7	NW (20.2)	2.9	–	
July	5.8	5.8	2.4	0.0	3.1	328.4	NW (21.9)	2.7	–	
Aug.	5.6	4.4	2.5	0.8	3.0	311.7	NW (20.3)	2.8	–	
Sept.	4.1	4.1	0.8	1.9	3.9	237.6	NW (19.2)	3.1	–	
Oct.	2.8	6.9	0.5	2.5	5.6	162.4	NW (21.2)	3.2	–	
Nov.	–	9.5	0.2	3.2	7.1	87.3	NW (25.2)	3.5	0.7	
Dec.	–	9.6	0.03	4.3	7.2	76.5	NW (29.1)	3.4	4.0	
Annual	–	93.7	14.3	32.7	5.6	2190.3		3.2	18.3	

[1] 1957–1962.

TABLE XXIV

CLIMATIC TABLE FOR SOFIA

Latitude 42°42′N, longitude 23°20′E, elevation 550 m

Month	Mean sta. press. (mbar)	Temperature (°C) mean daily	daily range	extreme max.	extreme min.	Mean vap. press. (mbar)	Precipitation (mm) mean	max. in 24 h	Days with snow (mean)	Snow depth (cm) mean	max.
Jan.	950.9	−1.7		16.9	−27.5	4.8	41.9	36.7		9.7	65
Feb.	949.3	0.6		20.7	−24.5	5.2	30.5	22.6		8.7	34
Mar.	950.0	4.6		30.9	−14.9	6.4	37.1	32.2		4.0	31
Apr.	949.0	10.6		28.5	−4.8	8.4	55.2	38.0		1.5	18
May	949.5	15.5		32.1	−1.5	11.6	70.8	54.9		–	–
June	950.3	19.0		34.0	2.5	14.4	89.8	71.7		–	–
July	950.0	21.3		36.7	6.9	14.6	59.2	60.4		–	–
Aug.	950.6	20.7		37.3	6.1	14.0	43.3	49.0		–	–
Sept.	953.0	17.0		37.5	−1.5	12.8	42.3	59.3		–	–
Oct.	953.1	11.1		33.2	−3.3	10.0	55.1	37.1		0.2	5
Nov.	952.7	5.5		24.2	−10.7	7.6	52.3	29.5		2.2	27
Dec.	951.9	0.7		20.0	−20.3	6.0	44.3	27.6		6.5	30
Annual	950.9	10.4		37.5	−27.5	9.6	621.8	71.7			65

Month	Mean evap. (mm)*¹	Number of days with precip. ≥0.1 mm	thunder-storm	fog	Mean cloud-iness (tenths)	Mean sun-shine (h)	Wind preval. direct.	mean speed (m/sec)	Days with snow cover mean	max.	Relat. humid. (%)
Jan.	–	14.0	0	11.5	7.7	54.5	W (15.5)	2.0	18.6		84
Feb.	–	13.1	0.1	7.0	6.9	91.1	W (15.8)	2.4	11.8		78
Mar.	–	13.9	0.3	5.3	6.5	137.7	W (16.3)	2.4	4.3		72
Apr.	3.0	14.6	1.0	2.5	6.1	186.6	W (15.3)	2.2	0.3		66
May	3.6	16.7	4.1	0.5	6.1	221.3	W (12.6)	2.1	–		68
June	4.3	14.8	5.1	0.6	5.2	260.5	W (14.4)	1.9	–		67
July	4.9	10.6	3.7	0.5	3.8	314.1	W (14.8)	1.9	–		62
Aug.	5.0	8.5	2.2	0.3	3.3	304.3	W (13.1)	1.8	–		61
Sept.	4.3	8.0	1.2	1.5	3.9	232.8	E, W (10.9)	1.6	–		68
Oct.	2.0	10.8	0.3	4.5	5.5	154.6	E (14.0)	1.7	0.0		75
Nov.	–	13.1	0.1	8.3	7.2	75.0	E (15.5)	1.8	2.1		83
Dec.	–	13.8	0.03	13.2	7.7	49.3	W (12.7)	1.6	9.7		85
Annual	–	151.9	18.1	55.7	5.8	2081.8		2.0	46.8		72

*¹ 1952–1962.

TABLE XXV

CLIMATIC TABLE FOR TITOGRAD

Latitude 42°26′N, longitude 19°16′E, elevation 40 m

Month	Mean sta. press. (mbar)	Temperature (°C) mean daily	daily range	extreme max.	extreme min.	Mean vap. press. (mbar)	Precipitation (mm) mean	max. in 24 h*¹	Days with snow (mean)*¹
Jan.	1015	5.6					179	109.7	1.2
Feb.	1015	6.2					195	91.5	0.9
Mar.	1015	9.5					135	74.9	0.8
Apr.	1014	14.0					98	93.6	0.1
May	1014	18.6					105	87.5	–
June	1014	23.5					60	148.0	–
July	1013	26.4					40	90.1	–
Aug.	1013	26.3					63	104.4	–
Sept.	1016	21.6					113	98.2	–
Oct.	1017	15.3					202	176.1	–
Nov.	1017	10.6					213	118.2	–
Dec.	1017	7.6					229	85.0	1.1
Annual	1015	15.4					1632	176.1	4.1

Month	Mean evap. (mm)	Number of days with precip. ≥0.1 mm*¹	thunder-storm	fog	Mean cloud-iness (tenths)*¹	Mean sun-shine (h)	Wind preval. direct.*¹	mean speed (m/sec)	Days with snow cover mean*¹	max.	Relat. humid. (%)
Jan.		10.4	0.6		5.7		N (21)		3.1		73
Feb.		9.1	0.9		5.5		N (17)		0.2		74
Mar.		12.1	1.0		6.2		N (20)		0.2		66
Apr.		11.4	1.4		6.0		N (13)		–		63
May		12.3	3.5		5.6		S (12)		–		63
June		8.0	4.4		4.5		N,S (14)		–		56
July		3.4	3.4		3.0		N (21)		–		48
Aug.		4.9	3.9		3.2		N (19)		–		48
Sept.		6.9	2.5		3.9		N (17)		–		58
Oct.		13.8	3.3		5.3		N (16)		–		68
Nov.		12.1	2.1		6.3		N (13)		–		77
Dec.		14.0	1.1		6.6		N (15)		1.0		75
Annual		118.4	28.1		5.2		N (196)		4.5		64

*¹ 1925–1940.

TABLE XXVI

CLIMATIC TABLE FOR PALAGRUŽA

Latitude 42°23′N, longitude 16°16′E, elevation 92 m

Month	Mean sta. press. (mbar)	Temperature (°C)				Mean vap. press. (mbar)*1	Precipitation (mm)		Days with snow (mean)*1
		mean daily*1	daily range	extreme			mean	max. in 24 h*1	
				max.	min.				
Jan.		9.1				8			0.6
Feb.		9.3				9			0.2
Mar.		10.7				10			0.1
Apr.		13.1				11			–
May		16.6				15			–
June		20.8				19			–
July		23.8				22			–
Aug.		23.8				22			–
Sept.		20.8				20			–
Oct.		17.2				16			–
Nov.		14.0				12			–
Dec.		10.6				10			–
Annual		15.8				15			0.9

Month	Mean evap. (mm)	Number of days with			Mean cloud-iness (tenths)*1	Mean sun-shine (h)	Wind	
		precip. ≥0.1 mm*1	thunder-storm*1	fog			preval. direct.*1	mean speed (m/sec)*1
Jan.		8	0.7		6.0		NW (33)	6.6
Feb.		7	0.6		5.9		NW (30)	6.2
Mar.		7	0.6		5.4		SE (36)	6.2
Apr.		6	1.1		5.0		SE (37)	5.6
May		5	3.0		4.5		NW (35)	5.3
June		3	2.6		3.8		NW (40)	4.3
July		2	3.1		2.4		NW (53)	5.0
Aug.		3	2.9		2.2		NW (49)	4.6
Sept.		5	2.8		3.3		NW (41)	5.2
Oct.		7	2.4		5.1		SE (36)	5.6
Nov.		10	1.4		5.4		SE (29)	6.6
Dec.		10	1.4		5.8		NW (33)	6.6
Annual		73	22.6		4.6			5.6

*1 After Biel, 1944.

TABLE XXVII

CLIMATIC TABLE FOR PLOVDIV

Latitude 42°09′N, longitude 24°45′E, elevation 160 m

Month	Mean sta. press. (mbar)	Temperature (°C)				Mean vap. press. (mbar)	Precipitation (mm)		Days with snow (mean)	Snow depth (cm)	
		mean daily	daily range	extreme			mean	max. in 24 h		mean	max.
				max.	min.						
Jan.	998.9	–0.4		19.4	–31.5	5.2	41.4	24.6		8.4	36
Feb.	997.3	2.1		23.6	–29.1	5.6	33.9	37.2		8.1	46
Mar.	997.6	6.0		28.4	–17.5	6.8	39.7	31.3		3.9	32
Apr.	995.4	12.2		30.7	–4.0	9.6	42.8	32.3		0.5	13
May	995.0	17.2		35.3	–0.3	13.6	55.4	39.0		–	–
June	995.1	21.0		38.5	6.0	16.8	66.8	79.6		–	–
July	994.4	23.4		39.8	8.2	17.6	47.1	49.7		–	–
Aug.	995.1	22.8		41.3	5.6	16.8	30.6	53.2		--	–
Sept.	998.4	18.3		36.5	0.2	14.4	32.5	42.2		–	–
Oct.	999.5	12.6		32.7	–5.8	11.2	44.5	52.5		–	–
Nov.	1000.2	7.1		23.9	–9.1	8.8	53.3	28.6		1.8	24
Dec.	999.2	2.3		22.1	–18.0	6.4	52.4	57.6		6.6	27
Annual	997.2	12.0		41.3	–31.5	11.1	540.4	79.6			46

Month	Mean evap. (mm)*1	Number of days with			Mean cloud-iness (oktas)	Mean sun-shine (h)	Wind		Days with snow cover	
		precip. ≥0.1 mm	thunder-storm	fog			preval. direct.	mean speed (m/sec)	mean	max.
Jan.	–	7.2	0.3	7.3	7.0	80.7	W (28.2)		10.8	
Feb.	–	5.2	0.1	3.7	6.5	104.8	W (30.3)		6.6	
Mar.	–	5.5	0.3	2.5	6.2	148.9	W (24.2)		2.2	
Apr.	2.6	6.5	1.5	0.6	5.6	202.9	W (23.3)		0.1	
May	3.0	8.5	5.5	0.7	5.7	234.4	W (21.4)		–	
June	3.5	8.0	8.6	0.3	4.9	270.8	W (27.9)		–	
July	4.5	5.9	5.2	0.1	3.6	327.5	W (26.6)		–	
Aug.	4.7	3.8	3.6	0.01	3.0	320.7	W (23.2)		–	
Sept.	3.0	3.8	1.7	0.9	3.8	241.4	W (18.5)		–	
Oct.	1.4	5.3	0.4	3.2	5.2	167.0	W (17.9)		–	
Nov.	–	7.2	0.2	5.7	6.9	87.0	W (20.5)		0.8	
Dec.	–	6.1	0.1	8.2	7.1	75.3	W (24.1)		4.7	
Annual	–	73.0	27.5	33.2	5.5	2261.4			25.2	

*1 1958–1962.

TABLE XXVIII

CLIMATIC TABLE FOR SHKODER (Shkodra)

Latitude 42°04'N, longitude 19°22'E, elevation 22 m

Month	Mean sta. press. (mbar)[1]	Temperature (°C) mean daily	daily range[2]	extreme[2] max.	min.	Mean vap. press. (mbar)	Precipitation (mm) mean	max. in 24 h	Days with snow (mean)[3]
Jan.	1011.0	5.3	5.7	17.4	−15.5		216		1.1
Feb.	1010.5	7.1	6.6	20.8	−14.5		173		1.2
Mar.	1010.7	10.1	7.7	23.5	−3.2		149		0.6
Apr.	1009.9	14.7	8.7	31.5	1.4		101		−
May	1009.9	19.1	9.3	34.5	6.2		114		−
June	1010.2	23.2	9.8	35.8	10.7		58		−
July	1008.7	26.4	10.3	40.0	13.4		30		−
Aug.	1009.0	26.0	10.2	40.0	10.5		44		−
Sept.	1011.8	22.2	9.5	39.7	10.5		128		−
Oct.	1013.3	16.5	7.9	30.2	2.7		262		0.0
Nov.	1012.9	11.6	5.8	22.4	−0.3		241		0.1
Dec.	1012.9	7.4	7.0	20.0	−10.6		227		0.6
Annual	1010.9	15.8	8.2	40.0	−15.5		1741		3.6

Month	Mean evap. (mm)	Number of days with precip. ⩾0.1 mm[3]	thunder- storm[3]	fog	Mean cloud- iness (tenths)[3]	Mean sun- shine (h)	Wind preval. direct.[2]	mean speed (m/sec)[2]	Days with snow cover mean	max.	Relat. humid. (%)[1]
Jan.		10	0.7		4.6		E (26.6)	2.6			73
Feb.		11	1.9		5.1		E (21.8)	2.7			73
Mar.		11	2.3		4.7		NW (21.4)	2.8			64
Apr.		10	1.7		5.1		NW (19.9)	2.6			64
May		8	1.6		4.3		SW (20.7)	2.4			65
June		6	2.6		3.4		W (20.8)	2.5			59
July		3	2.1		1.6		W (21.9)	2.5			54
Aug.		3	1.4		1.5		NW (19.6)	2.5			54
Sept.		7	3.1		2.7		NW (19.5)	2.5			61
Oct.		10	2.7		4.5		NW (19.0)	2.2			66
Nov.		11	1.2		4.6		E (24.1)	2.4			73
Dec.		11	1.3		5.1		E (27.8)	2.5			73
Annual		100	22.6		3.9		NW (18.4)	2.5			65

[1] 1951–1960; [2] after Isidorov, 1955; [3] after Biel, 1944.

TABLE XXIX

CLIMATIC TABLE FOR SKOPJE

Latitude 42°00'N, longitude 21°06'E, elevation 245 m

Month	Mean sta. press. (mbar)	Temperature (°C) mean daily	daily range	extreme[1] max.	min.	Mean vap. press. (mbar)	Precipitation (mm) mean	max. in 24 h[1]	Days with snow (mean)[1]
Jan.	1019	1.1		19.0	−23.0		46	31.5	3.6
Feb.	1018	2.9		18.1	−23.9		41	22.8	2.4
Mar.	1017	6.5		25.0	−18.4		38	60.0	1.8
Apr.	1015	12.1		30.2	−4.5		34	25.3	0.1
May	1015	17.0		33.5	−0.6		52	29.4	−
June	1015	21.6		37.0	3.3		49	67.3	−
July	1014	23.8		41.2	5.5		35	21.5	−
Aug.	1014	23.7		40.5	5.7		37	32.0	−
Sept.	1018	18.6		38.5	−3.2		42	30.0	−
Oct.	1020	11.9		34.2	−4.4		58	40.7	0.1
Nov.	1020	7.2		24.8	−9.7		71	37.2	0.2
Dec.	1020	2.9		19.4	−21.8		43	58.8	4.0
Annual	1017	12.4		41.2	−23.9		546	67.3	12.2

Month	Mean evap. (mm)	Number of days with precip. ⩾0.1 mm[1]	thunder- storm[1]	fog	Mean cloud- iness (tenths)[1]	Mean sun- shine (h)	Wind preval. direct.[1]	mean speed (m/sec)	Days with snow cover mean[1]	max.	Relat. humid. (%)
Jan.		7.4	0.0		7.2		W (11)		7.1		85
Feb.		6.4	0.1		6.6		W (10)		4.7		79
Mar.		7.6	0.2		6.5		SE,W (10)		1.0		73
Apr.		8.3	0.9		6.1		SE (11)		−		66
May		10.8	3.5		6.2		SE (12)		−		69
June		8.1	4.2		4.7		W (10)		−		63
July		4.4	3.3		3.1		W (12)		−		58
Aug.		4.7	3.2		3.1		W (12)		−		57
Sept.		4.3	1.2		3.9		W (9)		−		67
Oct.		8.1	0.6		5.3		W (9)		−		80
Nov.		6.1	0.1		6.4		W (7)		0.1		85
Dec.		11.0	0.1		8.1		W (13)		7.2		87
Annual		87.2	17.4		5.6		W (121)		20.1		72

[1] 1925–1940.

TABLE XXX

CLIMATIC TABLE FOR TIRANA

Latitude 41°18'N, longitude 19°48'E, elevation 114 m

Month	Mean sta. press. (mbar)[*1]	Temperature (°C)		extreme[*2]		Mean vap. press. (mbar)	Precipitation (mm)	
		mean daily	daily range[*2]	max.	min.		mean	max. in 24 h
Jan.	1003.5	7.3	9.8	21.4	−10.0		132	
Feb.	1003.3	8.3	10.5	23.7	−9.0		120	
Mar.	1003.7	10.6	10.0	27.2	−10.5		100	
Apr.	1003.3	14.4	11.4	31.2	−1.0		87	
May	1003.5	18.4	11.7	35.3	3.1		99	
June	1003.9	22.4	12.7	38.0	–		60	
July	1002.6	25.0	14.1	42.0	8.0		28	
Aug.	1003.0	24.9	14.7	39.8	8.5		39	
Sept.	1005.9	21.8	13.0	39.7	3.0		73	
Oct.	1006.3	17.4	11.1	35.5	0.1		157	
Nov.	1005.7	12.9	9.5	28.4	−5.8		152	
Dec.	1005.8	9.2	8.8	21.9	−8.0		142	
Annual	1004.1	16.0	11.5	42.0	−10.5		1189	

Month	Mean evap. (mm)	Number of days with			Mean cloud- iness (tenths)[*2]	Mean sun- shine (h)	Wind		Days with snow cover		Relat. humid. (%)[*1]
		precip. ≥0.1 mm[*2]	thunder- storm	fog			preval. direct.[*2]	mean speed (m/sec)[*2]	mean	max.	
Jan.		12			6.4		SE (34.1)	1.5			70
Feb.		10			6.1		SE (29.3)	1.7			70
Mar.		11			5.6		NW (27.8)	1.6			68
Apr.		11			5.4		NW (37.2)	1.5			67
May		10			5.4		NW (37.1)	1.5			69
June		6			3.9		NW (40.9)	1.6			62
July		4			2.2		NW (49.7)	1.7			58
Aug.		4			2.1		NW (50.3)	1.5			58
Sept.		6			3.4		NW (44.2)	1.4			64
Oct.		11			5.3		NW (29.5)	1.4			70
Nov.		13			6.1		SE (32.1)	1.4			74
Dec.		12			6.1		SE (36.6)	1.4			70
Annual		110			4.8		NW (31.7)	1.5			67

[*1] 1950–1960; [*2] after Isidorov, 1955.

TABLE XXXI

CLIMATIC TABLE FOR THESSALONIKI

Latitude 40°39'N, longitude 23°07'E, elevation 2 m

Month	Mean sta. press.[*2] (mbar)	Temperature (°C)		extreme		Mean vap. press. (mbar)	Precipitation (mm)		Days with snow (mean)[*1]
		mean daily	daily range	max.	min.		mean	max. in 24 h	
Jan.	1016.6	5.5					44		1.4
Feb.	1016.1	7.1					34		0.7
Mar.	1016.7	9.6					35		0.4
Apr.	1015.0	14.5					36		0.1
May	1014.1	19.6					40		–
June	1014.0	24.7					33		–
July	1012.5	27.3					20		–
Aug.	1012.9	26.8					14		–
Sept.	1016.8	22.5					28		–
Oct.	1018.9	17.1					55		0.1
Nov.	1018.9	12.0					56		0.3
Dec.	1018.3	7.5					54		0.7
Annual	1015.9	16.1					449		3.7

Month	Mean evap. (mm)	Number of days with			Mean cloud- iness (tenths)[*1]	Relat. humid. (%)
		precip. ≥0.1 mm[*1]	thunder- storm[*1]	fog		
Jan.		6	0.3		5.0	78
Feb.		6	0.3		5.4	71
Mar.		7	0.3		5.5	69
Apr.		7	1.2		5.2	67
May		6	5.6		4.5	66
June		6	7.2		3.3	56
July		4	4.8		2.2	51
Aug.		3	3.4		2.2	52
Sept.		4	2.5		3.0	60
Oct.		6	1.5		4.5	69
Nov.		7	0.7		5.8	76
Dec.		8	0.3		5.9	78
Annual		70	28.1		4.4	66

[*1] After Biel, 1944; 1951–1960.

TABLE XXXII

CLIMATIC TABLE FOR VLORE

Latitude 40°29'N, longitude 19°30'E, elevation 10 m

Month	Mean sta. press. (mbar)[1]	Temperature (°C)		extreme[2]		Mean vap. press. (mbar)	Precipitation (mm)	
		mean daily	daily range[2]	max.	min.		mean	max. in 24 h
Jan.	1014.2	9.2	8.1	24.0	−7.2		148	
Feb.	1014.2	10.2	8.9	26.0	−3.5		102	
Mar.	1014.5	11.8	9.1	28.1	−3.0		73	
Apr.	1013.5	14.9	9.4	30.0	3.0		60	
May	1014.3	18.7	10.1	36.0	5.0		49	
June	1014.6	22.4	11.0	38.8			29	
July	1013.1	24.7	11.6	39.C	12.5		9	
Aug.	1013.4	24.6	11.9	41.0	12.8		25	
Sept.	1015.9	22.0	10.8	38.0	4.0		65	
Oct.	1017.0	18.4	9.8	35.0	1.0		133	
Nov.	1016.2	14.6	8.3	32.0	−2.6		167	
Dec.	1016.2	11.0	8.4	23.5	−4.5		170	
Annual	1014.7	16.9	9.8	41.0	−7.2		1028	

Month	Mean evap (mm)	Number of days with			Mean cloud- iness (tenths)[3]	Mean sun- shine (h)	Wind		Days with snow cover		Relat. humid. (%)[1]
		precip. ≥0.1 mm[2]	thunder- storm	fog			preval. direct.[2]	mean speed (m/sec)[2]	mean	max.	
Jan.		11			5.5		E (24.4)	3.1			68
Feb.		8			4.6		N (26.1)	3.2			68
Mar.		8			4.8		N (31.3)	3.0			67
Apr.		8			3.6		N (31.7)	3.2			69
May		6			3.8		N (25.2)	2.5			70
June		3			2.8		N (30.8)	2.4			63
July		1			1.4		N (36.5)	2.1			62
Aug.		2			1.9		N (25.1)	1.9			63
Sept.		4			3.1		N (26.1)	2.3			67
Oct.		8			4.6		N (31.8)	2.5			70
Nov.		8			6.2		E (24.2)	3.1			71
Dec.		12			5.7		E (24.8)	3.8			67
Annual		79			4.0		N (27.8)	2.8			67

[1] 1951–1960; [2] after Isidorov, 1955; [3] after Biel, 1944.

TABLE XXXIII

CLIMATIC TABLE FOR LARISA

Latitude 39°37'N, longitude 22°15'E, elevation 76 m

Month	Mean sta. press.[1] (mbar)	Temperature (°C)		extreme		Mean vap. press. (mbar)	Precipitation (mm)	
		mean daily[2]	daily range	max.	min.		mean[2]	max. in 24 h
Jan.	1016.3	5.8					51	
Feb.	1016.0	7.4					40	
Mar.	1015.4	9.2					49	
Apr.	1013.2	13.7					35	
May	1013.2	19.7					45	
June	1012.7	25.4					30	
July	1010.5	28.0					15	
Aug.	1011.3	27.8					13	
Sept.	1015.5	22.4					31	
Oct.	1017.6	16.2					88	
Nov.	1017.8	11.4					64	
Dec.	1016.3	7.4					61	
Annual	1014.7	16.2					522	

Month	Mean evap. (mm)	Number of days with			Mean cloud- iness (tenths)	Mean sun- shine (h)	Wind		Days with snow cover		Relat. humid.[2] (%)
		precip. ≥0.1 mm	thunder- storm	fog			preval. direct.	mean speed (m/sec)[3]	mean	max.	
Jan.								0.7			81
Feb.								0.7			74
Mar.								1.0			74
Apr.								1.0			71
May								1.0			65
June								1.0			52
July								1.0			47
Aug.								1.0			45
Sept.								0.7			58
Oct.								0.5			73
Nov.								0.6			82
Dec.								0.5			82
Annual								0.7			67

[1] 1954–1960; [2] 1951–1960; [3] after Biel, 1944.

TABLE XXXIV

CLIMATIC TABLE FOR PATRAS (Patrai)

Latitude 38°15'N, longtitude 21°45'E, elevation 43 m

Month	Mean sta. press.*1 (mbar)	Temperature (°C)				Mean vap. press. (mbar)*2	Precipitation (mm)	
		mean daily	daily range	extreme			mean	max. in 24 h
				max.	min.			
Jan.	1015.5	9.7				9	123	
Feb.	1015.0	10.4				10	87	
Mar.	1015.3	11.9				11	72	
Apr.	1014.4	15.6				13	50	
May	1013.8	19.6				17	27	
June	1013.8	23.8				21	13	
July	1012.0	26.3				22	1	
Aug.	1012.3	26.4				23	6	
Sept.	1015.4	23.1				19	27	
Oct.	1016.4	18.7				17	82	
Nov.	1016.8	14.4				13	113	
Dec.	1015.8	10.9				11	148	
Annual	1014.7	17.6				16	749	

Month	Mean evap. (mm)	Number of days with			Mean cloud-iness (tenths)*2	Relat. humid. (%)
		precip. ⩾0.1 mm	thunder-storm*1	fog		
Jan.		1.2			5.7	74
Feb.		1.4			6.2	71
Mar.		1.6			5.8	69
Apr.		1.7			5.8	68
May		1.4			4.7	67
June		2.8			2.9	64
July		0.7			1.1	61
Aug.		0.8			1.0	62
Sept.		1.6			2.2	67
Oct.		1.9			4.3	71
Nov.		2.8			5.3	74
Dec.		1.4			6.2	74
Annual		19.3			4.3	69

*1 1932–1960; *2 after Biel, 1944.

TABLE XXXV

CLIMATIC TABLE FOR ATHENS (Athinai)

Latitude 37°58'N, longitude 23°43'E, elevation 107 m

Month	Mean sta. press. (mbar)	Temperature (°C)				Mean vap. press. (mbar)	Precipitation (mm)		Days with snow (mean)*1
		mean daily	daily range	extreme			mean	max. in 24 h	
				max.	min.				
Jan.	1016.1	9.3					62		1.7
Feb.	1015.8	9.9					36		1.2
Mar.	1015.6	11.3					38		0.6
Apr.	1014.3	15.3					23		0.1
May	1013.5	20.0					23		–
June	1012.9	24.6					14		–
July	1011.2	27.6					6		–
Aug.	1011.7	27.4					7		–
Sept.	1015.3	23.5					15		–
Oct.	1017.0	19.0					51		–
Nov.	1017.5	14.7					56		0.3
Dec.	1017.1	11.0					71		0.7
Annual	1014.8	17.8					402		4.6

Month	Mean evap. (mm)	Number of days with			Mean cloud-iness (tenths)*1	Mean sun shine (h)*1	Wind		Days with snow cover		Relat. humid. (%)
		precip. ⩾0.1 mm*1	thunder-storm*1	fog			preval. direct.*1	mean speed (m/sec)*1	mean	max.	
Jan.		12	0.6		5.5	149	NE (21)	1.9			74
Feb.		11	1.0		5.7	156	NE,SW (21)	2.2			70
Mar.		10	0.4		5.3	190	NE (19)	2.7			67
Apr.		8	0.4		4.8	215	NE (18)	1.8			63
May		7	2.0		4.0	232	S (19)	1.8			59
June		5	2.2		2.5	292	SW (20)	1.8			53
July		2	1.1		1.1	364	NE (26)	2.2			47
Aug.		3	0.6		1.2	340	NE (30)	2.2			47
Sept.		4	1.2		2.2	272	NE (30)	1.9			56
Oct.		8	1.8		4.0	210	NE (23)	1.8			67
Nov.		12	1.0		5.7	129	NE (32)	2.3			73
Dec.		12	0.9		5.9	108	NE (20)	2.1			75
Annual		93	13.2		4.0	2655		2.0			63

*1 After Biel, 1944.

Reference Index

Geographical Index

Aachen, 65
Abruzzi, 148
Adige, 147
– Valley, 38
Adriatic Sea, 15, 76, 130, 137, 142, 191, 193, 194, 198, 202, 206, 210
Aegean Islands, 191, 192, 203, 205–210, 212
– Sea, 193, 194, 198, 203, 205, 219
Africa, 6
–, North, 192
Ajdovščina, 218
Alassio, 176
Albania, 186
Alghero, 162, 163, 164
Alpine Basin, 188
Alps, 4, 5, 9, 10, 15, 17, 18, 21, 27, 37–57, 81, 135, 136, 150, 192
–, Bernese, 39
–, Dinaric, 186, 191, 203, 220
–, Julian, 196, 205, 209–219
Altdorf, 38, 56
Ancona, 127, 162, 163, 164, 178
Aosta, 147, 162, 163, 164
Apennines, 128, 129, 142, 150, 191
–, Ligurian, 137
Apulia, 147
Arlberg, 141
Arosa, 145
Asia Minor, 193, 202, 212
Athens (Athinai), 185, 186, 187, 188, 189, 198, 206, 213, 215, 215, 220, 235
Atlantic Ocean, 42, 130, 131, 192
Attica, 216
Augsburg, 67
Augusta, 162, 163, 164
Austria, 4–6, 37
Azores, 17, 80, 81, 85, 131

Badgastein, 45
Bad Reichenhall, 28
Bagnara, 150
Baikal, Lake, 80
Bakonya Forest, 99, 106
Balaton, Lake, 100
Balkan Mountains, 186, 194, 204, 205, 217, 219
– Peninsula, 131, 186–220
Baltic Sea, 10, 29–31, 42, 76, 88, 100, 112
Bamberg, 66
Bardonecchia, 149
Bari, 144, 162, 163, 164
Basel, 3, 26, 45
Basilicata, 144, 147
Bavaria, 36
Bavarian Forest, 5

Bay of Biscay, 131
Belgrade, 187–190, 193–196, 201, 217, 219, 220, 226
Benevento, 162, 163, 164
Berg, 56
Bergamo, 162, 163, 164
Berlin, 3, 8, 10, 20, 56
Bern, 36
Bernese Alps, 39
Bernina, 5
Bever, 22, 45
Bight of Cologne, 31, 33
Bight of Graz, 56
Bjelašnica, 187, 201, 208, 209, 212, 213
Black Forest, 5, 15, 27, 28, 36, 56
Black Sea, 131, 187, 193, 194, 197, 198, 202–220
Bled, 203
Bohemian Forest, 5, 37
Bohinjski greben, 209
Boka Kotorska, 203
Bologna, 127, 162, 163, 164, 170, 175
Bolzano, 162, 163, 164, 166
Bosna Valley, 220
Bosporus, 187, 200
Bothnia, Gulf of, 88
Bratislava, 78, 84, 93, 103, 104, 106, 108, 111, 112, 122
Bremen, 22, 29
Brenner Pass, 37
Brest, 10
Brindisi, 162, 163, 164
British Isles, 7, 15, 18, 192
Brno, 76, 78, 92, 103, 104, 106, 112, 121
Brocken, 34, 63
Brunico, 162, 163, 164
Bucharest, 185–190, 198–209, 215, 216, 220, 227
Budapest, 78–80, 86–96, 103–109, 112, 124
Bug, 78
Bukovina, 194, 198, 201, 203, 210
Bulgaria, 186, 194, 201, 202
Burgas, 194
Bydgoszcz, 78

Cagliari, 162, 163, 164, 182
Calabria, 130, 138, 142–148
– Mountains, 130
Calopezzati, 162, 163, 164
Candia, 201
Cape Bellavista, 162, 163, 164
– Caccia, 162, 163, 164
– Carbonara, 162, 163, 164
– Comino, 162, 163, 164
– Palinuro, 162, 163, 164
– Spartivento, 162, 163, 164
Capracotta, 149

Subject Index